Integrating Geographic Information Systems into Library Services:
A Guide for Academic Libraries

John Abresch
University of South Florida Libraries, USA

Ardis Hanson
University of South Florida Libraries, USA

Susan Heron
University of South Florida Libraries, USA

Pete Reehling
University of South Florida Libraries, USA

 Information Science Publishing

Hershey • New York

Acquisition Editor:	Kristin Klinger
Senior Managing Editor:	Jennifer Neidig
Managing Editor:	Jamie Snavely
Assistant Managing Editor:	Carole Coulson
Development Editor:	Kristin Roth
Copy Editor:	Angela Thor
Typesetter:	Larissa Vinci
Cover Design:	Lisa Tosheff
Printed at:	Yurchak Printing Inc.

Published in the United States of America by
Information Science Publishing (an imprint of IGI Global)
701 E. Chocolate Avenue
Hershey PA 17033
Tel: 717-533-8845
Fax: 717-533-8661
E-mail: cust@igi-global.com
Web site: http://www.igi-global.com

and in the United Kingdom by
Information Science Publishing (an imprint of IGI Global)
3 Henrietta Street
Covent Garden
London WC2E 8LU
Tel: 44 20 7240 0856
Fax: 44 20 7379 3313
Web site: http://www.eurospanbookstore.com

Library of Congress Cataloging-in-Publication Data

Integrating geographic information systems into library services : a guide for academic libraries / John Abresch ... [et al.].
 p. cm.
 Summary: "This book integrates traditional map librarianship and contemporary issues in digital librarianship within a framework of a global embedded information infrastructure, addressing technical, legal, and institutional factors such as collection development, reference and research services, and cataloging/metadata, as well as issues in accessibility and standards"--Provided by publisher.
 Includes bibliographical references and index.
 ISBN-13: 978-1-59904-726-3
 ISBN-13: 978-1-59904-728-7 (ebook)
 1. Libraries--Special collections--Geographic information systems. 2. Geographic information systems. I. Abresch, John.
 Z688.G33158 2008
 025.06'91--dc22
                                              2007036857

British Cataloguing in Publication Data
A Cataloguing in Publication record for this book is available from the British Library.

All work contributed to this book is new, previously-unpublished material. The views expressed in this book are those of the authors, but not necessarily of the publisher.

Integrating Geographic Information Systems into Library Services:
A Guide for Academic Libraries

Table of Contents

Chapter II
Information Economy and Geospatial Information 22

Chapter III
Spatial Databases and Data Infrastructure 53

Foreword

Maps seduce with color and design and often with grace and style. Maps represent adventure, potential, plans, and hope. Cartography is a symbology that transcends languages and time. Maps are stores of coded spatial data, coordinate observations of coastlines and depths, transportation classification and networks, street name and address ranges. Maps represent the current and the past states of geography. These cartographically encoded geographies originated as numbers and are frozen in paper as maps.

Libraries define collections by their storage needs; map libraries are stacks of flat metal cases for large sheets of paper. The map library typically sits alone, off to the side, rows of grey, metal, map cases housing map sheets. However, the map library is actually a store of numeric spatial information in symbolic, viz. cartographic, form. Until the mid-1960s the map was the exclusive method for storing symbolic spatial information. Nineteen sixty four marked the beginning of geographic information systems or GIS. A GIS is a computer-based database for capturing, storing, analyzing, and managing data and associated attributes that are spatially referenced to the earth.

Twenty years ago, debate raged over the definition of cartography and maps. The International Cartographic Association (ICA) invited redefinitions of cartography in light of innovations in computer technology. Two camps emerged, stressing the importance of the map on one hand, and the spatial database on the other. M. Visvalingum articulated a middle ground, focusing not on product, but on content:

"If cartography is concerned with the making and use of maps, then it is not just concerned with visual products: it is equally concerned with the processes of mapping, from data collection, transformation and simplification through to symbolism and with map reading, analysis and interpretation. These intellectual processes are expressed in terms of prevailing technologies and computer-based information technology is fast becoming the dominant technology of the day" (Visvalingum, 1989).

Today the technology is shifting yet again. To paraphrase Visvalingum, as computers have become ubiquitous, network-based IT has become the dominant technology of the day. Web 2.0 is the move to the Internet as a platform. Spatial data has been networked almost from its inception. Now, with Google Map™, Google Earth™ has rapidly become the poster child of Web 2.0. Librarians have been slow to engage the requirements of managing datasets in libraries much less the unique requirements of spatial data. Libraries have been more effective at building digital surrogate collections than collecting, describing, and providing access to very large, complex, born-digital spatial data. This book also provides a much-needed text to challenge the dialogue of spatial data and information in libraries, and to teach the management of spatial information in library and information science programs.

This book provides a vocabulary for discussing how to build and manage digital spatial data collections in libraries, integrating traditional map librarianship and contemporary issues in digital librarianship. Augmenting the services of the map library, GIS, a geospatial database management system, has uses that transcend the paper map.

These uses have created expectations: "[m]aps and GIS are important sources for the production of geographic knowledge. What are the power-knowledge relations of mapping as they occur against the historical horizon of possibilities and how can that horizon be enlarged?" (Crampton, 2003, p. 53). These types of discussions have created a continuum of thought on what *is* critical GIS. Pickles (1995, p. 4), for example, describes it as a "part of a contemporary network of knowledge, ideology, and practice that defines, inscribes, and represents environmental and social patterns within a broader economy of signification that calls forth new ways of thinking, acting, and writing." How far GIS can redefine how we look at populations, location, and natural resources is still unknown. However, redefinition continues and affects use and user.

In *Ground Truth*, GIS was seen as a way to create "new visual imaginaries, new conceptions of earth, new modalities of commodity and consumer, and new visions of what constitutes market, territory and empire" (Pickles 1995, p. viii). *Integrating Geographic Information Systems into Library Services: A Guide for Academic Libraries* will create new ways of viewing geographic and library and information sciences within the academic setting.

References

Crampton, J. W. (2003). *The political mapping of cyberspace*. Chicago, IL: University of Chicago Press.

Pickles, J. (1995). Representations in an electronic age: Geography, GIS, and democracy. In J. Pickles (Ed.), *Ground truth: The social implications of geographic information systems* (pp. 1-30). New York, NY: The Guildford Press.

Visvalingam, M. (1989). Cartography, GIS and maps in perspective. *Cartographic Journal, 26*(1), 26-32.

Patrick McGlamery

Information & Technology Services Area Head
University of Connecticut Libraries

Patrick McGlamery, MLS, has over 17 years of experience as map librarian at the University of Connecticut Libraries. Before that he had worked at the Library of Congress Geography and Map Division for 5 years. He has been involved with computer mapping in a library environment since 1987. Mr. McGlamery is active in the American Library Association's Map and Geography Roundtable (MAGERT) and Library Information Technology Association, and in the International Federation of Library Association's Geography and Map Section. Librarian of MAGIC, UConn's Map and Geographic Information Center, Mr. McGlamery is participating in Simmons College initiative, funded by the Atlantic Philanthropies, to train Vietnamese librarians. He is currently teaching "Academic Libraries" and "Digital Libraries" in Hue, Vietnam from June to August, 2007. The final Fall semester will be in both Danang and Thai Nguyen.

Preface

Unfolding landscapes evoke all sorts of feelings and memories. Maps allow us to visit or revisit areas of the world that fascinate us. They allow us to travel across continents, explore hidden cities, understand the planning of medieval walled towns, and escape to exotic locales that may no longer exist. The power of place is indescribable. The need for us as humans to understand "place," as well as our place in the world, is essential. Geography gives us those skills and concepts to understand the physical, human, political, historical, economic, and cultural factors that affect the human and natural environments.

Libraries are part of the human environment. They represent our attempts to understand, to wonder, and to reflect on the myriad wonderfulness of our universes, local and far away, real and imagined. Libraries house riches, from books to journals to maps to globes to pictures in all sorts of two- and three-dimensional formats. Libraries also provide ways of knowing and understanding a topic or place or person through classification and naming. Libraries provide ways to access and acquire those materials that can give us a deeper understanding of all those factors that make us human and that help create societies. Libraries are also places of instruction, of learning how to find that bit of knowledge that keeps us wondering or lying awake at night trying to figure it out.

Both geography and librarianship have evolved significantly in their breadth of understanding their respective universes, including the emergence of exciting conceptual and theoretical models, innovative methodologies, cutting-edge technologies, and application of these technologies. We cover the waterfront, so to speak, from the tangible, such as paper, photographs, and maps, to the intangible, such as digital objects, numeric/spatial data, and streaming media. We have also evolved from being "geography" and "library" to "geographic information science" and "library/information science."

In writing this monograph, we have tried to address these new forms of geography and library. Without a doubt, technology and the attendant uses of technology affect everyone one of us. In the world of information it is impossible to dissociate oneself from the use of technology. Further, the growth of the online and digital environments have ensured that technology is here to stay. Fifty years ago, we would have been hard-pressed to imagine ourselves pulling up a map or a book on a cellular phone or a personal digital assistant. Today, instant messaging and digital books are intrinsic to, if not our lives, to the lives of the children and teenagers who are immersed in the digital world.

Examining how academic libraries and geographic information science intersect must begin with a review of the information-based economy we now live in. Certainly, the convergence of computer technologies and communication technology in the past two decades has revolutionized business organizations in how they operate, especially with the rapid and efficient transmission of information on a global scale. This economic restructuring is driven by an information economy that continues to value knowledge work as commodity. Geospatial data and libraries have become important components of socioeconomic processes, political activities, and academic research within the emerging information economy.

The social milieu is another aspect of this new economic structure that cannot be ignored. What information is available affects *how* individuals participate, as well as *who* participates. Libraries offer digital services and digital resources to increase access to information to a wider community of online users, both in the physical library as well as to remote users. Chapters I and II attempt to place geospatial information science and library/information science in the context of the information economy and the digital infrastructure we know as the Internet.

To create a holistic view of the "landscape of information," librarians and geographers use classification schemes and measures relevant to the phenomena in the landscape under study. Analytic and statistical tools continue to enhance the use and display of spatial information, providing linkages to previously undiscovered and unknown relationships between factors. Research into the structure and interconnectedness of databases, data structures, and indexing methods have resulted in new data frameworks and typologies in both geographic and library information science. Both fields are still faced with challenges in the cataloging and mining of digital data. To do so will require us to address the challenges in describing geospatial works, such as quality and relevance of metadata, record formats, intellectual analysis of works, and search and retrieval frameworks to meet the different uses of geospatial information. These interrelated topics are integrated throughout Chapters II through VI.

Since the 1990s, digital geospatial data interoperability has been the target of major efforts by standardization bodies and the research community. With the rise of new digital models, applications, and networks, we suggest that libraries can better organize and increase the resource discovery of digital geospatial data. For some, a "geolibrary" that results from the intersection of the library and the spatial data infrastructure would extend the use of geographic information far beyond scope of a traditional map library. As remote access to digital resources increases, how libraries will address the information tasks performed by users is critical. First, users will have to create effective search criteria to gather materials, determine if the items they found actually can meet their information need, hone in on specific items that are "perfect," and then retrieve the actual item online. It sounds simple, however, in an online environment, access, discovery, and retrieval are more complicated. What will be important is that legacy materials, in print and superseded digital formats, are not lost to researchers and users, rather that they remain findable and usable through library catalogs and other digital frameworks. This is discussed in Chapters III, IV, and V as we delve into the design and development of databases, metadata frameworks, and standards to ensure interoperability and access.

To make things findable and retrievable requires compatibility between hardware and physical facilities; software applications and software; and network standards and transmission codes. It also requires that persons who produce and provide access to resources work within standards to ensure interoperability between *my* system and *your* system, our interfaces, and our respective products. Standards exist for cartography, hardware and software, telecommunications, and information technology at national and international levels. It also requires a common language to ensure availability, access, integration, and sharing of geographic information. How language is used in the discipline of geographic information science, as well as those disciplines using its methodologies and data, will have the user looking at semantics, which change as one moves across and within disciplines. It is exciting to see new forms of linguistic and semantic relationships emerge across fields and among researchers. Chapters IV, V, and VI address these issues from the perspective of cataloging, metadata, and ontology development.

For librarians, the opportunity to work with geospatial data and its users offers a world of exciting possibilities. There will be new services, new resources, new research collaborations, and possibly new business ventures, should libraries also become producers of data or other geographic information products. This means, of course, more sources, more options for sources, higher patron expectations, and, of course, more reliance on new technologies. Accordingly, the most remarkable opportunities and challenges emerge within academic libraries with regard to the incorporation of technology and services into our daily work lives. Both affect how libraries operate and how librarians keep up with ever-changing technology, user needs, and user expectations. It also affects the instruction and training we provide to our users, from the undergraduate student new to maps, much less complex data sets, to the researcher who is looking for assistance in managing a literature review or gathering background information on a topic that is inevitably squirreled away in thousands of places, none of them obvious. It also affects how we teach. Geospatial data requires us to rethink how questions are asked *and* answered. It also requires us to rethink how we teach users to navigate the foreign and highly mathematical territory of geospatial information. Chapters VI, VII, and VIII address these issues from the perspective of accessibility, reference services, and collection development.

Those of us who run libraries now have opportunities to support the scientific research infrastructure at our universities and colleges. GIS also allows us to increase our market of services and resources as geospatial data users are in every college, in every department, in every school, and throughout administrative units, such as facilities planning and building maintenance. It creates further opportunities for collaboration in large, distributed, and often international partnerships and consortia, as we house, share, and produce product. Most importantly, it allows us to keep current with innovative practices and technologies that can make the world a better place, or at least allow us to better understand it.

Education will also have to change to encompass GIS. Programs must be designed to best meet the information needs of library students and library professionals to acquire the necessary technical knowledge and computer skills to handle geospatial information. Even the most basic of GIS services requires significant investment in training programs and resources for librarians and staff. A more holistic, transdisciplinary approach to training and working with other disciplines will provide a richer, in-depth education for librarians with geospatial information. This is discussed further in Chapter IX.

What does the future hold for geographic information science and library/information science? Forecasting the future is always fraught with the possibility of being wrong. What we do suggest in Chapter X is that GIS applications will become easier to use and more intuitive for the user. As with computing, there will be accompanying increases in analytic capacity. Further, GIS software will become more embedded within current and emerging applications and technologies, much as word processing, spreadsheets, and databases are now found in computer "office" suites.

Can GIS help us assess, evaluate, and interpret trends of mutual influences across society? How will the digital divide, literacy, and economic disparities influence future applications and their use? Data integrity and privacy will continue to be a concern as data is misrepresented or misused. What will be the effects on social organizations, groups, and places affected by uses and outcomes of GIS, such as communities, business monopolies, or political hegemony?

For the four of us, all librarians in an academic environment, this book has allowed us to explore some of the larger, and smaller, issues that are at work in our interactions with students, researchers, community users, and other librarians. It has also permitted us to explore less obvious connections, such as social constructionism and the issues of trust in a distributed data-sharing environment. Most importantly, it has given us an opportunity to take questions that we have had with descriptive and semantic concerns and explore them more fully within the framework of geographic and library information sciences. After all, if semantics are enmeshed in philosophy and perception, a map and geospatial data are then also sites of critical inquiry.

It is our hope that the reader of this monograph will be intrigued, provoked, and reflective as he or she works their way through this attempt to tie geographic information science and library science, theory and practice, together in a coherent being, with applications in the real world for practitioners, students, educators, and those individuals fascinated with the world of maps and landscapes, real or imagined.

Chapter I

Geography and Librarianship

John Abresch, University of South Florida Libraries, USA

Ardis Hanson, University of South Florida Libraries, USA

Susan Heron, University of South Florida Libraries, USA

Peter Reehling, University of South Florida Libraries, USA

Introduction

There are many definitions of the study of geography. Most scholars define the discipline of geography as broadly concerned with the study of the earth's environment and interpretation of the different natural and man-made phenomena that occur across it. Geographers are interested in the interrelationships between phenomena across the earth's landscape in individual locations and across different regions. Though considered a social science by scholars, the field of geography incorporates methods and techniques that relate the study of geography to a variety of disciplines, such as anthropology, geology, ecology, political science, transportation, health, engineering, and library and information science. The multidisciplinary nature of geography provides opportunities for scholars in the discipline to apply these geographic concepts to many areas of study. The application of geographic techniques to new areas of study has provided the impetus for proposing new hypotheses and testing theories in different disciplines. The research has advanced geographic thought beyond established paradigms, as scholars use computer applications and remotely sensed data to redefine concepts of geographic space and to study the phenomena that occur in them.

Libraries have been an important facilitator in the development of techniques and methodologies in the discipline of geography, cartography, and related subjects. In the United States, for example, geographically themed collections have been integral to federal, academic, and public libraries, such as the Library of Congress, the Library of the American Geographical Society at the University of Wisconsin-Milwaukee, and the New York Public Library. These collections include maps, aerial photos, gazetteers, globes, three-dimensional models, atlases, remote-sensing data, geospatial data, and other materials that describe the spatial, geographic, geologic, and chronological aspects of the Earth.

Librarians have important roles in facilitating academic and scientific research and instruction. They apply their knowledge and skills in the collecting, archiving, and cataloging of geographic materials and participate in the scholarly publication process associated with geographic thought. This volume is intended to provide the librarian in academic library settings a guide to identify concepts and accepted guidelines in collecting, cataloging, and making accessible geographic data. The emphasis will be on digital geographic data or geospatial data since contemporary methodologies in geographic analysis are mostly computer based. The volume also examines other issues such as educational, user, and future issues.

Geographic Study and Maps

During the years following the Second World War, librarians in the United States adapted their information management techniques and operational frameworks to deal with the increased production and use of cartographic materials in both the public and private sector. An estimated 60,000 to 100,000 map sheets were being produced annually on a global scale and, in the United States, over 80% of map production was being generated by federal, state, and local agencies (Ristow, 1980). Many of the maps that were being produced were being acquired by academic libraries and public libraries. Published guidelines reflected current technology, such as the ubiquitous use of metal flat files used for the safekeeping of maps.

The early adoption of computer technology in librarianship was mirrored in geography and cartography with the advent of automated cartographic systems. Geographers and cartographers could quickly convert and transform map data into different projections and facilitate new methods and techniques in spatial analysis. The production of maps and cartographic materials also increased with new capabilities of the automated cartographic systems. The cartographic automation of the 1960s evolved into today's contemporary geographic information systems that are in widespread use throughout the private and public sectors and in academia. Indicative of the interdisciplinary nature of geographic study, many universities and colleges in the United States have experienced growth in the application of spatial analysis and geographic information systems. Digital geospatial data is found in many departments outside of geography, geology, and related disciplines. However, the challenges of managing and providing access to the spatial information, whether analyzed in a geographic information system or computer-generated map, have also increased since the 1960s. The discussion that follows introduces the relationships between geography maps, spatial thought, and map librarianship.

Maps and Map Libraries

The development of map librarianship paralleled the establishment of modern centers of geographic thought by associations in Europe and the United States, such as the Association of American Geographers and the British Geographical Society. Mid-19th century concepts of geography were characterized by description of locations by detailed narratives and by graphical means.

A key component of research in the discipline of geography is the depiction of natural or manmade features of a particular surface area of the earth on flat sheets of paper or related material, otherwise known as maps. Reflective of the diverse nature of geography, maps may depict a variety of characteristics or themes, such as transportation routes, population distributions, income levels, and even climatic regions, or depict the topography of an area, for example, the natural shape and form of the landscape. Often, maps would have differing levels of data graphically depicted across their surfaces to illustrate the interrelationships shown on the maps. A topographic map could depict the natural contours of a particular area and the location of other features, such as cities, towns, or political divisions.

An essential component in the creation of maps is the concept of scale. Scale refers to the amount of abstraction or variation in the map as it depicts the data relative to the landscape it is based upon (DeMars, 1997). Most maps are either large- or small-scale in their composition. Large-scale maps can show many features of a particular landscape, using a 1:1 ratio, that is, 1 inch on the map represents 1 mile of area. Small-scale maps represent much less detail using a 1:10 ratio, that is, 1 inch on a map represents 10 miles (DeMars, 1997). Nonetheless, there can be variation of data represented in maps. The individual mapmaker will choose the data elements to be depicted as well as the accuracy of their illustration. Mapmaker or cartographer A may emphasize the accuracy in the depiction of data set, while mapmaker or cartographer B may emphasize the accuracy in the depiction of data set Y, depending upon the purpose of the map (DeMars, 1997).

The advent of more precise measuring and surveying instruments in the 19th century increased the map's importance as a spatial research tool. Advances in instrumentation allowed the production of more accurate maps than in previous centuries. The rapid economic and political expansion of the United States was also a significant factor in the production and use of maps, for example, government-sponsored surveys of the American West, such as the United States and Mexican boundary survey, Pacific Railroad surveys, and the Northwest boundary survey, provided new data for maps. With maps being produced in ever-increasing quantities, additional collections of national scope were being compiled at the Newberry Library in Chicago and at Harvard University in Cambridge. These map collections displayed a wide array of socioeconomic, political, and physical information about the various landscapes being depicted, and were often arranged according by geographic area.

By the 1900s, maps were regarded as essential for geographic study by graduate departments at universities in the United States. Contemporary approaches to the study of geography were characterized by rich description of regions and advanced studies were regional in nature. Since maps portrayed different aspects of particular regions, comparison of maps could detect new patterns in, or allow interpretation of, the information depicted on the maps. Recognizing the need for more efficient organization of library cartographic collections, librarians and geographers began issuing classification rules and general advice about the cataloging and disposition of maps. Early attempts at publishing guidelines argued the distinction between the cataloging of maps and that of books (Ristow, 1980). Cataloging rules for maps would need to emphasize concepts such as geographic region as well as the thematic and physical characteristics of the map itself. By the 1940s, a number of pamphlets had been published that issued advice on the processing and handling of maps and the preferred methods for storage, such as in flat drawers in cabinets. A reassessment of map use in geographic scholarship began in the period after World War II, with the publication of *The Look of Maps* (Robinson, 1952). Robinson asserted that a scientific approach to cartography would base map production on a map's actual use rather than on aesthetics (Davenport & Prusak, 1997). He also felt that several steps in the production of maps, such as color, lettering, and symbols, could use objective guidelines. The consideration of objective guidelines in map symbolization and the map production process widened the scope of cartographic research, incorporating research from other disciplines, such as psychological research in human perception and cognition. Robinson's research in cartography was the beginning of an objective approach in map production and its relationship to scholarship in geography.

Communicating Geospatial Information

In the following decades, scholars attempted to apply a more rigorous scientific method to cartography and related geographic research by proposing new communication paradigms for cartography. In the 1960s, the new paradigm was "advocating a process" (Davenport & Prusak, 1997). Map making was essentially a process of communicating spatial data from a particular part of reality to the end user. Within the model, cartography was comprised of both internal and external filters. Internal filters were those of the mapmaker and included personal knowledge, experiences, and objectives. External objectives would include reasons for making a map, or even the process of map design such as classification, symbolization, and production. Use of the map would depend on the perceptual and cognitive abilities of the map user, and even the setting from which the map was being referenced, most likely in a library environment (Arms, 2001).

The communication paradigm in cartography was based on research into behavioral psychology. While the attempt to define the map production process within a communication model was an effort to apply objective analysis to cartography, the approach was criticized by scholars as being restrictive, the approach was too narrow in its consideration of the scope of a map's possible function (MacEachren, Buttenfield, et al., 1992). Application of the paradigm to a map assumed a specific message was being communicated by the cartographer. For example, regardless of the type of map (topographic or thematic), if that map's particular function is to simply depict location, the map has no real message to communicate (Davenport & Prusak, 1997). Further, the spatial data on the map could be interpreted in any given number of ways by different individuals. Davenport and Prusak (1997) suggest that the attempt to codify map production and interpretation within a narrowly defined process also did not take into account the use of technology that would be used for spatial analysis and data visualization. Through the use of technology, such as spatial data visualization software, the researcher could create maps for any stage of a research process. The map image could be part of an early stage of research and not reflect the final conclusions of a case study. One aspect of the communication paradigm that MacEachren (1995) considers applicable to cartography is in investigating the "perceptual and cognitive processes involved in both map reading and spatial information processing" (p. 8). The research approach would help determine the limitations of map symbolization.

Other scholars integrate map symbolization and other graphics into a type of cartographic language. Comparing maps and other graphics with words and numbers, Borchert (1987) defined maps as "diagrams of geographic systems and their evolution ... conveying cartographically communicated theories about global or regional geographical systems of resources and settlement" (p. 388) and the cartographic language of maps as being "a visual statement of geographic analysis, portraying various geographic phenomena based on spatial and temporal attributes" (p. 388).

In recent years, the integration of computer imaging technology into the cartographic process has facilitated the production of multiple geographic perspectives and maps (Borchert, 1987). Crampton (2001) uses the term *geographic visualization* for maps, graphics, and images that portray visible spatial relationships between data and to the "added capabilities of interactive mapping software, which can create transform, rotate, and skew map data into different projections" (p. 244). Crampton agrees with MacEachren in that cartographic visualization is based mostly on human ability to develop mental representations of the landscape and to arrange what is present into patterns.

In creating new graphic visualizations and cartographic representations with mapping software, cartographers and geographers are building upon a rich tradition of geographic analysis using maps found in library collections. Before the advent of computerized mapping software, geographers would collaborate with map librarians in determining better cataloging and classification procedures for paper map collections, thereby improving access to geographic information. Librarians can use the same collaborative approach with geographers in managing geospatial data in the electronic medium of geographic information systems.

Components of the digital economy, such as computer hardware and telecommunication systems, assisted cartographers and geographers in developing electronic mapping and digital geospatial data. Similar components of data communication infrastructure inspired the emergence of digital libraries (Brodeur, Bédard, Edwards, & Moulin, 2003). Arms (2001) defined digital libraries as "managed collections of information, with associated services, where information is stored in digital formats and accessible over a network" (p. 2). By applying the themes of efficient accessibility and classification to geospatial information in the online environment, librarians can be viable contributors to the geographic research process in both academic settings and in the private sector.

The development of new forms of geographic representation as a result of the use of new computer technologies and analytic techniques is indicative of the challenges faced by librarians in academic research environments. Traditionally, research librarians have been concerned with providing support to a community of users by focusing on services such as collection management, information literacy, and access, based on a static collection housed in a specific location. The migration of materials to digital formats and the prevalence of distributed collections accessible over complex telecommunications networks have altered the organization and physical structure of libraries. The production and distribution of information resources in digital formats have also introduced a number of social and economic factors that can affect traditional modes of the delivery of information in academic libraries. The following section discusses the emergence of what some scholars define as geographic information science, including elaboration on some of the disciplines' research foci. Discussion of the research areas of geographic information science not only defines the parameters of the research areas, but also identifies concepts and agendas that can be integrated into research library services.

Libraries: Facilitating Research in Geographic Information Science

In supporting research and instruction endeavors, academic libraries have a long history of managing information resources. Although academic libraries may differ in their missions by serving different user communities, their fundamental functions are similar. Most academic library functions include selecting, collecting, and cataloging information sources to ensure access to a community of users. Libraries also maintain an archival function, preserving materials for future users. Libraries serve important social functions to their communities of users. Ram, Park, and Lee (1999) suggest that there are three roles for libraries in teaching and learning: a practical role in sharing resources, a cultural role in preserving and organizing information resources, and a social/intellectual role by bringing together people and ideas. These traditional roles still remain primary characteristics of digital libraries (Ram et al., 1999). Digital libraries can provide enhanced and value-added services that traditional libraries could not (Chowdhury & Chowdhury, 2003). A digital library can be available 24 hours a day via a network that allows users to access materials no matter where they live. In this sense, digital libraries are global. Digital libraries are also dynamic in that materials can be added and updated by many contributors from any place at any time, although quality control procedures for resource collection enforced by the digital library may limit the dynamics of digital libraries (Chowdhury, 2002). Characteristics of a digital library include an open architecture, distributed information repositories, multiple formats, metadata and data, seamless and transparent access to resources, interoperability, user-friendly easy-to-use interfaces and query facilities, and service orientation to both information seekers and providers (Ram et al., 1999).

In recent years, the use of geographic information systems has facilitated the study of spatial phenomena in both human and natural environments in a number of disciplines in the sciences and social sciences. Some scholars view the varied applications of GIS in research as leading to a new direction of scientific inquiry or geographic information science. In 1994, representatives of 34 U.S. universities and other research organizations met in Boulder, Colorado and decided to establish an organization "dedicated to the development and use of theories, methods, technology, and data for understanding geographic processes, relationships, and pattern" (Mark, 1999b, p.15). The group referred themselves as the University Consortium for Geographic Information Science (UCGIS). In defining their mission and goals, they prepared a framework for the definition of the field that stated:

UCGIS Mission

- To serve as an effective, unified voice for the geographic information science research community;
- To foster multidisciplinary research and education; and
- To promote the informed and responsible use of geographic information science and geographic analysis for the benefit of society.

Goals

Unify:

Provide ongoing research priorities for advancing theory and methods in geographic information science.

Assess the current and potential contributions of GIS to national scientific and public policy issues.

Facilitate:

Expand and strengthen geographic information science education at all levels.

Provide the organizational infrastructure to foster collaborative interdisciplinary research in geographic information science.

Benefit society

Promote the ethical use of and access to geographic information.

Foster geographic information science and analysis in support of national needs.

Source: http://www.ucgis.org/

A more definitive of geographic information science arose out of a workshop sponsored by the National Science Foundation:

"Geographic Information Science (GIScience) is the basic research field that seeks to redefine geographic concepts and their use in the context of geographic information systems. GIScience also examines the impacts of GIS on individuals and society, and the influences of society on GIS. GIScience re-examines some of the most fundamental themes in traditional spatially oriented fields such as geography, cartography, and geodesy, while incorporating more recent developments in cognitive and information science. It also overlaps with and draws from more specialized research fields such as computer science, statistics, mathematics, and psychology,

and contributes to progress in those fields. It supports research in political science and anthropology, and draws on those fields in studies of geographic information and society. (Mark, 1999a, p. 7)

Although, the traditional academic community did not fully endorse the definition of geographic information science, academics working in research centers, such as the National Center for Geographic Information and Analysis (NCGIA) and the University Consortium for Geographic Information Science (UCGIS), did work to identify research themes and priorities that helped geographic information science develop as a discipline. In 1996, the UCGIS established 10 research priorities for the field of research (Mark, 2003), p. 7):

1.   Spatial Data Acquisition and Integration
2.   Distributed Computing
3.   Extensions to Geographic Representation
4.   Cognition of Geographic Information
5.   Interoperability of Geographic Information
6.   Scale
7.   Spatial Analysis in a GIS Environment
8.   The Future of the Spatial Information Infrastructure
9.   Uncertainty in Spatial Data and GIS-based Analyses
10. GIS and Society

From these research priorities, a discussion of geographic information research within the emerging discipline of geographic information science is more than discussing different applications of geographic information systems. Within the discussion of the different applications, one can draw relations between geographic information science and library and information science. One function of data that is important to both disciplines involves ontology and representation (Guarino, 1998; Schuurman, 2006; Schuurman & Leszczynski, 2006).

Ontology deals with what exists and what may possibly exist. This part of geographic information science looks at geographic concepts that are utilized by scientists in their research. Ontology has been described as "seeking to provide a consistent formal theory of tokens (instances) and types (kinds) in the real world, their relationships, and the processes that modify them (Mark, 2003, p. 8). Ontology in both information science and knowledge representation refers to "the specifications of the conceptualizations employed by different groups of users to domains of entities of different types" that " involves the laying down of a conceptually tractable taxonomy of the objects in the given domain of a sort that can support automatic translation from one data context to another. The representations are types or kinds in the digital domain, to be instantiated through data to become digital tokens (instances) that correspond to geographic things in reality" (Mark, 2003, p. 8). This representation is also referred to as data modeling. The different research areas in representation research are also relevant to the discipline of library and information science (Frank, 2003).

Concerning spatial data, efficient indexing of multidimensional data is an important problem in database research in computer science (Luo, Liu, Wang, Wang, & Xu, 2004; Wang, Wang, Luo, Wang, & Xu, 2004; Wu, Chen, & Yu, 2006). The study of cognitive models of geographic phenomena is another area of research that is important to geographic information science (Barkowsky, 2001; Jefferies & Yeap, 2001; Solomon, 2002; Steinhauer, Wiese, Freksa, & Barkowsky, 2001). This research area involves the "study of human perception, learning, memory, reasoning, and communication of and about geographic phenomena" (Mark, 2003, p. 10). The research area has been characterized by the study of human computer interactions with the intent to gain more understanding about geographic ontology. The research also benefits from related research in spatial cognition and perception. Another aspect of the study of human computer interactions is with the study of the design of user interfaces. This has relevance for user issues within the discipline of geographic information science. Other areas of important research areas within GIS include acquisition of geographic data, quality of geographic information, spatial analysis, and the social aspects of geographic information.

Library and Information Science

According to Saracevic (1999, p. 1052), "Information science has three general characteristics that are the leitmotif of its evolution and existence." However, these are shared characteristics, with attendant problems, across many fields.

First, information science is interdisciplinary in nature; however, the relations with various disciplines are changing. The interdisciplinary evolution is far from over.

Second, information science is inexorably connected to information technology. A technological imperative is compelling and constraining the evolution of information science, as is the evolution of a number of other fields, and moreover, of the information society as a whole.

Third, information science is, with many other fields, an active participant in the evolution of the information society. Information science has a strong social and human dimension, above and beyond technology (Saracevic, 1999, p. 1072)

This third piece is particularly relevant to libraries and library practice. Libraries have been described as hubs of social and intellectual interactions in communities and organizations as far back as the library in Alexandria, Egypt. In addition to providing information in the form of books, journals, videos, CDs, and other media, the library serves an important community building purpose. Since librarians are often closely integrated into the social fabric of their communities, libraries often serve as information hubs and librarians become "information networkers" (Robertson, 2000). Robertson (2000) identifies a number of community building aspects that libraries can facilitate:

- **The places:** It is important to have an environment where structure and meaning overlap. For example, location and topic are related in libraries.
- **The perspectives and identities:** It is important to have the ability to see what other people are doing. For example, in a library it matters what sections of a library other people are in, how often they do certain things, and what reading materials they are selecting.
- **The interaction opportunities:** Meeting places such as conference rooms, chat areas, hallways, and so forth, make it possible for people who notice each other to interact.
- **The facilitators:** Individuals such as library staff members notice what people are doing over time and can become community facilitators. (Robertson, 2000, p. 247)

Besides making available the physical spaces needed for community building, libraries offer the expertise and organizational skills to manage and interpret information that comprise globalization and the information economy. Libraries are well positioned to provide social inclusion to the information economy to groups that have been marginalized by the digital divide (Hull, 2003). Public libraries, for example, are substantially reducing disparities in access to computers and the Internet (Heuertz, Gordon, Moore, & Gordon, 2002). Patrons use public library computers for a wide array of purposes, ranging from business and education to entertainment. By providing classes and one-on-one assistance, libraries help their patrons gain the skills necessary to use technology and to obtain desired information. Geographic information is just one type of information that librarians provide instruction on how to use and how to find through both traditional and interactive tools, such as maps, globes, and gazetteers.

Libraries and geographic information are ubiquitous in everyday life. The purpose of this volume is to provide how these two areas are woven inextricably together, especially with the development of the online phenomenon known as the Internet and the emergence of the information economy.

Geospatial Data and the Information Economy

Chapter II of this volume outlines the framework of the information economy, and elaborates on the role of geographic information systems in facilitating research and instruction using digital geospatial data. The convergence of computer and communication technologies in the past two decades has caused private sector firms to reorganize their functional operations around the rapid and efficient transmission of information. The economic restructuring has been identified as comprising a new information economy and geospatial data has become an important part of socio economic processes, political activities, and academic research within this economy (Boxall, 2002; Carlsson, 2004; Grubesic & Murray, 2005; Jorgenson & Vu, 2005; Larsgaard, 1998).

Although the framework of the information economy may have been built by technological innovations and capital investments, cultural and political factors about the social milieu in which information is processed affect how individuals participate in the economy. Clearly, the emergence the new information economy, characterized by a convergence of telecommunication functions, is creating a class of singular information users. Digital libraries can offer a more varied informational experience to the community of online users.

Concepts in contemporary geographic research are built on earlier concepts defined by Hyland, Robinson, and others who felt that the geographer would identify inter-relationships and variations of the phenomena on the earth's surface to determine a holistic perspective of the landscape (Hyland, 2002; Robinson, 1952; Robinson & Petchenick, 1975; Tobler, 1970). Just as librarians use descriptive procedures (classification schemes and cataloging rules), geographers use classification schemes and measures empirically relevant to the phenomena in the landscape under study. Hartshorne, for example, tended to organize similar phenomena in either physical or human regions (Hartshorne, 1959). Later geographers also advocated the use of empirical analysis and statistical procedures in the analysis of phenomena across the landscape (Berry, Marble, & Joint Comp, 1968; Bunge, 1966; Carlsson, 2004; Kainz, 2004).

Research in the spatial display of information was modeled after the graphic display of map referenced data (Kainz, 2004), with much of the pioneering research in digital spatial data performed at the Laboratory for Computer Graphics and Spatial Analysis of the Harvard Graduate School of Design and the Geographic Information Systems Laboratory at SUNY Buffalo, among others (Lutz, Riedemann, & Probst, 2003). Kainz (2004) notes how conceptual development in the virtual representation of spatial data was made with the introduction of typology into geographic research involving mapping software: "the search for a stable and consistent representation of map structures of map data led to the use of topology and related graph theory was effective in constructing two dimensional data representations" (p. 16). Afterwards, research in geographic information systems expanded into areas such as spatial databases, spatial data structures, and indexing methods (Lutz et al., 2003). Even with the advent of geographic information systems, geographers are still presented with the challenge of cataloging their data and in devising effective data mining methods for research purposes.

Spatial Databases and Data Infrastructures

Chapter III discusses the information network of the NSDI and its relationship to digital libraries and related geoportals. Three significant factors have impacted the pervasiveness of geographic information systems: (1) personal computing and the World Wide Web, (2) the architecture of geographic information systems and its data processing capabilities, and (3) the rapid increase in applications throughout the public and private sectors. We suggest that the new Internet-based communication applications and digital libraries can be used to organize and ease the data mining of digital geospatial data. As Brodeur et al. (2003) have indicated, digital geospatial data interoperability has been the target of major efforts by standardization bodies and the research community since the 1990s. The most significant interoperability issue is the computer architecture, often legacy or homegrown architectures, upon which the system is based (Strasser, 1995).

Researchers also envision an intersection of the concepts of the digital library and that of the spatial data infrastructure in a sort of "geolibrary" (Chiles & Delfiner, 1999), a library filled with "georeferenced information that can have a geographic footprint." Including multimedia, images, and music that could be assigned a location attribute, geolibraries would thus extend beyond the scope of a traditional map library (Nogueras-Iso, Zarazaga-Soria, & Muro-Medrano, 2005, p. 6), and provide new services and resources for users to discover.

Describing Geospatial Information

Chapter IV provides an overview of current academic cataloging principles, issues in handling evolving formats, and challenges for academic catalogs and the issues involved in adequately describing geospatial works. Issues surrounding the quality and relevance of metadata (bibliographic access) become more critical in online venues, especially with geospatial data. It also addresses the kind of bibliographic records or metadata that will be required to meet the different uses of geospatial information and user needs and the organization and structure of these bibliographic data or metadata for intellectual and physical access to the works.

Increasingly, libraries are seen more as remote resources rather than as walk-in facilities. To meet this need, libraries must have sustainable systems of access and databases and durable objects that fulfill the three R's for users: reliability, redundancy, and replication of results. There are four generic information tasks users perform: "finding materials that correspond to the user's stated search criteria (e.g., in the context of a search for all documents on a given subject, or a search for a recording issued under a particular title); using the data retrieved to identify an entity (e.g., to confirm that the document, described in a record corresponds to the document sought by the user, or to distinguish between two texts or recordings that have the same title); using the data to select an entity that is appropriate to the user's needs (e.g., to select a text in a language the user understands, or to choose a version of a computer program that is compatible with the hardware and operating system available to the user); [and] using the data in order to acquire or obtain access to the entity described (e.g., to place a purchase order for a publication, to submit a request for the loan of a copy of a book in a library's collection, or to access online an electronic document stored on a remote computer)" (IFLA Study Group on the Functional Requirements for Bibliographic Records, 1998). Cataloging and classification provide information for resource discovery and selection of the appropriate work. Further, the rich legacy of maps can be connected to the digital world of data through the use of online catalogs and other digital frameworks.

Describing GIS Data Standards

Chapter V addresses the new methods of description developed to allow users, often with very different information needs, to find and retrieve relevant resources across different platforms and software systems. The success of a national spatial data infrastructure (addressed in Chapter III) depends on the development of a series of standards for that infrastructure. Infrastructure components encompass a variety of elements: hardware and physical facilities; software applications and software; and network standards and transmission codes (Hanson, 2006). When reviewing standards for geospatial data, one must look at standards for cartography, hardware and software, telecommunications, and information technology at national and international levels. This chapter will also address standards in geospatial data, interoperability and transferability, mark-up languages, and the development of the federal metadata standard for geospatial information. With the increased use of digital computation, data, information, and networks to replace and extend traditional research, description for digital data, applications, and services becomes increasingly more complex. User access can become a major issue in the provision of services, from both a library and data provider perspective.

Access Issues in Using Geospatial Data

With the creation of the Internet and the continued evolution of technologies in GIS, networking, and knowledge management, access to geospatial information is a critical component of research and practice. As discussed in Chapter V, standards increase the "understanding and usage of geographic information," the "availability, access, integration, and sharing of geographic information," and the "efficient, effective, and economic use of digital geographic information and associated hardware and software systems" (Albrecht, 1999, p. 151). All three are affected by different aspects of interoperability. Data interoperability is defined as the ability to access multiple, heterogeneous geoprocessing environments, either local or remote, by means of a single unchanging software interface. Syntactical interoperability assures that there is a technical connection, that is, that the data can be transferred between systems. Semantic interoperability assures that the content is understood in the same way in both systems. Research in geospatial interoperability must take into account not only data or structural issues but also semantics, which are enmeshed in philosophy and perception. Semantic interoperability requires new descriptive classifications to identify functionalities, creating new ontologies for GIS. Ontologies not only support query disambiguation and query term expansions, they allow the creation of the spatial indexes to support the search and the annotation of Web resources, Web documents, and geographic data sets. Other issues emerge in the sharing of data across organizations. Trust among participants as to data quality, integrity, and intellectual property is essential. The use of GIS needs to mitigate organizational and political factors that may have serious implications for end users in their use of geospatial information.

Reference Services, GIS, and Academic Libraries

Chapter VII will examine the provision of reference services in academic libraries using geospatial information. Since there are varying levels of GIS services that libraries offer, reference staff will require the skills and knowledge to provide service from running a full GIS lab to the provision of static maps, which may actually be created by the librarian in response to a specific reference or research question. Reference librarians field questions from undergraduates, graduate students, teaching faculty, and research faculty. In many areas, the academic library supports a wide range of community users as well, with varying degrees of knowledge and skills in the use of geographic information. More sources, more options for sources, higher patron expectations, and, of course, more reliance on new technologies creates a constantly changing environment.

This chapter addresses some of the issues in the provision of services across different academic libraries and provides a look at the skills necessary to conduct appropriate reference interviews and to instruct the patron in how to access geospatial information. A number of issues are still of concern in offering reference GIS services, such as complex question differentiation, evaluation of GIS hardware and software, and pedagogical skills necessary to teach complicated GIS systems to a heterogeneous group of users. In addition to establishing standards for GIS centers and services to reduce variance in user needs and expectations across academic libraries (Badurak, 2000), reference librarians have several other key opportunities in collection development, instruction on accessing spatial information and software tools, and facilitating user knowledge of GIS concepts and applications (Abbott & Argentati, 1995).

Collection Management Issues in GIS

Chapter VIII outlines collection management issues in academic libraries. Librarians can rely on an extensive tradition of collecting geographic materials in both public and academic librarians as described by Ristow (1980) and Larsgaard (1998) for outlining strategies to build new collections. Managing a digital geospatial data collection in an academic library is a process that involves decision making in a dynamic and changing environment that has many challenges, ranging from the academic to the technological. Librarians face new demands in the acquisition, cataloging, and disposition of digital spatial data as their library collections migrate from a primarily print format to a primarily digital format. Other challenges include building a digital geographic data collection from scratch, acquiring computer software and hardware systems, managing and displaying the geographic information, and training in the use of the technology and the materials.

Strategies for Integrating GIS in Library and Information Science Education

Chapter IX examines issues in designing educational programs for library students and for library professionals to best meet *their* information needs as they acquire the technical knowledge and computer skills necessary to use GIS software in addition to other library services. It can be daunting considering the range of services that may be required. A library professional may be collaborating and assisting academic departments in using GIS software, implementing and operating a GIS service, facilitating access to digital geospatial data collections, framing the research question, or providing a range of research support. A tiered approach in delivering GIS information services is considered best, depending on need, from the most basic to more intense research support (Martin, James, James, & James, 1993). Even the most basic of GIS services requires significant investment in training programs and resources for librarians and staff involved in the service delivery to acquire GIS related skills (Hartshorne, 1959; Martin et al., 1993). With the establishment of geolibraries and data infrastructures, there will be changes in the field of map librarianship. Librarians will need to take a more holistic approach to training and to work with other disciplines in creating development opportunities for librarians with GIS technologies (Grubesic & Murray, 2005).

Prognostication: Trends in GIS, Geospatial Data, and Academic Libraries Services

Chapter X explores future trends in using geospatial data. The literature clearly suggests that GIS development will continue to become easier to use, more intuitive for the user, increase in analytic capacity, and, without a doubt, become more embedded within current and emerging applications and technologies. However, in addition to the technology, there are larger societal issues that need to be reviewed. For example, how will we assess, evaluate, and interpret trends of mutual influences between GIS and society? The digital divide, literacy, and economic disparities will influence what is developed for use in GIS and how it is used. Data integrity and privacy will continue to be a concern as the effects of worms, Trojan horses, and other hacker activity can cause data to be misrepresented or used for purposes not conducive to the well-being and safety of societies. What will be the effects on social organizations, groups, and places affected by uses and outcomes of GIS, such as communities, business monopolies, or political hegemony? Finally, Chapter X will also look at some of the research and practice trends in librarianship and how those connect with the larger discussion of GIS and society.

It is our hope that the reader of this monograph will be intrigued, provoked, and reflective as he or she works their way through this attempt to tie geographic information science and library science, theory and practice, together in a coherent being, with applications in the real world for practitioners, students, educators, and those individuals fascinated with the world of maps and landscapes, real or imagined.

References

Abbott, L. T., & Argentati, C. D. (1995). GIS: A new component of public services. *The Journal of Academic Librarianship, 21*(4), 251-256.

Albrecht, J. (1999). Towards interoperable geo-information standards: A comparison of reference models for geo-spatial information. *The Annals of Regional Science, 33*(2), 151-169.

Arms, W. Y. (2001). *Digital libraries*. Cambridge, MA: MIT Press.

Badurak, C. (2000). Managing GIS in academic libraries. *WAML Information Bulletin, 31*(2), 110-114.

Barkowsky, T. (2001). Mental processing of geographic knowledge. In D. R. Montello (Ed.), *Spatial information theory: Foundations of geographic information science: International Conference, COSIT 2001 Morro Bay, CA, USA, September 19-23, 2001, Proceedings* (pp. 371-386). Berlin, Germany: Springer-Verlag GmbH.

Berry, B. J. L., Marble, D. F., & Joint Comp . (1968). *Spatial analysis: A reader in statistical geography*. Englewood Cliffs, N.J: Prentice-Hall.

Borchert, J. R. (1987). Maps, geography, and geographers. *The Professional Geographer, 39*(4), 387-389.

Boxall, J. (2002). Geolibraries, the global spatial data infrastructure and digital Earth: A time for map librarians to reflect upon the moonshot. *INSPEL, 36*(1), 1-21.

Brodeur, J., Bédard, Y., Edwards, G., & Moulin, B. (2003). Revisiting the concept of geospatial data interoperability within the scope of human communication processes. *Transactions in GIS, 7*(2), 243-265.

Bunge, W. (1966). *Theoretical geography*. Lund, Sweden: CWK Gleerup, Publishers.

Carlsson, B. (2004). The digital economy: What is new and what is not? *Structural Change and Economic Dynamics, 15*(3), 245-264. [Special issue: New and old economy: The role of ICT in structural change and economic dynamics].

Chilès, J.-P., & Delfiner, P. (1999). *Geostatistics: Modeling spatial uncertainty*. New York, NY: Wiley.

Chowdhury, G. G. (2002). Digital libraries and reference services: Present and future. *Journal of Documentation, 58*(3), 258-283 .

Chowdhury, G. G., & Chowdhury, S. (2003). *Introduction to digital libraries*. London: Facet.

Crampton, J. (2001). Maps as social constructions: Power, comunication, and visualization. *Progress in Human Geography 25*(2), 235-252.

Davenport, T. H., & Prusak, L. (1997). *Information ecology: Mastering the information and knowledge environment*. London: Oxford University Press.

DeMars, M.N. (1997). *Fundamentals of geographic information systems*. New York, NY: Wiley & Sons.

Frank, A. U. (2003). Ontology for spatio-temporal databases. In G. Goos, J. Hartmanis, & J. van Leeuwen (Eds.), *Spatio-temporal databases: The CHOROCHRONOS approach* (pp. 9-77). Berlin, Germany: Springer-Verlag GmbH.

Grubesic, T. H., & Murray, A. T. (2005). Geographies of imperfection in telecommunication analysis. *Telecommunications Policy, 29*(1), 69-94.

Guarino, N. (1998). Formal ontology and information systems. In N. Guarino (Ed.), *Formal ontology in information systems* (pp. 3-15). Amsterdam, Netherlands: IOS Press.

Hanson, A. (2006). Organization and access to electronic resources. In V. L. Gregory *Selecting and managing electronic resources* (rev. ed.) (pp. 45-64). New York , NY: Neal-Schuman.

Hartshorne, R. (1959). *Perspective on the nature of geography* (Monograph series of the Association of American Geographers No. 1). Chicago, IL: Published for the Association of American Geographers by Rand McNally.

Heuertz, L., Gordon, A., Moore, E., & Gordon, M. (2002). *Public libraries and the digital divide: How libraries help: A report to the Bill & Melinda Gates Foundation.* Seattle: University of Washington, Evans School of Public Affairs. Retrieved December 2006, from http://www.gatesfoundation.org/NR/Downloads/libraries/eval_docs/pdf/PLDD_%20020729.pdf

Hull, B. (2003). ICT and social exclusion: The role of libraries. *Telematics and Informatics, 20,* 131-142.

Hyland, N. C. (2002). GIS and data sharing in libraries: Considerations for digital libraries. *INSPEL, 36*(3), 207-215.

IFLA Study Group on the Functional Requirements for Bibliographic Records. (1998). *Functional requirements for bibliographic records: Final Report: IFLA Study Group on the Functional Requirements for Bibliographic Records.* Frankfurt, Germany: Deutsche Bibliothek: Frankfurt am Main. Retrieved 09/23/2001, from http://www.ifla.org/VII/s13/frbr/frbr.pdf

Jefferies, M. E., & Yeap, W. K. (2001). The utility of global representations in a cognitive map. In D. R. Montello (Ed.), *Spatial information theory: Foundations of geographic information science: International Conference, COSIT 2001 Morro Bay, CA, USA, September 19-23, 2001, Proceedings* (pp. 233-247). Berlin, Germany: Springer-Verlag GmbH.

Jorgenson, D. W., & Vu, K. (2005). Information technology and the world economy. *Scandinavian Journal of Economics, 107*(4), 631-650.

Kainz, W. (2004). *Geographic information science* (Version 2.0 ed.). Vienna: Institut für Geographie und RegionalforschungUniversität Wien Universitätsstraße. Retrieved from http://www.geografie.webzdarma.cz/GIS-skriptum.pdf

Larsgaard, M. L. (1998). *Map librarianship: An introduction* (3rd ed.). Englewood, CO: Libraries Unlimited.

Luo, Y., Liu, X., Wang, X., Wang, W., & Xu, Z. (2004). Design open sharing framework for spatial information in semantic Web. In H. Jin, Y. Pan, N. Xiao, & J.

Sun (Eds.), *Grid and cooperative computing: GCC 2004 third international conference, Wuhan, China, October 21-24, 2004, proceedings* (pp. 145-152). Berlin, Germany: Springer-Verlag GmbH.

Lutz, M., Riedemann, C. R., & Probst, F. (2003). A classification framework for approaches to achieving semantic interoperability between GI Web services. In W. Kuhn, M. F. Worboys, & S. Timpf (Eds.), *Spatial information theory* (pp. 186-203). Berlin,Germany: Springer.

MacEachren, A.M. (1995). *How maps work representations, visualization, and design.* New York: Guilford Press.

MacEachren, A. M., Buttenfield, B. P., et. al. (1992). Visualization. In R. F. Abler, M. G. Marcus, & J. M. Olson *Geography's inner worlds: Pervasive themes in contemporary American geography* (1st ed., pp. 99-137). New Brunswick, N.J.: Rutgers University Press.

Mark, D. M. (1999a). *Geographic information science: Critical issues in an emerging cross-disciplinary research domain: Workshop report* . Workshop on Geographic Information Science and Geospatial Activities at NSF January 14-15, 1999. Retrieved December 2006, from http://www.geog.buffalo.edu/ncgia/workshopreport.html

Mark, D. M. (1999b). *Geographic information science: Critical issues in an emerging cross-disciplinary research domain.* Paper presented at the meeting of the NSF workshop to assess the needs for basic research in Geographic Information Science and Technology, January 14-15, 1999. Retrieved December 2006, from http://www.ncgia.buffalo.edu/GIScienceReport.html

Mark, D. M. (2003). Geographic information science: Defining the field. In M. Duckham, M. F. Goodchild, & M. Worboys (Eds.), *Foundations of geographic information science* (pp. 3-18). New York: Taylor & Francis.

Martin, G. J., James, P. E., James, E. W., & James, P. E. (1993). *All possible worlds: A history of geographical ideas* (3rd ed.). New York, NY: Wiley & Sons.

Nogueras-Iso, H., Muro-Medrano, P.R., & Zarazaga-Soria, F. J. (2005). *Geographic information metadata for spatial data infrastructures resources, interoperability, and information retrieval* (1st ed.). BerlinL New York Springer.

Ram, S., Park, J., & Lee, D. (1999). Digital libraries for the next millennium: Challenges and research directions. *Information Systems Frontiers, 1*(1), 75-94.

Ristow, W. W. (1980). *The emergence of maps in libraries.* London: Mansell Publishing.

Robertson, S. (2000). The digital city's public library: Support for community building and knowledge sharing. In T. Ishida, & K. Isbister (Eds.), *Digital cities* (pp. 246-260). New York, NY : Springer.

Robinson, A. H. (1952). *The look of maps: An examination of cartographic design.* Madison, WI: University of Wisconsin Press.

Robinson, A. H., & Petchenick, B. B. (1975). The map as a communication system. *The Cartographic Journal, 12*, 7-14.

Saracevic, T. (1999). Information science. *Journal of the American Society for Information Science, 50*(12), 1051-1063.

Schuurman, N. (2006). Formalization matters: Critical GIS and ontology research. *Annals of the Association of American Geographers, 96*(4), 726-739.

Schuurman, N., & Leszczynski, A. (2006). Ontology-based metadata. *Transactions in GIS, 10*(5), 709-726.

Solomon, P. (2002). Discovering information in context. *Annual Review of Information Science and Technology, 36*, 229-264.

Steinhauer, J. H., Wiese, T. o. m., Freksa, C., & Barkowsky, T. (2001). Recognition of abstract regions in cartographic maps. In D. R. Montello (Ed.), *Spatial information theory: Foundations of geographic information science: International Conference, COSIT 2001 Morro Bay, CA, USA, September 19-23, 2001, Proceedings* (pp. 306-321). Berlin, Germany: Springer-Verlag GmbH.

Strasser, T. C. (1995). Desktop GIS in libraries, technology and costs: A view from New York State. *Journal of Academic Librarianship, 21*(4), 278-282.

Tobler, W. R. (1970). *Selected computer programs* (Michigan geographical publications No. 1). Ann Arbor, MI: University of Michigan. Dept. of Geography.

Wang, Y., Wang, W., Luo, Y., Wang, X., & Xu, Z. (2004). A metadata framework for distributed geo-spatial databases in grid environment. In H. Jin, Y. Pan, N. Xiao, & J. Sun (Eds.), *Grid and cooperative computing: GCC 2004 third international conference, Wuhan, China, October 21-24, 2004, proceedings* (pp. 153-160). Berlin, Germany: Springer.

Wu, K.-L., Chen, S.-K., & Yu, P. (2006). Query indexing with containment-encoded intervals for efficient stream processing. *Knowledge and Information Systems, 9*(1), 62-90.

Chapter II

Information Economy and Geospatial Information

John Abresch, University of South Florida Libraries, USA

Peter Reehling, University of South Florida Libraries, USA

Ardis Hanson, University of South Florida Libraries, USA

Introduction

The recent socioeconomic trends, convergence of telecommunication technologies and the emergence of information as an integral component of the contemporary economy, have had significant effects on individuals and on wider social groups in the population. The current information node infrastructure of the telecommunications industry, which has facilitated that convergence of the telecommunications technology, is comprised of a variety of links. These links include data clearinghouses, data providers, and data warehouses, which themselves combine to form complex information networks as well as individual links, or single participants. All of these links affect how information flows across the network. Libraries, as participants in the information network infrastructure, are well suited to affect the nature of data processes in the current information economy.

Although the framework of the information economy may have been built by technological innovations and capital investments, there are cultural and political factors about the social milieu in which information is processed that affect how individuals participate in the economy. The composition of the information networks creates a variety of challenges to the successful searching, discovery, and mining of data for users in a variety of situations. Libraries can facilitate the interactions of individuals with different types of information that are integral to successful participation in the information economy. A key informational component of the information economy is how libraries can facilitate participation of their users with geospatial information.

This chapter will explore different socioeconomic aspects of the information economy and the role of libraries. The development of geographic information systems, the importance of "value added" services and an examination of how information is being increasingly commoditized is also included. Public aspects of geospatial information, such as government-produced GIS, will be discussed. How libraries can play a role in facilitating some of the social aspects of the distribution of the information economy, such as the digital divide, will also be examined.

Role of Information in Contemporary Economy

In the information economy, many private sector firms and government agencies have become consolidated around a framework of telecommunication networks and related information technologies. However, integrating communications and information technologies into their organizations have affected organizational processes, including production, distribution, and administration of products and services (Fincham, 2006; Fors & Moreno, 2002; Vyhmeister, Mondelo, & Novella, 2006; Xu, Wang, Luo, & Shi, 2006). Researchers have noted that organizational processes in both the public and private sectors have become characterized by applications, such as electronic data exchange, distributed databases, computer-based communication, and client server computing (Bieberstein, Bose, Walker, & Lynch, 2005; Pilkington & Fitzgerald, 2006; Shah & Mehta, 1998; Strnadl, 2006; Versteeg & Bouwman, 2006). Other researchers have indicated that geographic concepts, such as space and location, are significant factors in the distribution of data as it flows between individuals and organizations across communication networks, therefore, the use of geographic information systems in the facilitating and analysis of data has also increased greatly (Grubesic & Murray, 2005; Malecki, 2002; Van Gorp, Maitland, & Hanekop, 2006; Warf & Grimes, 1997; Zook, 2006)

The development and integration of geospatial information in a variety of administrative, production, and service functions within organizations in both the public and private sectors can have an effect on the role that libraries play in the information economy. The geospatial information that is used by individuals and organizations in the information economy is often produced by a mix of private developers and government agencies. The data used by geographers and geographic information systems analysts in their operations is diverse and distributed across a wide network of server locations, over a complex series of information nodes. Libraries can provide a unique entry point to this multitude of sources.

Economic activities during the past few decades have increasingly been characterized by different functional uses of information. The establishment and growth of diverse information industries, such as statistical bureaus, marketing associations, trade journals, and consulting agencies, have integrated information in their operations. The development builds on early studies of how information not only impacted, but helped define, different economic processes (Braman, 2006; Mccarthy, 1956). During the decades following World War II, individuals in the banking and accounting sectors began defining information as having value and affecting production (American Bankers Association, 1970). Statements, such as the "Statement of Basic Accounting Theory," asserted that accounting systems could be conceptualized as an "application of general theories of information to the problem of efficient economic operations" (American Accounting Association, 1966, p. 9). The National Science Foundation began to fund research about the economics of information (United States Senate Committee on Labor and Public Welfare, 1968). Other organizations, such as the Organization for Economic Co-operation and Development (OECD) and the United Nations Economic, Scientific, and Cultural Organization (UNESCO), were also debating policy issues about the economics of information. However, to develop an information policy, one must first understand that policy is "… the generic name of any formulation, simple or complex, vague or exact, general or special, discretionary or detailed, of guidance for action in the face of circumstances which, lying necessarily in the future, can be approached only by conjecture and imagination" (Shackle, 1961, p. ix).

The first bibliography of information economics was published in 1971 (Olsen, 1971). By the mid 1970s, developments in the literature illustrated how the general theories of information were being investigated by economists (Lamberton, 1974; Lamberton, 1975; Marschak, 1971) and other researchers in the sciences and social sciences (Barzakay, 1971; Cawkell & Garfield, 1975; Cooper, 1973; Hindle & Raper, 1976; Marschak, 1974; Mick, 1979; Regan, 1970; Wilson, 1972). Clearly, "… the existence of learning processes and likely variation in policy criteria in a business organization imply that the decision-making unit is undergoing continual change" (Lamberton, 1965, p. 74). Further, "those responsible for shaping information policy must determine the appropriate mix of information inputs to achieve social objectives, while at the same time they must have regard for equity considerations" (Lamberton, 1974, p. 145).

By 1976, the American Economic Association officially recognized information as an economic topic and incorporated it into core ideas. By the 1980s, research into this area was generating much debate among researchers about what comprised information economics (Benson & Parker, 1987; Cruise O'Brien, 1983; Jussawalla & Ebenfield, 1984; Jussawalla & Lamberton, 1982; Lamberton, 1990; Lebas, 1980;

Parker & Benson, 1987; Parker, Benson, & Trainor, 1988; Repo, 1987; Rubin, 1983; Seyhun, 1986).

Among the ideas being debated was the emergent issue of how to incorporate information into modes of production and thereby study its effects in different sectors of the economy. Early economists described three factors of production that included land, labor, and capital (Marx & Engels, 1970; Mill & Ashley, 1965; Ricardo, 1911; Smith, 1937). The competing theories on defining just how these three factors relate, especially capital, in a system is contentious and open to debate. Economists differ on how to define capital. It can be described concretely as something of value as in money or more abstractly as an association with labor and consumption. In examining the value of transformative functions of an economic activity, the definition of capital can be more broadly interpreted, in some cases to include information. For example, the definition of human capital is the "sum total of skills embodied within an individual: education, intelligence, charisma, creativity, work experience, entrepreneurial vigor, ... it is what you would be left with if someone stripped away all of your assets-your job, your money, your possessions" (Wheelan, 2002, p. 99). A number of articles and other publications that were produced in the 1990s discussed different forms of intangible capital including information, cultural capital, linguistic capital, and social capital. Social capital would involve "networks of communication and communication-based institutions and their rules, norms of social practice, and relationships of trust" (Braman, 2006, p.14). Networks themselves would be considered as facilitating a sum of knowledge that would have value as a factor in production (Braman, 2006, p.14).

Another perspective on integrating information into economic theory was in examining how change in the economy is affected by information. Braman (2006) suggests that "the production of knowledge and its transformation into technologies and other applied mechanisms such as competition can alter the order of different sectors of the economy and thus its equilibrium" (p. 15). Since the "defining trend of the modern economy is the shift to the intangible. The economic landscape is ...moulded by intangible streams of data, images and symbols. The source of economic value and wealth is ... the creation and manipulation of dematerialised content ...Non-linear and non-deterministic, the intangible economy raises a whole series of measurement issues. More fundamentally, it changes the role, the function, and the perception of economic measurement data. Because information is its key resource and output, the intangible economy is highly data-sensitive and intrinsically self reflective" (Goldfinger, 1997, p. 191). However, as Goldfinger suggests, the effects of the applications of information and technology are only apparent over a period of time when it is possible to perform analysis with systematic data. Recent studies have focused on defining the role of technology in causing economic growth. Some researchers consider economic change as being influenced by aspects of the innovation process, aspects that are characterized as being "information intense" or more developed technologically than what had currently existed. There has been much speculation on the correlation between technological development and economic growth and the social consequences of economic activities. In Marshallian neoclassical economics (Marshall, 1920), externalities refer to by-products of activities that affect the well-being of people, where those impacts are not reflected in market prices; the costs (or benefits) associated with externalities do not enter standard cost accounting schemes.

The backbone of the information economy is in the linkages that combine to form a network. If the network is a combination of technology and human activities that processes information as they traverse the network, then the uneven flow of information across the network can cause externalities to manifest themselves. These externalities can be caused by incomplete information that can result in the fluctuating prices of a commodity. The effect of content of information on the economy is up to debate. Are the contextual aspects of information processed in the economy irrelevant or do the contextual aspects of information in the economy have distinct sociopolitical value? For those who argue the latter, the contextual aspects of information do shape the perception and understanding of economic processes themselves. The cultural variability of information creates difficulties in determining the locational aspects of data. Data may be associated with a physical carrier, but it is not bound by the location of that carrier. This aspect of information has created challenges to economists who have attempted to illustrate Braman's (2006) concept that would "distinguish between the costs of information at different stages of an information production chain" (p. 20). Further, the value of the information chain hinges upon relevance that "is a subjective question of mapping an utterance on the conceptual map of a given user seeking information for a particular purpose defined by that individual" and the "utility of a piece of information will depend on a combined valuation of its credibility and relevance" (Benkler, 2002, p. 383).

Most of the economic models that attempt to illustrate an information production chain have the elements of information creating, processing, flows, and use incorporated into their structure. A number of organizations have created models that have the elements of information acquisition, production, assembly, storage, monitoring, interpretation, and exchange (Braman, 2006). United States federal documents describe an information cycle that includes information creation, collection, processing, and distribution (Blumenthal & Inouye, 1997a; Blumenthal & Inouye, 1997b; National Archives and Records Administration, 2005) as do international organizations (Centre on Transnational Corporations, 1983;; Gassmann, 1985; Hamelink, 1984; Jussawalla & Cheah, 1987; Organisation for Economic Co-operation and Development & Committee for Information, Computer, and Communications Policy, 1983).

Creators, or individuals who are involved in the innovation of new concepts and new industries, are highly valued. However, the idea of an information sector is not new. In 1962, over 50 industries were identified that comprised an information sector (Machlup, 1962; Machlup & Leeson, 1978-1980). Other economists preferred to identify information sector industries by using existing industry classifications, as identified by Standard Industrial Classification (SIC) codes (Porat, 1977; Porat & Rubin, 1977; Rubin & Porat, 1977a; Rubin & Porat, 1977b). However, critics of the approach questioned the validity of defining information on so specific a criteria, fearing that existing industrial classifications "ignore a major part of the processes of investment in information in the economy" and that "most significant information activities do not produce goods in tangible form and thus not included in the SIC system" (Braman, 2006, p.22). The debate over how to classify industries in the information sector illustrates the variability of trying to define information for econometric purposes. The debate is further magnified when trying to define the economic aspects of information in an international sense. Cultural context is a factor in defining what comprises the information sector, with definitions varying between countries and methods that describe material and information related production. Even though legal negotiations about the nature of information and production between countries, such as with the General Agreements on Tariffs and Trade (GATT), have attempted to arrive at some definitional parameters for information, the process of trying to understand how information works in an international trade environment is problematic due to the great variability production across different countries. For example, the North American Industry Classification System (NAICS) was adopted in 1997, replacing the old Standard Industrial Classification (SIC) system. Developed jointly by the U.S., Canada, and Mexico to provide new comparability in statistics about business activity across North America, the NAICS also better distinguishes between manufacturing and information sectors (United States Office of Management and Budget, 1998; United States Office of Management and Budget, 2002). Although imperfect, the NAICS is an essential tool to determine service and product providers in the information sector, to remain current on industry analyses, statistics, and leading companies (Darnay & Simkin, 2006).

The international economic environment is indeed a complex one, and in defining how information operates within economic systems, one can examine the redefining of economic components such as the agent, the firm, and the market (Clarke & McGuinness, 1987; Nagel, Shubik, & Strauss, 2004; Vickers, 1968). As discussed in classical economic theory, agents tend to act in ways that put themselves in a better position than previously held. Braman (2006) questions the decision-making processes of economic agents as they improve their position according to economic theory. Processes including "rationality, perfect information, and individualistic independence" (p. 26) are subject to myriad influences and the amount of information available. Further, there are limits to the amount of information that an individual agent can have access to or even process in order to maximize his position. Research in the fields of psychology, that is, cognitive miserliness (Fiske & Taylor, 1991), has indicated that there are cognitive and neurological limitations that affect how information is perceived and understood. Technology limitations can also affect the capture, recording, and transmission of information, thereby, affecting its cognition by an agent or organization (Saade, & Otrakji, 2007). The cultural context of information, as well, can have an affect on how information is understood (Irrmann, 2005). The social organization of society can have an significant impact on the flow of information creating barriers that can slow or impede the finding of information by individual agents (Tyler & Gnyawali, 2002; Workman, 2005). The concept of social organization and community is also important in understanding how information is processed in regards to economic function.

In classical economic theory, economic agents act purposively. Since they have an end in mind and find means to attain those ends, the actions of individual agents bear a causal relationship to overall market outcomes. Further, an economic agent uses information in order to "obtain satisfaction or benefits gained from consuming a particular good or service according to a hierarchical ranking of preferences" (Braman, 2006, p. 28). The social context of how information can be pursued and utilized also affects the actions of an individual economic agent. An individual may use information and acquire goods and services beyond what is necessary for the maximizing of their position and thus, may not make the best use of information in a classical economic sense. Cultural processes can affect preferences for a particular commodity adding or detracting from its value. The decision-making process is interconnected among various individuals in society, which can affect the utility factor in economic activities. Societies and communities can be conceptualized as components of a larger structure or market in which economic activities are performed. The larger structure can operate as an "informational mechanism" with different types of information arrayed spatially across it. How economic agents operate within markets is subject to much debate as economists advocate different operational models about the functioning of markets (Malone, Weill, Lai, D'Urso, Herman, Apel, et al., 2006). One model centers on the prices of commodities and services that are distributed across a market. Economic agents enter the market with a certain knowledge level and assumptions about prices that influence their interaction with the market. The search for further price information for an eventual outcome process frames their activities in the market. Social aspects and cultural ties that are part of the market structure affect how information interacts with markets. The distribution of information across markets has an impact on the economic decision process of the firm as well as the individual.

If a market is a social arrangement that allows buyers and sellers to discover information and carry out a voluntary exchange of goods or services, the firm is an alternative system of allocation to the market that exists to organize production in a non-price environment. Although there is a distinction between a market and a firm, most economists admit that the two shade into each other. A business model describes the elements and relationships that express the business logic of a firm (Osterwalder, Pigneur, & Tucci, preprint). Just as NAICS classes businesses into sectors, firms are often classed into "internally consistent sets of firms," referred to as strategic groups or configurations, allowing typologies and taxonomies to explore the determinants of performance (Malone et al., 2006). Early studies of information flow within existing business entities usually focused on individual managers. In a market environment in which reliable information was at a premium, business managers made decisions based on an individualized perspective that made use of personal social networks as well as market information. The growth of larger business enterprises with extensive production and distribution linkages increased the need for accurate market information, which led to the eventual progression toward extensive record keeping overseen by professional managers. The availability of accurate market information and information on other firms in related enterprises began to affect business decisions. An increased efficiency in record management usually decreases what some economists have labeled transaction costs, or the searching for information needed for a particular type of economic activity. Early studies in the nature of transaction costs within business processes determined that firms usually invested in those transaction costs that were more expensive, while leaving the more inexpensive transaction costs to the marketplace.

The application of advanced telecommunication technology by firms has encouraged commercial enterprises to distribute their transaction costs on an even wider basis for more efficiency. Contemporary studies in the nature of firms and different economic sectors have expanded the focus of study beyond the single firm to a hierarchy of firms or even larger social constructs, such as networks.

Research into the role of information and economic theory bridges the mathematical boundaries of idealized econometric models to account for information sector variability and the irrationality of human behavior. Research in the social dynamics of industries that comprise the information sector should take a wider, holistic approach. Studies on the information sector should integrate concepts and analytical methods from other disciplines. One such discipline is political science, which has often focused on the effect of uneven flows of information and of the impact of imperfect information on decisions (Braman, 2006). In expanding their descriptive scope of the components of the information sector, economists expanded their idea of the firm and the market by including variables with wider spatial structures. Such structures include telecommunication networks and social communication networks. It is with the physical network of the telecommunications sector that some researchers, especially in the field of geography, have sought to combine spatial concepts of space, place, and landscape to provide a rigorous and quantifiable platform in which to measure the complex socioeconomic phenomena that comprises the information sector.

Geographies of the Internet

The study of contemporary geographic spaces and technology inevitably focuses on the ubiquitous telecommunications networks that comprise the Internet. The complex interlocking nodes of the Internet create a variety of physical spaces bounded by technology such as computer hardware, fiber optics, and communication software. Upon the network of the Internet, geographers have identified a great many physical and virtual places as well. Geographers view the concept of space as more of an abstract expression, but the concept of place is bounded in the social milieu of a given location and can evolve as the local society that identified the space changes, and as new technologies alter the unique expression of the location. Places can be a unique experience of an individual person, but more often, they are the social constructs of many people with shared experiences. What is unique about places on the Internet is they can alter the bounded physical locations of places as identified by society and encourage the defining of new concepts of place that use elements of technology combined with historical notions of place, or "cyberspaces coexist with geographic spaces, providing a new layer of virtual sites superimposed over geographic spaces" (Kitchin, 1988, p. 403). Using such logic, geographers identify places on the Internet with a variety of parameters, including socioeconomic, political, and cultural factors. In defining their research about the geographic nature of telecommunications and of the Internet, geographers tend to focus on two broad aspects of research. On one hand, geographers are interested in the technological aspect of telecommunications and the Internet and how society and interacts with the technology. Research by geographers on the Internet "developed lines of research focused on the technology and infrastructure of the Internet (bandwidth, fiber networks, etc.) and how the use of this technology has blended with existing cultural, political, and economic structures manifest in physical places (virtual communities, e-commerce, etc.)" (Zook, 2006, p. 58).

Geographers intent on researching the technical aspects of the Internet tend to focus on analyzing locationally referenced Internet infrastructure data, essentially, the locations of nodes, communications networks. Factors in locating the infrastructure across the landscape include: "amounts of bandwidth coming from a populated area, Internet fiber backbone, and points of presence (POPs) or broadband development" (Zook, 2006, p. 59). Many of the studies in the locational aspects of the Internet infrastructure depict an uneven distribution of its composition. Developmental trends of the Internet mirror what Zook (2006) describes as patterns of agglomeration. The most densely developed Internet hubs are located in larger urbanized areas, while less densely developed rural areas lack high bandwidth rates. The distribution of the Internet has a significant effect on how information flows across the Internet and how the information is used by people. The physical distribution of the Internet is especially relevant to the delivery of library information and resources.

In the dynamic environment of the information sector, the library can be an important cornerstone in facilitating information flow between individuals and organizations. The stable presence of the publicly funded institutions can offer access to information as well as guidance in interpreting information for users. Libraries can also offer other services such as archival functions in an environment wherein information durability can be problematic. The following section examines more closely the mechanisms that enable information to be transmitted from place to place and person to person that libraries use.

Convergence of Communication Technologies

Telecommunication convergence in the information economy has had a significant effect on the delivery of library information resources and services. Convergence promises a "clean slate" approach, where things are reengineered to provide better, more flexible service to the user (Fowler, 2002, p. 11). Convergence, according to Fowler (2002), can be defined in four major categories of the telecommunications industry that correspond to layers, or sets of layers, in the Open Systems Interconnect (OSI) network model. The categories are transport (comprising the *physical* layer), switching (comprising the *data link* and *network* layers), and applications (comprising the *application* layer) (Fowler, 2002, p. 12).

- **Transport:** The same physical pipes (optical fiber, microwave, copper) and transport technology (usually Synchronous Optical Network [SONET]) carry multiple services, usually of different customers, for example, multiple T1 or T3 links. Convergence at this level is primarily used by carriers to provision their infrastructure; it is largely transparent to users as they continue to see and pay for separate services.
- **Switching:** The same cable plant carries different types of traffic and does appropriate switching. Content and presentation to the user is unchanged, with the possible exception of new features. Historically, this has been the layer about which most discussion has centered. The distinction between services becomes less distinct or disappears entirely under network layer convergence; at the present time, Internet protocol (IP), a switching technology, is envisioned as the common medium for all (or many) types of telecommunications traffic, especially voice and Internet traffic.
- **Application (Content):** The same end-user device or type of device and network handles and delivers all content; the user does not have separate network interface devices (e.g., television receiver, radio, VCR, computer, etc.)
- **Telecommunications/IT:** There is also a fourth meaning of convergence, which will be considered here, that refers to *blurring of the distinction between*

telecommunications and information processing. Examples are use of Applications Service Providers (ASPs) and network computing. (Fowler, 2002, pp. 11-12)

With convergence, traditionally separate functions are now available through one source or channel, such as an ISP or through a single vendor. Of particular interest to libraries is the technological convergence at the application level. Multiple resources in varying formats, including text, data, images, graphics, audiovisual media, streaming media, games and simulations, and so forth, are now available over one transmission network, through the same user equipment, and through standard, ubiquitous software applications. Previous channels would have included, and perhaps required, delivery via postal mail service or broadcast media, carrying a variety of formats and media. Enterprise systems, interactive video-on-demand, and telecalls/teleconferencing are becoming the norm (Ayres & Williams, 2004; Bieberstein et al., 2005; Xu et al., 2006). What will limit the rapid expansion of these applications will be intellectual property rights issues (Ayres & Williams, 2004; Braman & Roberts, 2003).

As the United States continues to build infrastructure and to recreate how the infrastructure is policed, broadband will undergo a number of changes (Federal Communications Commission, 2004; Ferguson, 2002). However, increased broadband access will continue to build convergence. The United States Congress has been urged to recognize and "encourage the convergence of voice, data, image and video information into bit streams" to "[e]nsure the greatest possible regulatory flexibility, to allow for unpredictable future service needs, market developments and technological innovation" (Committee on Communications and Information Policy , 2005, p. 5). Congress has also been urged to reduce barriers to competition, restructure the market in the public interest, and increase spectrum efficiency with both licensed and unlicensed models of spectrum use (Committee on Communications and Information Policy , 2005, p. 5). Convergence is seen as an evolutionary, rather than a revolutionary process. Initiatives to ensure broadband (100 Mbps) to the majority of homes in the United States by 2010, if realized, may speed up these trends (as well as many at the other levels of convergence). Initiatives include the UTOPIA (Utah Telecommunication Open Infrastructure Agency, 2003), the National LambdaRail Network (National LambdaRail, 2007), the NorthEast Education and Research Network (NEREN) (Northeast Research and Education Network, 2004), the Third Frontier Network (TFN) (Ohio Supercomputer Center, 2006), and the hundreds of other fiber communities in the United States (Ross, 2004).

Integrated media systems that seamlessly combine video, audio, computer animation, text, and graphics into a common digital display medium were noted as a new area of convergence (Mihram & Mihram, 1995). Further, Mihram and Mihram (1995) predicted that infrastructure installation, product creation, and commercialization relating to integrated media systems would become areas of research and development for both the public and private sectors. Today, geographic information systems are seen as the next pivotal technology in this convergence, again due to the advances in interoperability and data standards (Camarata, 2005, ¶3). Again, we see infrastructure installation, product creation, and commercialization in GIS becoming areas of research for the public and private sectors. Standards development, in particular, is a crucial area for GIS convergence. Many organizations work collaboratively to promote the advancement of open geospatial standards and specifications, especially as the market expands into domains associated with Internet and telecommunications communications technologies. These organizations include the Open Geospatial Consortium, W3C, and OASIS, as well as national and international standards associations. Since much of the data in the contemporary information economy have geospatial or geotemporal components, no applications, systems, or technologies are unaffected. Further, "[i]t is through the efficient use of GIS/geospatial technologies that we are able to understand and leverage the value of spatial/location based information and processes in the broader context of ICT and enterprise information systems" (Camarata, 2005, ¶4).

Convergence continues through business processes and architectures. Business architectures are the starting point from which to develop related and integrated functional, information, process, and application architectures (Versteeg & Bouwman, 2006). These architectures are aided by the rapid development and deployment of open, standards-based service-oriented architectures. Examples include the development of XML and XML Schema (a simple data format and a logical data description mechanism); SOAP (a simple object access protocol used as a remote invocation facility); WSDL (a service interface description mechanism); and XQuery (a declarative query language). In addition, more Web service tiers are integrated with business logic and workflow (Alonso, Casati, & Machiraju, 2004; Tatemura, Hsiung, & Li, 2003). There is increased deployment of mobile devices, communications, and services with even newer new micromobility management schema and technologies (Langar, Tohme, & Bouabdallah, 2006; Lo, Lee, Chen, & Liu, 2004). There are also smarter and richer client tools (especially in the GIS world) with more robust and flexible secure information exchange solutions. Certainly, enterprise systems will "serve as a 'unifying' element through their capacity to manage and utilize geospatial content, capabilities and services in enterprise environments" (Camarata, 2005, ¶ 6).

As the landscape of information technology changes through the convergence in the industry, the public sector, including libraries, is affected as well. As academic disciplines transition to newer forms of working and educational environment, the situation necessitates a commitment that encompasses several objectives: to change basic educational tools, to retrofit installations of school- and campus-wide data networks, and to create affordable networks that would link schools, homes, and communities (Mihram & Mihram, 1995). The development of new media technologies tends to be spatially uneven, often concentrating in one sector of society while marginalizing other sectors. The process of being marginalized by technology can often be through cultural, socioeconomic, geographic, technological and political means (Cullen, 2001). The uneven distribution of technology, computer networks, and of the information they transmit, has been characterized as contributing to a "digital divide" between those who do have access to the Internet and those who do not (Norris, 2001).

The Social Structure of the Information Economy

The digital divide has been described by researchers in a variety of dichotomies in both contemporary American and international settings (Burkett, 2000). In contemporary American society, researchers apply econometrics to socioeconomic data to portray how particular income groups have sufficient capital to purchase computer hardware and software for Internet access, while others do not. Researchers also look at factors such as race, ethnicity, and educational attainment to further illustrate differences in technology access (Ferrigno-Stack, Robinson, Kestnbaum, Neustadtl, & Alvarez, 2003). Different technological standards are identified, within contemporary American society, that can contribute to uneven access to information technologies. In international settings, the literature on the digital divide includes many of the factors described in articles about technology and information access in America, though other factors are emphasized, for example, political ones (Fahmi, 2002).

Many research and policy papers that examine different aspects of the digital divide identify minority ethnic groups, indigenous peoples, and specific groups of people as being disadvantaged in participating in the information economy. Factors include low incomes, few educational qualifications, low literacy levels, unemployment, age or disabled status, and single parent households (Cullen, 2001). Diverse minority groups often live in large urban centers that have complex telecommunication networks, unlike persons living in rural and frontier areas. However, both rural and urban areas may have older telecommunication infrastructures, affecting access. In the 2002 report, *A Nation Online*, the Department of Commerce noted the increasing rates of Internet usage among traditionally underserved groups:

In every income bracket, at every level of education, in every age group, for people of every race and among people of Hispanic origin, among both men and women, many more people use computers and the Internet now than did so in the recent past. Some people are still more likely to be Internet users than others are. Individuals living in low-income households or having little education, still trail the national average. However, broad measures of Internet use in the United States suggest that over time Internet use has become more equitable. (pp. 10-11)

However, there still exists a significant gap in the number of computers in low-income schools and communities (Prieger, 2003; Wilbon, 2003). Prieger (2003) also did not find that language was a statistically significant factor in Internet access, though many research reports indicate that English language ability is an important factor in participation in the information economy. Households with a record of higher educational attainment would be more inclined to use computers, have Internet access, and place a high value on literacy (Wilbon, 2003).

An important factor influencing participation in the information economy is the acquisition of information and communication technology (ICT) skills. Individuals who are members of the various groups that have been marginalized from computers and of the Internet often lack the necessary skills to use ICT applications. Cullen (2001) argues that the interaction of factors, such as cost, restricting access to equipment, low educational achievement, and cultural, age or gender-based exclusion from literacy and computing skills, counteracts against the spread of such skills in disadvantaged communities. Efforts to improve access to computers and programs to improve ICT technology skills among marginalized groups have been incorporated in a variety of community outreach programs by both public and private agencies. Many such programs try to establish and create a culture that is more conducive to ICT technologies.

Community Internet Initiatives

The combination of municipal and commercial computer networks in some large metropolitan areas has led to the creation of community information networks. Community information networks are often built around existing social networks involved in neighborhood activities like employment and economic opportunity centers, youth and family centers, health, education, and affordable housing initiatives (Borgida, Sullivan, Oxendine, Jackson, Riedel, & Gangl, 2002; Zielstra, 1999). One such project in Chicago, the Chicago Area Northside Neighborhood Online Network, is an organization that offers training and Internet access to community based entities throughout the city (Light, 1999). Building on existing human networks in the communitiés, the project has successfully trained residents and staff of over 60 community organizations, creating a unique multiracial, mixed-economic, and mixed-gender pool of community users. The NeighborTech program works with Chicago's inner city neighborhoods. NeighborTech uses a variety of methods (training classes, informational meetings, newsletters, and seminars) to draw in neighborhood residents, small businesses, and non-profit organizations located in disadvantaged neighborhoods. The Erie Neighborhood House, a non-profit, multiservice agency in

Erie, New York, operates the Erie Technology Center. The mission of the Technology Center is to provide computer and information literacy to West Town residents with limited English proficiency and low educational attainment (Light, 1999). By working with other educational programs in the area, the Technology Center integrates traditional teaching/learning methods with current technology applications for students ranging from prekindergarten to senior citizens.

Grants from federal agencies, such as the U.S. Department of Commerce, are also used to build community information networks outside of urban areas in rural locations (Borgida et al., 2002). The ItascaNet network, established in the town of Grand Rapids in north central Minnesota, is mostly rural, with a shrinking population base. Local community leaders used the grant funds to build an online network to increase the community's access to, and use of, the national information infrastructure, reduce disparities in access levels among community residents, increase information available to community members, and facilitate the sharing of data and information among partner organizations (Borgida et al., 2002). The ItaskaNet network involved five partner agencies that oversaw the purchasing of a server and cable for connections between the agencies. A significant member agency of the partnership was the public library. Internet-linked computers were made available to students in the public schools and to citizens in the public library, and free computer training classes were offered to the community. Researchers have identified libraries as an important cultural resource in contemporary American society. They can be an important node in the contemporary computer network and provide computer access and information service to the user community (Borgida et al., 2002).

The Role of Libraries in the Information Economy

The emergence of the new information economy comprised of computer networks that link users to one another or to larger organizations is creating a class of singular information users. To aid users in participating in the information economy, libraries designed a virtual presence to deliver information services, usually in the form of a "digital library." Digital libraries can offer a more varied informational experience to the community of online users. Libraries, with their traditional strengths of information collection, description, organization, and dissemination, can provide a more holistic learning experience for these new user communities. In commenting on the structure and environment of digital libraries, the DLF (Digital Library Federation) has offered the following definition: "Digital libraries are organizations that provide the resources, including specialized staff, to select, structure, offer intellectual access to, distribute, interpret, preserve the integrity of, and ensure the persistence over time of collections of digital works" (Waters, 1998), ¶3). In elaborating on the concept, Hanani and Frank (2000) define six major characteristics that should be integral to digital libraries (pp. 212-213):

1. "Collection of data objects: A library holds together a collection of data objects, items and resources. The items can be books, journals and documents (e.g., HTML pages), multimedia objects (such as pictures or images, tapes or video files, etc.). The library objects can be available locally, or indirectly, by using a network to access them.
2. "Collection of metadata structures: A library contains a collection of metadata structures, such as catalogs, guides, dictionaries, thesauri, indices, summaries, annotations, glossaries, and so forth.
3. "Collection of services: A library provides a collection of services, such as various access methods (search, browse, etc.) for. as well as consultation for. different users; management of the library (purchase, shelf arranging, computerization, communication); logging/statistics and performance measurement evaluation (PME); selective dissemination of information (SDI) or push mode, as it is called on the Internet.
4. "Domain focus: A library has a domain focus and its collection has a domain focus; purpose. For example, art, science, or literature. Also, it is usually created to serve a community of users and therefore, is finely grained.
5. "Quality control: A library uses quality control in the sense that all its material is verified and consistent with the profile of the library. The material is filtered and its metadata is usually enriched (e.g., annotated).
6. "Preservation: The purpose of preservation is to ensure protection of information of enduring value for access by present and future generations. Preservation includes the allocation of resources for preservation, preventive measures, and remedial measures to restore the usability of selected materials." (Hanani & Frank, 2000, pp. 212-213)

As librarians compile digital collections of materials, the networked environment of which they are a part offers access to many more information resources. In describing libraries and the new online environment in which they operate, libraries are placing lesser emphasis on the materials they collect and house, and more emphasis on the kind of material they are able to obtain in response to user requests (Berry, 1996). The trend includes libraries forming partnerships to deliver material from elsewhere in time to answer a user's information needs. The shift to on-demand delivery of material from elsewhere is an effect of recent growth in digital networking in an environment where standards for description were established and refined over the past 35 years. Librarians are also moving primarily away from being caretakers of physical collections to people who identify resources that exist.

The earlier survey of the information economy has illustrated a variety of information sector industries and organizations that utilize information, such as geographic data, in their varied production and distribution functions. The flexibility of spatially referenced information, and its applicability to a variety of research techniques and spatially referenced applications, create a powerful tool in which librarians can raise the profile of their institution's involvement in the information sector. In defining the parameters of geographic information librarianship, the varied applications of geographic information in government functions, community initiatives, and private sector development can offer strategies in dealing with the social effects of trends like the digital divide and information marginalization. The following section provides an overview of the development of geographic information systems and related applications.

Development of Geographic Information Systems (GIS)

During the 1960s and 1970s, geographers and cartographers began adapting computer analytical methods to capturing graphical data portrayed on maps. Previously, the emphasis of cartographic research was based on the "idea of storing graphical features that were displayed on maps in computer files" (Kainz, 2004). Only with the later use of mathematical models and structures based on theory in topology were researchers able to apply "logically consistent two dimensional data representations" (Kainz, 2004). By the 1980s, the ability to create stable and consistent representations of map data was integral to research and development in geographic information systems. Kainz (2004) describes the impact of the micro- and personal computer in the development of powerful desktop software packages in word processing, database management, and statistical analysis and the development of desktop mapping software (MapInfo, ESRI, and Intergraph) that could integrate visualizations of map data with corresponding data in other databases. He further describes the rapid progression of research on spatial data structures, indexing methods, and spatial databases. A workspace in a GIS software package could contain graphic representations of a spatial dataset, integrate data from relational databases, and have other corresponding information in text formats or numeric formats. The convergence of computer hardware innovations with research into mathematical spatial modeling, and mapping software during the 1980s clearly contributed to better-defined geographic information systems.

A geographic information system (GIS) can be defined as "a computer-based technology and methodology for collecting, managing, analyzing, modeling, and presenting geographic data for a wide range of applications" (Davis, 2001, p. 13). A GIS essentially combines five components, people, data, hardware, software, and methods, for the purpose of finding solutions to issues that have a spatial context.

The fundamental operations of a GIS application are capturing, storing, querying, analyzing, displaying, and outputting data. The nature of geographic data can be best understood by three major concepts: *feature geometry*, *attributes*, and *topology*. *Feature geometry* represents features, such as houses, roads, or property boundaries, and establishes where these features are located in the real world. All features are symbolized by points, lines, or polygons. *Attributes* provide a description of the features and are stored in an associated table that is linked to the features. An attribute field can contain an address, street name, or land-use code, and so forth. *Topology*, the most abstract concept of geographic data, defines either the behavior or the spatial relationships that exist between features. For example, a GIS layer representing a transportation network requires topology rules to accurately depict one-way streets, overpasses, and right of way scenarios. Without topology, data integrity associated with sound editing, display, or analysis is not maintained.

Geographic knowledge is represented in five data formats: *maps* and *globes*, *geographic datasets*, *data models*, *processing and workflow models*, and *metadata*. Interactive *digital maps and globes* can query information and present it to the user. Since digital maps and globes have limited analytical functionality, they are generally used to resolve location and directional questions. *Geographic datasets* contain feature geometry, attributes, and topology. *Data models* are templates with defined topology schemas, and are used in the data creation process to ensure data standardization and integrity are maintained. *Processing and workflow models* are necessary when managing GIS projects to visually depict and duplicate the geoprocessing procedures associated with the spatial analysis. *Metadata* documents the four previous data formats, and is the key to organizing, discovering, and evaluating GIS data resources. With the proliferation of desktop computing and numerous GIS mapping applications, geospatial data has become an important part of various socioeconomic processes, political activities, and academic research that comprise the information economy. The following discussion outlines some applications of spatial data and geographic information systems in the information economy.

Applications of Spatial Data and GIS

The literature of the information economy is rich with descriptions of how computer technology and the Internet are altering socioeconomic processes within organizations and throughout political and commercial networks in which they are integrated. Some researchers assert that the use of information communication technologies (e.g., e-mail) can facilitate closer contact between members of management and other employees of an organization, thus rearranging lines of command for long-distance interactions, resulting in a direct savings on transportation costs, and improving the decision-making process in the organizational structure (Kokuryo & Takeda, 2005). Other scholars have discussed how the integration of information systems into government organizations' infrastructure have affected resource management (Heeks, 1999), such as the decentralization of decision making.

As part of extensive commercial and political networks, organizations exchange information through complex optical and cable networks that comprise the Internet. In summarizing research about the information economy, Grubesic and Murray (2005) describe the Internet as "a complex mesh of interconnected computers, fiber optic cables, routers and human users" (p. 70). They further describe the Internet as "a series of smaller networks linked together by hardware, software, and many peering agreements between Internet Service Providers. The distribution of smaller networks across the United States alone amounts to over 166 million users, 7,000 Internet Service Providers, and a large grouping interlinked cable and telephone companies that offer information services" (Grubesic & Murray, 2005, p. 70). The reliance of organizations upon a telecommunications network infrastructure to transmit different forms of electronic information adds a spatial component to the data. Grubesic and Murray (2005) note that a "unique element of many datasets is the geographical or spatial entities they represent. The geographic component of the data could correspond to a location of a point of presence, central office, or the path a fiber optic cable traverses" (p. 72). Geographic information systems can analyze, store, and process the spatial data associated with an organization's datasets. Since a GIS is often used to visually organize geographic data in order to facilitate different types of analysis for the data, geographic information systems are described as a "computer program for acquiring, storing, interpreting, and displaying spatially organized information" (Green & Bossomaier, 2002, p. 1). Besides recording the locational aspects of a particular data set such as neighborhood, street, or country, GIS data may have additional information characteristics, such as temperature, cost, color, or even detailed demographic information or public health data. The uses of GIS in the information economy are many, and often GIS databases are built with data from different types of sources in both the public and private sector, such as from ESRI, Inc., the U.S. Census Bureau, and even the Centers for Disease Control and Prevention.

Definitions of geospatial data vary within the research literature in geography. Some researchers define spatial data as being "anything dealing with the concept of space, in the geographic context, primarily dealing with the distribution of things on the surface of the earth " (DeMers, 1997, p. 474) while other researchers define spatial data as "data that occupies geographic space" (Davis, 2001). There are many types and characteristics of geospatial data. However, most spatial or geographic data have specific location according to a global geographic referencing system (e.g., latitude or longitude), and may be illustrated by other characteristics, such as size and shape of a particular dataset. Davis (2001) further elaborates that the size would be calculated by the amount of area, and shape would be defined by the position of the shape points of the dataset, for example, an administrative zone or economic zone.

Geospatial data collected by government agencies across different areas can include topographic data, hydrographic data, earth science data, and soil and forest survey inventories. Other types of information that can be geospatially referenced include social and economic data (e.g., population and industrial characteristics). An important aspect of geospatial data is its potential for multiple applications (Groot & McLaughlin, 2000). GIS technology facilitates the integration and comparison of different data sets, which allows for greater statistical analysis of the data content. It was suggested that the annual federal spending on geospatial data activities in the United States was over $4.4 billion (Koontz, 2003).

Since the 1960s, digital spatial data has been produced in a variety of formats and in a number of carriers. Digital spatial data can include information digitized and recorded by a single researcher, or be in the form of large-scale geographic coverages processed and packaged by private firms. Much digital spatial data is also generated by government agencies at the city, state, and federal level, and is issued in a variety of formats like on CD-ROMs with computer files with accompanying attribute data in relational databases (Decker, 2001).

The proliferation of spatial data and the rapidly evolving technological environment in which spatial data is being used in all manner of research has led some researchers to speculate on the nature of future applications of GIS (Sui, 2004). Future development in GIS will be in areas such as geocomputation, social informatics, information ecology, and a spatially integrated social science. Sui (2004) notes that the "diffusion of spatial analytical tools" and their integration with "visualization tools" will lead to the use of geographic metaphors important in describing political-economic activities across contemporary social and cultural regions (p. 66). One research area involving digital geospatial data has been the electronic space of the Internet, or cyberspace.

Recent trends involving spatial data and geographic research have focused on defining epistemologies and methodologies using GIS to better explain sociogeographic phenomena across different environments. Longan (2002) argues that geographers have only begun to explore social, cultural, and political aspects of cyberspace. In his examination of the community networking environment across the United States, he attempts to define a sense of place in the online environments that different neighborhoods and towns are setting up in their areas (Longan, 2002). Other researchers examine the Internet and the effects of telecommunications between communities and social classes and the distribution of wealth, arguing that Internet community networks may help disadvantaged groups overcome the problems of distance by using cyber communications to access needed resources (Warf & Grimes, 1997), or find that access to telecommunications has positive effects on a community's sense of identity (Albrow, 1997) Still others have used GIS and spatial data in examining urban cultural and political spaces (Páez & Scott, 2004) or to study economic production and social consumption (Kidner, Higgs, & White, 2002; Nyerges, Jankowski, & Drew, 2002). Still other researchers envision continued development of representation of real world data using computer languages (Goodchild & Haining, 2004).

Map Libraries in Transition

In attempting to put some order to the many types of geospatial data produced by government agencies, private firms, and other organizations, librarians can turn to map librarianship for guidelines and best practices in designing procedures for processing spatial data. As discussed in Chapter I, map libraries, especially in the United States, have benefited from lengthy depository programs wherein maps produced by the United States Geological Survey, Army Map Service, and Inter-American Geodetic Survey were deposited across many libraries. The experiences of processing the cartographic collections in libraries are beneficial in learning about the scope of map collections in libraries as well as to understand cataloging and classification schemes for maps and cartographic materials (Andrew & Larsgaard, 1999; Larsgaard, 1998). The rapid migration to digitally produced cartographic materials in a variety of formats (magnetic tape, floppy, CD ROM, and servers) presents librarians with new challenges in processing spatial data (Parry & Perkins, 2001).

The same technological changes that have transformed the production of cartographic materials have affected the functioning of library services and collections as well. The advent of the Internet has provided an opportunity for the traditional academic library to evolve its services and reposition its collection to take advantage of the communication possibilities that the World Wide Web presents. Applications of GIS technology and geographic information can also provide libraries with tools that can be used to overcome technological and social barriers that have come about due to the digital divide. The public aspect of GIS, especially in terms of government produced and facilitated data, create more of a sense of urgency in providing access to the information. The discussion that follows examines the convergence of communication technologies in the contemporary information economy and its effect on various segments of the private and public sector including libraries.

Conclusion/Summary

In 2005, the Pew Internet & American Life Project released a survey of 1,286 individuals. Roughly a third of the experts are affiliated with an academic institution, another third work for a company or consulting firm, and the rest are divided between non-profit organizations, publications, and the government (Fox, Anderson, & Rainie, 2005). Comparing the survey results with an earlier 1990-1995 predictions database, certain similar themes about the impact of technology and society emerged.

1. "Technological change is inevitable, and it will result in both beneficial and harmful outcomes. Those surveyed see the impact of the Internet as multidirectional and complex, as did predictors at the dawn of all other communications technologies.
2. "A technology is never totally isolated in its influence as a change agent. Many social trends commonly associated with the coming of the Internet are the result of changes spurred by multiple forces; some already were in motion as the Internet came into common use. We must not fall into the trap of technological determinism; the Internet should not be fully credited nor should it take all of the blame.
3. "Entrenched interests prefer the status quo and often work to block or delay innovations introduced by new technologies such as the Internet. Respondents see this happening in copyright clashes, education, health care, and other areas.
4. The business of projecting the future impact of a technology can be difficult and full of inconsistencies." (Fox et al., 2005, pp. 47-48)

Clearly, the emergence of the new information economy, characterized by a convergence of telecommunication functions, is creating a class of singular information users. Digital libraries can offer a more varied informational experience to the community of online users. Libraries with their traditional strengths of information collection, description, organization, and dissemination can provide a more holistic learning experience for the community of digital library users. Librarians can prepare a learning environment of Internet resources that is easier to navigate by classifying different online information sources.

The capabilities of digital libraries can be used to organize and ease the mining of a variety of data across the Internet, such as digital geospatial data. The Internet was described as an optimum medium for the sharing of GIS data files and related information (Pienaar & Brakel, 1999). With the continuing development of the Internet, and of data standards, data transfer protocols, and increasing interoperability of database systems, the framework is being put into place to easily facilitate GIS data transfer on the Internet. The following chapter contains a fuller discussion of the NSDI infrastructure, and introduces concepts of data standards that are essential in organizing numeric datasets such as digital geospatial datasets. The diverse nature of geospatial data presents challenges to libraries, such as the archiving, description, and accessing of the digital formats of information. By applying their traditional mission of identifying, cataloging, and providing access to information, libraries can ensure accessibility to GIS software and spatial data for reference, instruction, research, and commercial endeavors.

References

Albrow, M. (1997). Travelling beyond local cultures: Socioscapes in a global city. In J. Eade (Ed.), *Living the global city: Globalization as local process* (pp. 37-55). London: Routledge.

Alonso, G., Casati, F. K. H., & Machiraju, V. (2004). *Web services: Concepts, architectures and applications.* New York, NY: Springer.

American Accounting Association. (1966). *A statement of basic accounting theory.* Evanston, IL: The Association.

American Bankers Association. (1970). *Proceedings of a Symposium on Public Policy and Economic Understanding, November 17, 1969.* New York, NY: American Bankers Association.

Andrew, P. G., & Larsgaard, M. L. (1999). *Maps and related cartographic materials: Cataloging, classification, and bibliographic control.* Binghamton, NY: Haworth Information Press.

Ayres, R. U., & Williams, E. (2004). The digital economy: Where do we stand? *Technological Forecasting and Social Change, 71*(4), 315-339.

Barzakay, S. N. (1971). Policymaking and technology transfer: Need for national thinking laboratories. *Policy Sciences, 2*(3), 213-227.

Benkler, Y. (2002). Coase's penguin, or, Linux and the nature of the firm. *The Yale Law Journal, 112*(3), 369-446.

Benson, R. J., & Parker, M. M. (1987). *Advancing the state of the art in strategic planning.* St. Louis, MO: Washington University.

Berry, J. W. (1996). Digital libraries: New initiatives with worldwide implications. *IFLA Journal, 22*(1), 9-17.

Bieberstein, N., Bose, S., Walker, L., & Lynch, A. (2005). Impact of service-oriented architecture on enterprise systems, organizational structures, and individuals. *IBM Systems Journal, 44*(4), 691-708.

Blumenthal, M. S., & Inouye, A. S. (1997a). *Assessment of formats and standards for the creation, dissemination, and permanent accessibility of electronic government information products: Phase I deliverables.* Washington, DC: Computer Science and Telecommunications Board, National Research Council. Retrieved January 2007, from http://www.nclis.gov/govt/gpo1.html

Blumenthal, M. S., & Inouye, A. S. (1997b). *Statement of work for phase II.* Washington, DC: Computer Science and Telecommunications Board, National Research Council. Retrieved January 2007, from http://www.nclis.gov/govt/state.html

Borgida, E., Sullivan, J. L., Oxendine, A., Jackson, M. S., Riedel, E., & Gangl, A. (2002). Civic culture meets the digital divide: the role of community electronic networks. *Journal of Social Issues, 58*(1), 125-141.

Braman, S. (2006). The micro- and macroeconomics of information. *Annual Review of Information Science and Technology, 40*, 3-52.

Braman, S., & Roberts, S. (2003). Advantage ISP: Terms of service as media law. *New Media & Society, 5*(3), 422-448.

Burkett, I. (2000). Beyond the "information rich and poor": Futures understandings of inequality in globalising informational economies. *Futures, 32*, 679-694.

Camarata, S. J. (2005). GIS and geospatial technology in the information age: A new perspective. *Directions Magazine, May 08*. Retrieved January 2006, from http://www.directionsmag.com/article.php?article_id=856&trv=1

Cawkell, A., & Garfield, E. (1975). Cost-effectiveness and cost benefits of commercial information-services. *Current Contents, (34)*, 6-10.

Centre on Transnational Corporations. (1983). *Transborder data flows: Access to the international online data-base market: A technical paper*. New York, NY: United Nations.

Clarke, R., & McGuinness, T. (1987). *The economics of the firm*. Oxford, England: Cambridge.

Committee on Communications and Information Policy . (2005). *Providing ubiquitous gigabit networks in the United States: An IEEE-USA Committee on Communications and Information Policy (CCIP) white paper*. Washington, DC: CCIP. Retrieved December 2006, from http://www.ieeeusa.org/volunteers/committees/ccip/docs/Gigabit-WP.pdf

Cooper, M. D. (1973). Economics of information. *Annual Review of Information Science and Technology, 8*, 5-40.

Cruise O'Brien, R. (1983). *Information, economics, and power: The North-South dimension*. Boulder, CO: Westview Press.

Cullen, R. (2001). Addressing the digital divide. *Online Information Review, 25*(5), 311-320.

Darnay, A., & Simkin, J. P. (2006). *Manufacturing & distribution U.S.A.: Industry analyses, statistics, and leading companies* (4th.ed.). Detroit, MI: Thomson Gale.

Davis, B. E. (2001). *GIS: A visual approach* (2nd ed.). Albany, N.Y: Delmar Thomson Learning.

Decker, D. (2001). *GIS data sources*. New York, NY: John Wiley & Sons.

DeMers, M. N. (1997). *Fundamentals of geographic information systems*. New York: J. Wiley & Sons.

Fahmi, I. (2002). The Indonesian Digital Library Network is born to struggle with the digital divide. *International Information & Library Review, 34*(2), 153-174.

Federal Communications Commission. (2004). *Availability of advanced telecommunications capability in the United States, Fourth report to Congress*. Washington, DC: FCC.

Ferguson, C. H. (2002). *The United States broadband problem: Analysis and policy recommendations: A Brookings Working Paper*. Washington, DC: The Brookings Institution. Retrieved December 2006, from http://www.brookings.edu/views/papers/ferguson/working_paper_20020531.pdf

Ferrigno-Stack, J., Robinson, J. P., Kestnbaum, M., Neustadtl, A., & Alvarez, A. (2003). Internet and society: a summary of research reported at Webshop 2001. *Social Science Computer Review, 21*(1), 73-117.

Fincham, R. (2006). Knowledge work as occupational strategy: Comparing IT and management consulting. *New Technology Work and Employment, 21*(1), 16-28.

Fiske, S. T., & Taylor, S. E. (1991). *Social cognition* (2nd ed.). New York, NY: McGraw-Hill.

Fors, M., & Moreno, A. (2002). The benefits and obstacles of implementing ICTs strategies for development from a bottom-up approach. *Aslib Proceedings, 54*(3), 198-206.

Fowler, T. B. (2002). Convergence in the information technology and telecommunications world: Separating reality from hype. *The Telecommunications Review, 12*, 11-30.

Fox, S., Anderson, J. Q., & Rainie, L. (2005). *The future of the Internet: In a survey, technology experts and scholars evaluate where the network is headed in the next ten years*. Washington, DC: Pew Internet & American Life Project. Retrieved January 2007, from http://www.pewinternet.org/pdfs/PIP_Future_of_Internet.pdf

Gassmann, H.-P. (1985). *Transborder data flows: Proceedings of an OECD conference held December 1983*. Amsterdam, The Netherlands: North-Holland for the Organisation for Economic Co-operation and Development.

Goldfinger, C. (1997). Intangible economy and its implications for statistics and statisticians. *International Statistical Review/Revue Internationale De Statistique, 65*(2), 191-220.

Goodchild, M. F., & Haining, R. P. (2004). GIS and spatial data analysis: Converging perspectives. *Papers in Regional Science, 83*(1), 363-385.

Green, D., & Bossomaier, T. R. J. (2002). *Online GIS and spatial metadata*. London; New York: Taylor & Francis.

Groot, R., & McLaughlin, J. (2000). *Geospatial data infrastructure: Concepts, cases and good practice*. New York, NY: Oxford University Press.

Grubesic, T. H., & Murray, A. T. (2005). Geographies of imperfection in telecommunication analysis. *Telecommunications Policy, 29*(1), 69-94.

Hamelink, C. J. (1984). *Transnational data flows in the information age*. Lund, Sweden: Studentlitteratur AB.

Hanani, U., & Frank, A. J. (2000). The parallel evolution of search engines and digital libraries: Their convergence to the mega-portal. In Y. Kambayashi, G. Wiederhold, J. Klavans, W. Winiwarter, & H. Tarumi (Eds.), *2000 Kyoto International Conference on Digital Libraries: Research and Practice, November 13th-16th, 2000, Kyoto University, Kyoto, Japan* (pp.211-218). Los Alamitos, CA: IEEE Computer Society.

Heeks, R. (1999). *Reinventing government in the information age: International practice in IT-enabled public sector reform.* London: Routledge .

Hindle, A., & Raper, D. (1976). Economics of information. *Annual Review of Information Science and Technology, 11*, 27-54.

Irrmann, O. (2005). Communication dissonance and pragmatic failures in strategic processes: The case of cross-border acquisitions. *Advances in Strategic Management, 22*, 251-266.

Jussawalla, M., & Cheah, C. W. (1987). *The calculus of international communications: A study in the political economy of transborder data flows.* Littleton, CO: Libraries Unlimited.

Jussawalla, M., & Ebenfield, H. (1984). *Communication and information economics: New perspectives.* Amsterdam, The Netherlands: North Holland.

Jussawalla, M., & Lamberton, D. M. (1982). *Communication economics and development.* Honolulu, Hawaii; Elmsford, NY: East-West Center; Pergamon Press.

Kainz, W. (2004). *Geographic information science* (Version 2.0 ed.). Vienna, Austria: Institut für Geographie und RegionalforschungUniversität Wien Universitätsstraße. Retrieved from http://www.geografie.webzdarma.cz/GIS-skriptum.pdf

Kidner, D., Higgs, G., & White, S. (2002). *Socio-economic applications of geographic information science.* London: Taylor & Francis.

Kitchin, R. (1988). Towards geographies of cyberspace. *Progress in Human Geography, 22*(3), S385-S406.

Kokuryo, J., & Takeda, Y. (2005). Value creation on networks and in corporate activities. *Digital Economy and Social Design* (pp. 152-164). Tokyo: Springer.

Koontz, L. D. (2003). *Geographic information systems challenges to effective data sharing: testimony before the Subcommittee on Technology, Information Policy, Intergovernmental Relations and the Census, Committee on Government Reform, House of Representatives: Testimony* (GAO No. 03-874 T). Washington, D.C: U.S. General Accounting Office.

Lamberton, D. M. (1965). *The theory of profit.* Oxford, England: Blackwell.

Lamberton, D. M. (1974). *The information revolution.* Philadelphia, PA: American Academy of Political and Social Science.

Lamberton, D. M. (1975). *Who owns the unexpected? A perspective on the nation's information industry.* St. Lucia, Queensland, Australia: University of Queensland Press.

Lamberton, D. M. (1990). *Information economics: Threatened wreckage or new paradigm?* South Melbourne, Australia: Centre for International Research on Communication and Information Technologies (CIRCIT).

Langar, R., Tohme, S., & Bouabdallah, N. (2006). Mobility management support and performance analysis for wireless MPLS networks. *International Journal of Network Management, 16*(4), 279-294.

Larsgaard, M. L. (1998). *Map librarianship: An introduction* (3rd ed.). Englewood, CO: Libraries Unlimited.

Lebas, M. (1980). *Toward a theory of management control: Organizational process, information economics and behavioral approaches.* Jouy-en-Josas, France: CESA.

Light, J. S. (1999) From city space to cyberspace. In M. Crang, P. Crang, & J. May (Eds.), *Virtual geographies: Bodies, space and relations* (pp. 109-130). New York, NY: Routledge.

Lo, S. C., Lee, G., Chen, W. T., & Liu, J. C. (2004). Architecture for mobility and QoS support in All-IP wireless networks. *IEEE Journal on Selected Areas in Communications, 22*(4), 691-705.

Longan, M. W. (2002). Building a global sense of place: The community networking movement in the United States. *Urban Geography, 23*(3), 213-236.

Machlup, F. (1962). *The production and distribution of knowledge in the United States.* Princeton, NJ: Princeton University Press.

Machlup, F., & Leeson, K. (1978-1980). *Information through the printed word: The dissemination of scholarly, scientific, and intellectual knowledge* (No. 1-4). New York, NY: Praeger Publishers.

Malecki, E. J. (2002). The economic geography of the Internet's infrastructure. *Economic Geography, 78*(4), 399-424.

Malone, T. W., Weill, P., Lai, R. K., D'Urso, V. T., Herman, G., Apel, T. G., et al. (2006). *Do some business models perform better than others?* (MIT Sloan Research Paper No. 4615-06). Boston, MA: Massachusetts Institute of Technology (MIT), Sloan School of Management. Retrieved December 2006, from http://ccs.mit.edu/papers/pdf/wp226.pdf

Marschak, J. (1971). Economics of information systems. *Journal of the American Statistical Association, 66*(333), 192-219.

Marschak, J. (1974). Limited role of entropy in information economics. *Theory and Decision, 5*(1), 1-7.

Marshall, A. (1920). *Principles of economics* (8th ed.). London, England: Macmillan & Company.

Marx, K., & Engels, F. (1970). *Das Kapital: A critique of political economy.* Chicago, IL: H. Regnery.

Mccarthy, J. (1956). Measures of the value of information. *Proceedings of the National Academy of Sciences of the United States of America, 42*(9), 654-655.

Mick, C. K. (1979). Cost-analysis of information-systems and services. *Annual Review of Information Science and Technology, 14*, 37-64.

Mihram, D., & Mihram, G. A. (1995). The convergence of telecommunications and today's academic libraries. In H. Olson (Ed.), *Proceedings, Proceedings, Canadian Association for Information Science/L'Association canadienne des sciences de l'information: Connectedness: Information, Systems, People, Organizations, University of Alberta, Edmonton, Alberta. June 7 - 10, 1995.* Edmonton, Alberta, Canada: CAIS/ACSI. Retrieved November 2006, from http://www.cais-acsi.ca/proceedings/1995/mihram_1995.pdf

Mill, J. S., & Ashley, W. J., Sir. (1965). *Principles of political economy, with some of their applications to social philosophy.* New York, NY: A.M. Kelley, bookseller.

Nagel, K., Shubik, M., & Strauss, M. (2004). The importance of timescales: Simple models for economic markets. *Physica A: Statistical Mechanics and Its Applications, 340*(4), 668-677.

National Archives and Records Administration. (2005). *Benchmarking report on business process analysis and systems design for electronic recordkeeping.* Washington, DC: National Archives and Records Administration.

National LambdaRail. (2007). *National LambdaRail.* Retrieved January 2006, from http://www.nlr.net/

Norris, P. (2001). *Digital divide: Civic engagement, information poverty, and the Internet worldwide.* Cambridge, England: Cambridge University Press.

NorthEast Research and Education Network. (2004). *NEREN project overview.* [s.l.]: NEREN. Retrieved January 2006, from http://www.neren.org/documents/NEREN%20Overview%20for%20Members%20v2.8%2012-3-04AF.pdf

Nyerges, T., Jankowski, P., & Drew, C. (2002). Data-gathering strategies for social-behavioural research about participatory geographical information system use. *International Journal of Geographical Information Science, 16*(1), 1-22.

Ohio Supercomputer Center. (2006). *Third Frontier Network (TFN).* Columbus, OH: Ohio Board of Regents. Retrieved January 2006, from http://www.osc.edu/oarnet/tfn/

Olsen, H. A. (1971). *The economics of information: Bibliography and commentary on the literature.* Washington, DC: ERIC Clearinghouse on Library and Information Sciences.

Organisation for Economic Co-operation and Development, & Committee for Information, Computer, and Communications Policy. (1983). *An exploration of legal issues in information and communication technologies.* Paris, France: Organisation for Economic Co-operation and Development.

Osterwalder, A., Pigneur, Y., & Tucci, C. L. Clarifying business models: Origins, present, and future of the concept. *Communications of the Association for Information Systems, 15*(May), PrePrint. Retrieved December 2006, from http://www.businessmodeldesign.com/publications/Preprint%20Clarifying

%20Business%20Models%20Origins,%20Present,%20and%20Future%20o
f%20the%20Concept.pdf

Páez, A., & Scott, D. M. (2004). Spatial statistics for urban analysis: A review of techniques with examples. *GeoJournal, 61*(1), 53-67.

Parker, M. M., & Benson, R. J. (1987). *Information economics: An introduction.* St. Louis, MO: Center for the Study of Data Processing.

Parker, M. M., Benson, R. J., & Trainor, H. E. (1988). *Information economics: Linking business performance to information technology.* Englewood Cliffs, NJ: Prentice Hall.

Parry, R. B., & Perkins, C. R. (2001). *The map library in the new millennium.* Chicago; London: American Library Association; Library Association Pub.

Pienaar, M., & Brakel, P. (1999). The changing face of geographic information on the Web: A breakthrough in spatial data sharing. *Electronic Library, 17*(6), 365-371.

Pilkington, A., & Fitzgerald, R. (2006). Operations management themes, concepts and relationships: A forward retrospective of IJOPM. *International Journal of Operations & Production Management, 26*(11-12), 1255-1275.

Porat, M. U. (1977). *The information economy: Sources and methods for measuring the primary information sector (detailed industry reports).* Washington, DC: U.S. Dept. of Commerce, Office of Telecommunications.

Porat, M. U., & Rubin, M. R. (1977). *The input-output structure of the information economy.* Washington, DC: U.S. Dept. of Commerce, Office of Telecommunications.

Prieger, J. E. (2003). The supply side of the digital divide: Is there equal availability in the broadband Internet access market? *Economic Inquiry, 41*(2), 346-363.

Regan, J. E. (1970). Dynamic aspects of information flow within a society. *IEEE Transactions on Engineering Writing and Speech, EW13*(2), 65-73.

Repo, A. J. (1987). Economics of information. *Annual Review of Information Science and Technology, 22*, 3-35.

Ricardo, D. (1911). *The principles of political economy & taxation.* London, England: J.M. Dent & Sons, Ltd.

Ross, S. S. (2004). Where the broadband systems are. *Broadband Properties Magazine, December*, 42-45. Retrieved January 2006, from http://www.broadbandproperties.com/2004%20issues/dec04issues/Where_the_Broadband_Ross.pdf

Rubin, M. R. (1983). *Information economics and policy in the United States.* Littleton, CO: Libraries Unlimited.

Rubin, M. R., & Porat, M. U. (1977a). *Information economy: The interindustry transactions matrices.* Washington, DC: U.S. Dept. of Commerce, Office of Telecommunications.

Rubin, M. R., & Porat, M. U. (1977b). *Information economy: The technology matrices.* Washington, DC: U.S. Dept. of Commerce, Office of Telecommunications.

Saade, R. G., & Otrakji, C. A. (2007) First impressions last a lifetime: Effect of interface type on disorientation and cognitive load. *Computers in Human Behavior, 23*(1), 525-535.

Seyhun, H. N. (1986). *Empirical tests of information economics models: Relation between expected value, cost and variance of information, and market efficiency.* Ann Arbor, MI: University of Michigan, School of Business Administration.

Shackle, G. L. S. (1961). *Decision, order, and time in human affairs.* Cambridge, England: University Press.

Shah, V., & Mehta, K. T. (1998). Workforce, information technology and global unemployment. *Industrial Management & Data Systems, 98*(5), 226-231.

Smith, A. (1937). *An inquiry into the nature and causes of the wealth of nations.* New York, NY: Modern library 1937.

Strnadl, C. F. (2006). Aligning business and IT: The process-driven architecture model. *Information Systems Management, 23*(4), 67-77.

Sui, D. Z. (2004). GIS, cartography, and the "third culture": Geographic imaginations in the computer age. *The Professional Geographer, 56*(1), 62-72.

Tatemura, J., Hsiung, W.-P., & Li, W.-S. (2003). Acceleration of Web service workflow execution through edge computing. In *The Twelfth International World Wide Web Conference, 20-24 May 2003, Budapest, Hungary* (pp. 1-12). [s.l.]: WWW Consortium (W3C). Retrieved January 2006, from http://www2003.org/cdrom/papers/alternate/P172/p172-tatemura/p172-tatemura.html

Tyler, B. B., & Gnyawali, D. R. (2002). Mapping managers' market orientations regarding new product success. *Journal of Product Innovation Management, 19*(4), 259-276.

United States Office of Management and Budget. (1998). *North American industry classification system: United States, 1997.* Washington, DC: Executive Office of the President & Office of Management and Budget: For sale by the U.S. G.P.O., Supt. of Docs.

United States Office of Management and Budget. (2002). *North American industry classification system: United States, 2002* (2002 Rev ed.). Washington, DC: Executive Office of the President & Office of Management and Budget: For sale by the U.S. G.P.O., Supt. of Docs.

United States Senate Committee on Labor and Public Welfare, S. S. o. S. (1968). *National Science Foundation Act Amendments of 1968: Hearings before the United States Senate Committee on Labor and Public Welfare, Special Subcommittee on Science, Ninetieth Congress, first session, on Nov. 15, 16, 1967.* Washington, DC: Government Printing Office.

Utah Telecommunication Open Infrastructure Agency. (2003). *White paper: Utah's public-private fiber-to-the-premises initiative.* [s.l.]: UTOPIA.

Van Gorp, A. F., Maitland, C. F., & Hanekop, H. (2006). The broadband Internet access market: the changing role of ISPs. *Telecommunications Policy, 30*(2), 96-111.

Versteeg, G., & Bouwman, H. (2006). Business architecture: A new paradigm to relate business strategy to ICT. *Information Systems Frontiers, 8*(2), 91-102.

Vickers, D. (1968). *The theory of the firm: Production, capital, and finance.* New York, NY: McGraw-Hill.

Vyhmeister, R., Mondelo, P. R., & Novella, M. (2006). Towards a model for assessing workers' risks resulting from the implementation of information and communication systems and technologies. *Human Factors and Ergonomics in Manufacturing, 16*(1), 39-59.

Warf, B., & Grimes, J. (1997). Counterhegemonic discourses and the Internet. *The Geographical Review, 87*(2), 259-274. Retrieved July 2006, from http://www.amergeog.org/gr/Apr97/Apr97-Warf.html

Waters, D. J. (1998). What are digital libraries? *CLIR Issues, 4*(July/August). Retrieved December 2006, from http://www.clir.org/PUBS/issues/issues04.html

Wheelan, C. J. (2002). *Naked economics: Undressing the dismal science.* New York, NY: Norton.

Wilbon, A. D. (2003). Shrinking the digital divide: The moderating role of technology environments. *Technology in Society, 25*, 83-97.

Wilson, J. H. (1972). Costs, budgeting, and economics of information processing. *Annual Review of Information Science and Technology, 7*, 39-67.

Workman, M. (2005). Expert decision support system use, disuse, and misuse: A study using the theory of planned behavior. *Computers in Human Behavior, 21*(2), 211-231.

Xu, L. D., Wang, C. G., Luo, X. C., & Shi, Z. Z. (2006). Integrating knowledge management and ERP in enterprise information systems. *Systems Research and Behavioral Science, 23*(2), 147-156.

Zielstra, J. (1999). Building and testing a Web-based community network. *The Electronic Library, 17*(4), 231-238.

Zook, M. (2006). The geographies of the Internet. *Annual Review of Information Science and Technology, 40*, 53-78.

Chapter III

Spatial Databases and Data Infrastructure

John Abresch, University of South Florida Libraries, USA

Peter Reehling, University of South Florida Libraries, USA

Ardis Hanson, University of South Florida Libraries, USA

Introduction

The emergence, in recent years, of digital libraries and of Internet-based communication applications have led some researchers to propose that the emerging data infrastructure of the Internet and the capabilities of digital libraries can be used to organize and ease data-mining digital geospatial data across the Internet. Digital geospatial data interoperability, the target of major efforts by standardization bodies and the research community since the 1990s, "has been seen as a solution for sharing and integrating geospatial data, more specifically to solve the syntactic, schematic, and semantic as well as the spatial and temporal heterogeneities between various real world phenomena" (Brodeur, Bédard, Edwards, & Moulin, 2003, p. 243). Some researchers point to the problem that many GIS systems are singular in nature, are generally isolated, and lack interoperability, due in part to the computer architecture upon which they are based (Lutz, Riedemann, & Probst, 2003). This chapter will discuss the emergence of a national spatial digital infrastructure vis à vis the development of a national telecommunications infrastructure. Federal policies, standards, and procedures will be reviewed that assist in the management and production of geospatial data. Several examples of current geospatial libraries will be examined. The chapter will conclude with a short implications section on what are necessary next steps and future trends.

Characteristics of Spatial Data

As discussed earlier in Chapter II, geographic data is comprised of variables that represent real-world phenomena. These can be natural such as climate regions, topographic features, vegetation zones, and other natural processes. They can also refer to entities and objects that represent manmade activities such as buildings, roads, bridges, cable networks. In representing real-world phenomena, researchers use various abstract models that can represent some or many characteristics of the phenomena under consideration. Having been recorded by the individual researcher or captured by mechanical means, the data representing different aspects of the phenomena are often arranged in layers. The layering of information is representative of the cognitive process. Individuals tend to perceive information about the particular space they occupy by mentally processing inputs from a variety of senses, thereby building up a mental image or map of the area. The layering or thematic ordering of a particular place is thus rendered.

Other attributes of real-world phenomena are its spatial characteristics (geometry) and its temporal (time) characteristics. The definition of space is integrated with not only the cognitive processes associated with human perception, but also of cultural values as well. Culture affects the value and rendering of a conceptual map of particular place. Other factors helping to define space are found in various classical and contemporary concepts in mathematics, such as Euclidian notions of geometry and measurement. Contemporary ideas of quantum mechanics further add to the concepts of space in the environment by blurring the boundaries of Euclidean geometries. The concept of space in models representing real-world phenomena are also conceptualizations of the space of data features. It has been noted that "spatial information is always related to geographic space, that is, large-scale space. This is the space beyond the human body, space that represents the surrounding geographic world. Within such space, we constantly move around, we navigate in it, and we conceptualize it in different ways" (Kainz, 2004, p. 30).

The most common form of model used for representation of real-world phenomena is the map. Maps are two-dimensional depictions of a particular aspect usually rendered on paper or other print media. Maps can be general, thematic, and even topographic in nature. Statically depicted phenomena are bound within the parameters of scale and accuracy of the data captured or recorded for depiction. Map scale determines the spatial resolution of the information. The larger the scale, the more detail can be depicted on the map. Determining the correct scale for a particular rendering of spatial data or map generalization is part of the processes associated with cartography. Cartographic concepts of data representation and of the graphical layering of data form essential parts of digital geographic databases and geographic information systems. Creating digital representations of real-world phenomena in a database form are discussed next.

Conceptualizing Spatial Databases

Much like maps, spatial databases store representations of phenomena in the real world. Geographic information systems use spatial databases, aka GIS databases. Spatial databases comprise a system, or hierarchy, of data models. The representations of the data and its potential applications across a network area drive its definition. The data model closest to the level of the end user is referred to as a conceptual data model. To design an effective database, the developer must know how the spatial data will be used and what the intended product will be. The definition at this level will identify what possible applications the spatial data will have, such as flood zone analysis, voting district redistribution, or library user analysis. The "commonly used conceptual data model is the entity-relationship (ER) model; it uses primitives like entity type to describe independently existing entities, relationship type to define relationships between entities, and attributes to describe characteristic values of entities and relationships" (Kainz, 2004, p. 32). A conceptual database schema is the definition for the entire hierarchy of data models.

A central assumption for the design of a spatial database schema is that spatial phenomena in a real-world setting exist in Euclidean space. Complex relationships exist between all the phenomena in the real world. These relationships have a variety of characteristics that give them specific spatial and temporal attributes. The real-world phenomena depicted can also be classified thematically. The thematic classification of data in layers can depend on the purposes for the data depicted. The data can thus relate to items such as economic zones, library service areas, hydrographic areas, or physical features. Kainz (2004) suggests that "the representations of spatial phenomena (spatial features) are stored in a scale-less and seamless manner. Scale less means that all coordinates are world coordinates given in units that are normally used to reference features in the real world (geographic coordinates as latitude and longitude, or metric units in meters). From there, calculations can be easily performed and any (useful) scale can be chosen for visualization" (p. 32). As in print maps, the accuracy of the data being recorded or captured is significant in the composition of a spatial database. Information derived from direct observation of phenomena should have the geographic coordinates defined. The accuracy of spatial database attributes is also affected if the data was derived from a secondary source of cartographic information such as a map. The scale of the data from the map would shape the feature coordinates in the database. An advantage of digital spatial databases is that a database does not depict boundaries between spatial phenomena, such as in map sheet boundaries or other partitions of the geographic space, other than imposed by the spatial features themselves.

Spatial databases are real-world models in "that they are scale-less, potentially three-dimensional, dynamic, and seamless. It is easy to query a database, and to combine data from different layers (spatial join or overlay). Spatiotemporal data-bases consider not only the spatial and thematic but also the temporal extent of the features they represent" (Kainz, 2004, p. 33). (The number of spatial, temporal, and spatiotemporal data models that have been developed is too large to address in this volume. For more information, the reader is referred to Kainz). However, the basic elements of a spatial database schema provide a structure in which to design a system for spatial querying and retrieval of information.

Elements of Database Design

As discussed earlier, spatial data are representations of real-world phenomena. The digital representation of the real world is often referred to as a digital landscape model (DLM). With the flexibility to represent different characteristics of the features of data at different levels of scale, the DLM is a central component in the processing of spatial data and for analysis. Model generalization, or conceptual generalization, refers to the process that uses a geometric component in the model (Kainz, 2004). Generalization means the reduction in the complexity of information. Unlike in cartography, which means suppressing unnecessary detail, in database development, generalization means information abstraction, or the suppression of detail in order to widen the meaning of the information. Utilizing the digital landscape model, the user can derive graphic representations of different aspects of spatial data in either digital form or traditional cartographic processes.

A common model is the ANSI/SPARC layered model of database architecture. Three schema, a physical schema, a conceptual schema, and user views, comprise this model (Tsichritzis & Klug, 1979). For example, using this model in the building of a spatial database provides organization to a data set. A schema adds order to the variables in a database. An overall framework (or logical structure that defines the database) is identified and defined as the physical schema. Database variables are mapped and their attribute relationships identified. Concepts can be mapped as well and their relationships are also identified.

The first step in organizing a spatial database is defining a logical schema for a set of variables. This initial step recognizes data categories that outline the parameters

Table 1. ANSI/SPARC architecture data models and schemas

SCHEMAS	MODELS
Schema	Models used to derive the schema
External views	Based on different *n* user perspectives, a spatial model is created by defining and describing a subset of the real world.
Conceptual schema	External views synthesized to create conceptual schema using semantic data modeling techniques, for example, the entity-relationship approach.
Logical schema	Conceptual schema transformed into a logical schema using the relational model. Emphasis is on redundancy removal.
Physical schema	Logical schema mapped into data structures and algorithms. A "hidden" process not accessible or seen by user.

of groups of data. The framework allows for inserting attribute information, which further describes the attributes of the variables. Populating data categories creates a DLM. Since the DLM is an object-orientated topographic database, its data structure facilitates spatial analysis and linkage of geographic objects to external data. A DLM uses the vector as its primary geometric form, and often contains explicit or implicit topological information. The objects, their attributes, and the relations between the objects are referred to in terms of real-world entities (Kainz, 2004). Entities are comprised of type classification, attributes, and relationships. An entity may have one or more attributes, such as a building's (entity) attributes may be its characterizing material, such as block, brick, or frame. Attributes describe quantitative data ranked by three levels of accuracy: ordinal, interval, and ratio. Ordinal (ranked) may rank an entity from "worst to bad to good to better to best," interval (numeric) may address an entity's age or income, and ratio (scale) may address the length or area of an entity.

Another aspect of designing a database is in evaluating user perceptions of the data. Since a database generally serves multiple users or user groups, users may have very different perceptions of the attribute data collected. Each user (or group) receives his or her external view of the database to create a personalized conceptual database schema. Database designers then merge the external views of the data into a single conceptual schema of the database.

In designing a database, a conceptual schema is not determined by the parameters of a measurement tool, technique, or paradigm, but by its flexibility, which allows it to deal with the vagueness and uncertainty of defining different aspects of human-centered phenomena in the real world. After the phenomena types are defined, the conceptual schema is transformed into a logical schema using one of the logical data models, such as a relational data model. Since each fact should be stored only once in a database, the logical schema allows the development of a redundancy-free dataset. A physical schema is the result of the implementation of the logical schema with particular database management software.

Table 2. Data modeling

Physical Reality	Real World Model	Data Model	Database	Language
Phenomena	Entity	Object	Object	Symbols
Properties	Type	Type	Type	Lines Text
Connections	Attributes	Attributes	Attributes	Images
	Relationship	Relationship	Relationship	Tables
		Geometry	Geometry	Charts
		Quality	Quality	

The structure of spatial databases provides a method for libraries with digital geospatial data collections and services to create a system for the discovery and querying of geographic information across the online environment of the Internet. Many of the sources of digital geospatial data, such as private corporations and government agencies, reside in locations across the United States and the globe. Besides spatial databases, building a digital geospatial collection in an online environment will use a variety of software applications and hardware tools that will assist in the administration of information as it passes from host to user. Next is a brief discussion of the development of some of the applications and protocols utilized in data transfer in the online environment of the Internet.

Emergence of a Telecommunications Network

The contemporary information economy emerged as industries in established economic sectors, such as manufacturing and production services, incorporated computer technologies to their daily operations. The concentration of industrial, technological, and social capital of such industries in urban areas enabled the building of advanced telecommunication services. Firms were able to take advantage of existing telephone lines and exchanges to build new information networks that were quickly using new communication software applications.

In 1961, a researcher at RAND for the U.S. Department of Defense, working on how the U.S. telecommunications infrastructure could survive a "first strike," published a proposed digital data communications system based on a distributed network concept" (Baran, 1964). He introduced the concept of redundancy and the use of message-block (packet-switching) networks with no single outage point as a method of building communications systems to withstand outages. This became the underlying data communications technology for the Internet.

The development of ARPANET, created by the Advanced Projects Agency of the U.S. Department of Defense in 1969, was an early catalyst in the integration of computer and telephone technologies (Ayres & Williams, 2004), and enabled the exchange of information related to scientific study occurring in advanced computing centers. In the beginning, ARPANET's architecture consisted of 4 nodes (sites) located at the University of California at Los Angeles, the University of California at Santa Barbara, the University of Utah, and Stanford Research Institute.

Early experimentation with the system resulted in the creation of Telnet, an openly accessible public packet data service that allowed a computer operator at a terminal or PC to log onto a remote computer, run a program, and initiate FTP, an early file transfer protocol (FTP). The CCITT (International Consultative Committee on Telephony and Telegraphy) approved the first guidelines for X.25, a network protocol using virtual circuits that became the backbone of the TCP/IP (Transmission control protocol/Internet protocol) protocol.

Why is a communication protocol important? Communication between computers means sending messages from one machine to another. There are three types of communication. Simplex communication is message travel in only one direction. Half-duplex communication is asynchronous message travel in both directions. Half-duplex communication is not simultaneous, for example, much like using ham radio, the first person must say "over" at the end of his communication so the person at the other end knows that it is his turn to talk. Full-duplex communication is simultaneously sending and receiving messages in both directions simultaneously, with no lags or gaps in the transmission or receipt. Obviously, addressing mechanisms that allow unique identification of senders and receivers are very important. Other critical mechanisms are rules on how data travels, error-detection and error-correc-

Table 3. TCP/IP protocol

Layer	Function	Protocols Used
Process (Application) Layer	"Higher level" protocols, such as SMTP, FTP, SSH, HTTP, and so forth operate in this layer.	SMTP, FTP, SSH, HTTP, DHCP , IMAP4, IRC, MIME, POP3, SIP, SNMP, TELNET, TLS/SSL, RPC, RTP, SDP, SOAP
Host to Host (Transport) Level	Flow control and connection protocols live here. Opens and maintains connections. Ensures packets are actually received.	TCP, UDP, RSVP, DCCP, SCTP
Internet (Internetworking) Layer	Defines IP addresses and routing schemes for navigating packets from one IP address to another. Performs network routing, flow control, network segmentation/desegmentation, and error control functions.	IP, ARP, BGP, ICMP, IGMP, IGP, RARP
Data Link Layer	Responsible for node to node (hop to hop) packet delivery.	ATM, Bluetooth (PANs), DTM, Ethernet, FDDI, Frame Relay, GPRS, Modems, PPP, Wi-Fi
Physical Layer	Describes the physical equipment necessary for communications, such as twisted pair cables, the signaling used on level protocols using that signaling.	Bluetooth RF, Ethernet physical layer, ISDN, Modems, RS232, SONET/SDH, USB, Wi-Fi, Power line communication

Figure 1. The 14 fields that comprise an IP packet

Version	IHL	Type-of-service	Total Length	
Identification			Flags	Fragment offset
Time-to-Live		Protocol	Header Checksum	
Source Address				
Destination Address				
Options (+ padding)				
Data (variable)				

← 32 bits →

tion, disassembling and reassembling long messages, avoiding data overflow due to fast transmitters and slow receivers, and routing of messages.

TCP/IP is the basic communication language, or protocol, of the Internet and for private networks (e.g., intranets or extranets). TCP/IP communication is primarily a point-to-point protocol (PPP), meaning each communication is from one point (or host computer) in the network to another point (or host computer). There are five layers to the TCP/IP protocol. The higher layer, the transmission control protocol, disassembles a message or file into smaller packets that are transmitted over the Internet and received by a TCP layer that reassembles the packets into the original message. The third layer, the Internet protocol, handles the address part of each packet so that it gets to the right destination. TCP/IP and the higher-level applications that use it are collectively said to be "stateless" since every client request is considered a new request and unrelated to any previous request. Being stateless frees network paths so that everyone can use them continuously, unlike "dedicated" lines, such as plain old telephone service (POTS) or fax lines.

By the late 1980s and early 1990s, the Web was accelerating away from a dedicated mainframe environment to a distributed client-server system. This distributed system challenged the development of new protocols. Considering the variety of personal computers, simple terminals, servers as well as platforms, and operating systems, requirements for new protocols had to be simple, cross-platform, and non-computer specific. Based upon the idea of HyperCard, Berners-Lee of the Conseil Européen pour la Recherche Nucléaire (CERN; European Council for Nuclear Research) developed hypertext transfer protocol (HTTP) and its accompanying text format, hypertext mark-up language (HTML). HTML was based on SGML (standard generalized mark-up language), an internationally agreed upon method for marking up text into structural units, such as paragraphs, headings, and so forth, that was non-machine, non-platform specific. Two significant components of HTTP and HTML are the use of hypertext links to "anchor" items to each other inside and outside the "page" and the "www" naming protocol (URL) used for addressing Web sites. A uniform resource locator (URL) consists of three parts: the name of the protocol (http), the host name where the page resides (www.itc.nl), and the name of the hypertext document (home.htm). HTTP, TCP/IP, FTP, Telnet, and SMTP form a "suite" of protocols that Internet users use as a matter of course.

Building on a basic design of a document-sharing protocol, the Web eventually developed into a medium for the creation of a variety of Web sites that met myriad individual and organizational demands and needs. Although many of the top-level domains of the net were still educational, much of the early development of the Internet occurred in the private sector, with Internet service providers (ISPs), such as America Online (AOL). From its beginning as a specialized dial-up service for Apple Macintosh users in 1989, within 6 years AOL had a million subscribers. This was a substantial inroad in an environment where there were about 20 to 30 million users. The success of these early Internet service providers triggered tremendous growth in the number of Web sites during the second half of the 1990s. The availability of inexpensive access through local land telephone lines and relatively cheap cost of transmitting data. The number of commercial (.com or dot com) domains increased from around 1 million in 1994 to nearly 25 million at the beginning of 2000, while other categories of users (.net, .mil, .org, and .gov) added another 19 million domains, of which 15 million were cable-based net service providers (Ayres & Williams, 2004). The development of HTTP and HTML spawned thousands of new sites and inspired new information services, such as "browsers," "search engines," and "portals" to enable net users to find information of interest. More specialized and powerful search engines continue to emerge, including the current leader, Google. Additional Web-based services, and continued growth depended on the established telecoms or cable TV systems and their telephone wire infrastructure.

Rapid development of the Internet in the 1990s was accompanied by a simultaneous investment in construction of new telephone access lines in the United States. Investment in new lines by major carriers in the United States increased 32% between 1990 and 2000 from 119.8 million to 157.6 million lines (Gabe & Abel, 2002). The number of the new lines that incorporated ISDN (integrated services digital network) technology was 129.6 million, an increase of over 850% from the previous number of 13.6 million (Gabe & Abel, 2002). Transmitting data at speeds ranging from 128 kilobits per second to over 150 megabits per second, ISDN networks support a wide of range of simple to complex voice and non-voice services, and allow for the transmission of multiple channels of information, such as voice, data, fax, and video, over a single wire (Kessler, 1990). The spatial distribution of ISDN-technology-enabled networks across the United States is uneven, with a concentration of high-speed data lines in areas that have the necessary infrastructure for its development (Gabe & Abel, 2002).

Broadband, defined as 200 megabits per second of data output, is about four-times faster than a 56 Kbps dial-up modem and about eight-times faster than most people's actual download speeds, since many IPS's modems offer a maximum of 28.8 Kbps (Strover, 2001). Since many contemporary computer applications, for example, music clips, video clips, and streaming video, use considerable amounts of bandwidth, broadband access is growing as an essential Internet service expectation (Malecki, 2003; Prieger, 2003). These graphics intensive applications have many business and entertainment functions, and are not possible at lower dial up Internet connections. Broadband Internet access is usually provided through digital subscriber line (DSL) technology. Data on the Regional Bell Operating Companies indicate that more than 56% of all cities with populations above 100,000 had DSL available, but less than 5% of cities with populations less than 10,000 had DSL service (Malecki, 2003).

A number of ISPs have been building data lines in predominately rural areas. These ISPs tend to be small businesses that fill in gaps in data line access where no larger telecommunication firms are present (Malecki, 2003). Since the costs of establishing Internet service warrant a subscriber base of at least 200 households, some small telephone exchanges are too small to sustain an ISP of that size (Strover, 2001). The rural ISPs offer dial-up access in low bandwidth transmission formats. Most of the Internet services in rural areas tend to be related to e-mail and e-commerce transactions. If access to more services or larger bandwidth is required, costs are significantly higher for the user as well as the ISP, hence, Internet use in rural areas is held down by higher costs.

The economics of fiber optics can be expensive for urban connections as well. In situations where new trenches must be dug for the cable, installing fiber in metro areas can run in the hundreds of thousands of dollars per mile. Although installation cost can often be reduced by intensive use of existing infrastructure, it is difficult to justify the investment when existing phone lines can deliver adequate bandwidth via ADSL (Asymmetric Digital Subscriber Line) technology. ADSL is a modem technology that transforms POTS (plain old telephone service) lines into high-speed

Figure 2. Server farm configuration

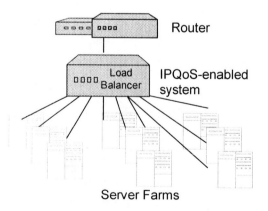

Server Farms

digital lines. By splitting an existing telephone line signal into two, one for voice and the other for data, ADSL technology can work at up to 8-Mbps download. Since the baseband is occupied by POTS, should ADSL fail, POTS service is guaranteed. Similar to telephone switching centers, the Internet requires specialized host computers called servers and routers. Servers are typically clustered in so-called "server farms" around the world, located in urban areas where most of the traffic originates. Server farm capacity is currently increasing at an annual rate of 50% (Ayres & Williams, 2004).

One of the biggest providers is Exodus Communications, with 9 server farms in Silicon Valley alone and 35 more in big cities around the world. Think of server farms as "wholesalers" in the Internet and the "data caches" as retailers, located around the world near consumers of content. Technically, the function of the data caches is to recombine the individual packets of data, dispatched by the routers by different routes (because of changing conditions from moment to moment), into coherent streams designated for final customers. The leader in this field is Akamai, with 11,000 caching servers in 62 countries. Akamai's customers are the content providers (including CNN and Yahoo!). The net result of all this investment was to create an oversupply of long-distance fiber-based carrier capacity, without a matching growth in local broadband access capacity for which the established telecoms retained their monopoly (except where cable TV was also available). As of early 2002, long-distance optical fiber channel capacity is said to be only 2%–5% utilized,

Figure 3. Putting it all together: Retrieving data from the Web

By opening a browser from his or her personal computer, a user can search, access, and view a variety of different media in a variety of formats (sound, video, or images) from a global selection. To view different types of media, the browsers need external viewers, or helper applications. Frequently these helper applications (also known as "plug-ins") are built into the browser. Machines accessing the Web through a browser must be directly connected to the Internet, or at least able to establish a PPP connection with an ISP. Web pages are addressed through a URL.

http://www.usf.edu

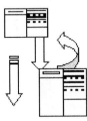
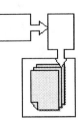

1. A user clicks a hyperlink or enters a URL.
2. The browser determines the URL (when a hyperlink was clicked).
3. The browser asks the domain name service (DNS) for the IP address of the host computer.
4. DNS resolves the name and returns the IP address.
5. The browser establishes a TCP connection to the host.
6. The browser requests the document.
7. The host server sends the requested file.
8. The TCP connection is closed.
9. The browser displays all the text of the document. The browser fetches and displays all images in the document.

whereas local access in many places outside the biggest cities, still dependent on copper wires, is badly congested. The limiting factor is electric power consumption, especially as the farms get larger (one new server farm, being built by iXguardian near London, will have its own gas-fired power plant).

By opening a browser from his or her personal computer, a user can search, access, and view a variety of different media in a variety of formats (sound, video, or images) from a global selection. To view different types of media, the browsers need external viewers, or helper applications. Frequently these helper applications (also known as "plug-ins") are built into the browser. Machines accessing the Web through a browser must be directly connected to the Internet, or at least able to establish a PPP connection with an ISP. Web pages are addressed through a URL.

1.   A user clicks a hyperlink or enters a URL.
2.   The browser determines the URL (when a hyperlink was clicked).
3.   The browser asks the domain name service (DNS) for the IP address of the host computer.
4.   DNS resolves the name and returns the IP address.
5.   The browser establishes a TCP connection to the host.
6.   The browser requests the document.
7.   The host server sends the requested file.
8.   The TCP connection is closed.
9.   The browser displays all the text of the document. The browser fetches and displays all images in the document.

Building on the data transfer capabilities of the Internet, also known as the World Wide Web, and its accompanying telecommunication infrastructure, the online environment of the Internet is rapidly becoming a standard platform for GIS as government agencies and private companies begin to exploit the data exchange capabilities of the Internet.

Characteristics of Distributed Spatial Databases

Spatial data infrastructures are based on large amounts of spatial data distributed over many agencies. The survey of the information economy indicates that industries in the information sector rely on efficient data transfer between organizations and users for their operations. Even though spatial databases, as described earlier in the chapter, may have useful socioeconomic purposes, a system linking many spatial databases would have far greater applications in the distributed environment of the

Figure 4. Visual representation of the ANSI/SPARC architecture

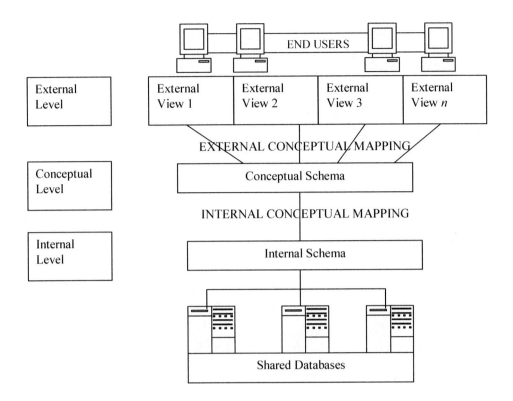

information sector. What is the structure of a distributed spatial database system? First, two components, a database and a database management system, comprise a database system. The database (DB) is an organized collection of stored data; the database management system (DBMS) is software for building and maintaining a database. The strength of a database system is in its design. It contains a number of features, such as "persistency, storage management, recovery, concurrency control, ad hoc queries, and data security," that allow for the efficient processing of data (Kainz, 2004, p. 84). A persistent storage capability allows data to exist independently of the application of a program.

The ability to recover data is important in the workings of a spatial database system. If in the execution of a program is not successful, the database management system can return the database to its former state before the attempted software application or to its former uncorrupted state. A concurrency control mechanism is also part of the normal operations of a distributed spatial database system. Its purpose is to avoid inconsistencies caused by concurrent read and write operations to the database and security access to the server on which the database is located. The ability of the database system to offer access to multiple users is also based on having data access security delineations.

The architecture of most databases is based on the standards of the ANSI/SPARC Study Group on Data Base Management Systems (Tsichritzis & Klug, 1979). These databases have three levels: internal, conceptual, and external level. The external level is closest to the user, the internal level is closest to the physical storage, and the conceptual "exists in the middle." Although individual users may have different external views on the database, there is only one conceptual view and one internal view. A data definition language (DDL) defines the views. A data manipulation language (DML) describes the processing of the database objects (Bobak, 1996).

Databases store models of conceptualized real-world phenomena. In the design process of a database, several data models are used to describe the various feature attributes of the variables in the database. A data model describes the contents and organization of the data. As discussed earlier, they can be classified into conceptual, logical, and physical data models. Proceeding with the variable feature descriptions as defined by the various data models, the data are then organized as conceptual model. The next step would be to construct a logical data model. A database management system can then implement the logical data model.

An example of a logical data model is the relational data model, which is in widespread use among desktop GIS users using commercial software (e.g., ArcGIS and MapInfo). A relational data model is characterized by a clear distinction between the graphical and descriptive data of features representing real-world elements. Graphical elements are stored in layers of the digital map file and attribute data is stored in the form of a relational database table. Spatial features on the digital map are linked through the identifier with the record of the relational database table containing further descriptive data of the feature. A commonly used language for the relational database model is structured query language (SQL). Supported by all major relational DBMS vendors, SQL is both a data definition and data manipulation language (Kainz, 2004).

In a spatial data infrastructure, many databases are distributed over many organizations across a network, and are interconnected through a communication network. Every site runs a unique database management system. Local applications operate on local hardware, while distributed or global applications can involve multiple sites on the data infrastructure. The design process for distributed databases builds on the design process for most database design, that is, it has a conceptual step, logical step, and physical design phase step. An additional design dimension is the distribution design step. The integration into a network introduces a number of design ideas to the building of distributed databases, such as distribution transparency, fragmentation, and replication. Distribution transparency indicates that most users do not know where the data in a database is actually located. Replication, where duplicate datasets may be stored in different locations, and fragmentation, when different parts of a dataset may be stored on different servers, are frequently performed for security and practical reasons. In both cases, the user always perceives a centrally located database.

Kainz (2004) notes that "there are two possible approaches to the design of distributed databases: top down and bottom up. The top-down design is applied to new databases that are designed from scratch" (p. 86). Many GIS data producers and providers participate in the United States Federal Geographic Data Committee (FGDC) data infrastructure initiative. A system of distributed spatial databases that provide users with access to myriad federal, state, and local datasets, the central component of the data infrastructure is the "clearinghouse." Originally defined as a "system of software and institutions to facilitate the discovery, evaluation, and downloading of digital geospatial data," today the FGDC defines a clearinghouse as "a community of distributed data providers who publish collections of metadata that describe their map and data resources within their areas of responsibility, documenting data quality, characteristics, and accessibility (http://www.fgdc.gov/dataandservices/). The FGDC clearinghouse(s) also address the descriptive component of the data through the use of metadata, which is discussed in more detail in chapters IV and V. The clearinghouse computer systems integrate many interoperating metadata servers, using a distributed, client-server architecture. On the client side, the most common software application is usually browser-based, such as Internet Explorer or Netscape Navigator. As described earlier in the chapter, a client residing on computer X can interact with server Y, located at another location, using a set of instructions called protocols. Since most traffic on the Internet uses the TCP/IP (transmission control protocol/Internet protocol), the TCP/IP software suite is frequently embedded in a computer's operating system (OS) software, such as in Microsoft Windows, MAC OS X, Unix, and Linux free ware. One protocol selected to provide search interoperability among different servers is the ISO 10163-1995 (or ANSI Z39.50-1995) search and retrieve protocol. The Z39.50, initially developed for the library community, contains client and server software that establishes a connection, relays a query, returns the query result, and presents retrieved documents in various formats.

Databases, Web Services, and Internet GIS

There are a number of advantages to a GIS that is available from the Internet including world wide accessibility, use of a standard interface, and cost-effective maintenance (Green & Bossomaier, 2002). The traditional model of GIS is a system that consists of a single software package, and accompanying data on a single machine. With the advent of distributed producers and consumers of digital geospatial data, this model is no longer valid. As Green and Bossomaier (2002) note "many GIS projects are often multi-agency, multi-disciplinary, multi-platform, and multi-software. Large numbers of contributors may be involved, and there may be a large pool of potential users" (p. 23). Further, "a central practical issue is how to provide widespread, device-independent access to GIS for large numbers of contributors and users" (p.23). The essential differences between a traditional GIS and an online GIS are the "separation of user interface, data storage, and processing" (p. 23). In stand-alone GIS, all of the elements are within a single machine; in an online environment, the elements are distributed across many computers.

During the 1980s, GIS was conceptualized strictly as a software application: "a computer based system that provides the following four sets of capabilities to handle georeferenced data: 1. input; 2. data management (data storage and retrieval); 3. manipulation and analysis; and 4. output" (Aronoff, 1993). This definition is accurate when conceptualizing GIS as an application; however, it does not address GIS as an interactive system where hardware, software, data, methods, and people come to create geographic knowledge for distribution. In the early 1990s, the concept of a GIS system evolved. ESRI, a privately held consulting firm, defined GIS as an "organized collection of computer hardware, software, geographic data, and personnel designed to efficiently capture, store, update, manipulate, analyze, and display all forms of geographically referenced information" (Rhind & Connolly, 1993).

Reid, Higgins, Medyckyj-Scott, and Robson (2004) observed, "The exploitation of geospatial data within diverse policy environments, allied with the increasing attention being afforded to cross-discipline social and environmental issues, has led to the demand for infrastructures to assist in the discovery, dissemination, and exploitation of geospatial data" (¶2). The authors also note that the structures are often referred to as spatial data infrastructures (SDIs). " Most SDIs will integrate geospatial data and metadata, which can provide the means to access the data and also establishes the needed licensing agreements between users to make use of the data" (Reid, Higgins, & Medyckyi-Scott, 2004, ¶4).

The diverse information services of SDIs mirror information services of digital libraries, the latter defined as a collection of services and the collection of information objects that support users in dealing with the information objects and the organization and presentation of those objects through electronic means. Three significant factors of SDIs are the provision, organization, and presentation of information and services to a specific group of users. By integrating with the digital library environment, the process of mining through digital geospatial data becomes more varied with additional data added to the original projection datum (Reid et al., 2004).

In delivering digital geospatial data over the Internet, the identification, retrieval, and delivery of information in a virtual environment is described as a "URL-addressable resource that performs functions and provides answers" (Seybold as cited in Parsons, 2003, p. 5). Web services are described as an "encapsulation of existing software functionality in a common form that allows the services it performs to be visible and accessible to other software applications" (Parsons, 2003, p. 5). The single Web services-based application can request services (three uses of services in one sentence) from other Web services, and can expect to receive the results or responses from those requests in an expected form. One advantage of Web services technology is its ability to interoperate in a loosely coupled manner, an "ask and wait" approach. The user can request a certain type of service across the network, and wait for responses. Web services can also be found and used by other applications, agents, and clients on the Internet (Parsons, 2003).

In a digital geospatial data retrieval system, a Web service can be published inside or outside the host's firewall by providing a document describing its operation and function. Using the Web services definition language (WSDL) markup language, the core of the document may illustrate a service when interpreted on a national grid reference responds with relevant coordinates on the host's mapping system. The document then described how the service is prompted and what the executables are. The document can also be published to a node, a clearinghouse, or other Internet site using the Universal Discovery, Description, and Integration (UDDI) registry. Parson (2003) also notes that communication between Web services pass "XML messages wrapped in an interoperable framework to allow the messages to cross different networks, use different application architectures and systems" (p. 6).

The widespread use of digital geospatial data by a number of different individuals and agencies in both the public and private sectors will continue to create large amounts of archived data. To make such data sets searchable across the environment of the Internet, librarians must be willing to take a varied approach in determining the appropriate metadata scheme to enable effective mining across the World Wide Web. A significant factor in the development of Web GIS services and in online services in digital libraries has been the National Spatial Data Infrastructure initiative by the Federal Geographic Data Committee.

Organizing a National Data Infrastructure

In 1994 the Federal Geographic Data Committee (FGDC) recommended that a National Spatial Data Infrastructure (NSDI) be organized. The NSDI would be an "umbrella of policies, standards, and procedures under which organizations and technologies interact to foster more efficient use, management, and production of geospatial data" (1994). The NSDI was conceptualized as fostering more cooperation and interaction between different organizations from the public and private sector. The recommendation was a response to alterations in the traditional flows of data across the government based on new convergences of information, computer technology, and communication. A significant component of the data being used in government transactions and commercial transactions was digital geospatial data.

The FGDC considered digital geospatial data as being critical to solving many "environmental, economic, and social problems." The FGDC had also recognized that "the use of GIS technologies for the digital analysis of spatial problems had become pervasive" (1994, p. 1). With the integration of GIS and increased use of geospatial data into problem-solving operations, a major goal of the NSDI was to make "an environment to respond to the current need of digital geospatial data." (FGDC, 1994, p. 1), The FGDC envisioned building the framework around the procedures, technology, and guidelines that enhance integration, sharing, and use of these data, and the institutional relationships and business practices that encourage the maintenance and use of data. The framework would represent the best available data for an area, and be certified, standardized, and described according to a common standard. Today, the FGDC standards provide a foundation on which organizations can build by adding their own detail and compiling other data sets.

Meeting Digital Geospatial Data Needs

The FGDC reported that many billions of dollars are spent by organizations on an annual basis in both the public and private sector in attempts to manage digital geospatial data (Federal Geographic Data Committee, 2006a). With digital geospatial data becoming a valuable asset within the business practices of many public and private sector organizations, most organizations can only afford to collect or purchase only a small portion of information they require. Often only the most basic digital geospatial data is acquired by organizations, usually a dataset that pertains to a specific geographic area or that has some unique attribute characteristics. The datasets are often incompatible with other datasets, due to differences in software platforms. Researchers at the FGDC found that data collected by differing organizations may be similar in geographic extent use different geographic bases and standards. Hence, "many of the resources organizations spend on geographic information systems (GIS) go toward duplicating other organizations' data collection efforts. The same geographic data themes for an area are collected again and again, at great expense." A data framework would improve the ability for organizations to share data. The framework would provide a "basic geographic data in a common format and an accessible environment that anyone can use and to which anyone can contribute. In this environment, users can perform cross-jurisdictional and cross-organizational analyses and operations, and organizations can funnel their resources into applications, rather than duplicating data production efforts" (Office of Management and Budget, 2000 as cited in Maryland State Geographic Information Committee, 2002, p. 17).

Researchers at the FGDC further noted in their findings that "geographic data users from many disciplines have a recurring need for a few themes of basic data: geodetic control, orthoimagery, elevation, transportation, hydrography, governmental units, and cadastral information" (Federal Geographic Data Committee, 2006b). Many organizations produce and use such data every day. The framework provides basic information for these data themes. By attaching their own geographic data, which can cover innumerable subjects and themes, to the common data in the framework, users can build their applications more easily and at less cost. The seven data themes provide basic data that can be used in applications, a base to which users can add or attach geographic details and attributes, a reference source for accurately registering and compiling participants' own data sets, and a reference map for displaying the locations and the results of an analysis of other data.

Researchers at the FGDC envision the information framework to be an evolving data resource to which geographic data producers can contribute. Integral components of the information network incorporate "procedures, guidelines, and technology to enable participants to build, integrate, maintain, distribute, and use framework data. These elements ensure that users can depend on accurate, detailed data that can be certified and integrated into the framework to create a trustworthy data source; users can update their data holdings from the framework data; and users can attach additional information to the framework" (Federal Geographic Data Committee, 2006b). These procedures would ensure the standardization of datasets and enable the efficient transfer of digital geospatial data across the Internet. The FGDC envisions the information framework integrating data from all types of organizations in both the public and private sector. The framework environment is designed to be responsive to the needs of the geographic data community in terms of "data creation and maintenance, and provides unrestricted access to data" (Federal Geographic Data Committee, 2006b).

A Cooperative Information Network

The FGDC also envisions organizations building the information framework by coordinating data development activities. Organizations can coordinate framework data along two dimensions. "The first dimension emphasizes opportunities for organizations with similar needs. An example is a metropolitan area in which local governments, their customers, state and federal agencies with facilities in the area, utilities, and others require high-resolution spatial data for their operations. In this case the framework provides a starting point for sharing the commonly needed geographic base information and allows each organization to add the unique information it requires to meet its business needs" (Federal Geographic Data Committee, 2006b).

The second dimension emphasizes opportunities for organizations needing different amounts of detail for an area. For example, a local government, a regional transportation planning organization, and a state transportation agency may require road data for an area, but at different levels of geography. For efficient data sharing the organizations would have to share the results of their individual efforts and would benefit from using a common geographic base and generalized data created from this base. The framework provides a starting point for a base and the data generalized from it, "providing the organizations with contemporary and consistent data for decision making and helping them avoid confusion caused by differences in the vintages, common attributes, and other characteristics of the base data" (Federal Geographic Data Committee, 2006b).

The framework is being developed by this entire community, with organizations from all areas playing roles. For some, the framework will supply the data they need to build applications. Others will contribute data, and some may provide services to maintain and distribute data. Some organizations will play several roles in framework development, operation, and use. An essential community partner in the information network can be libraries. Librarians have the skills and experience needed to help the information network to facilitate digital geospatial data and to aid in its search and mining procedures. Some researchers have advocated the "geolibrary" as a possible model for a library to facilitate digital geospatial data.

Digital Geolibraries and Digital Collections

One approach used by librarians in making digital geospatial data available to their users is to incorporate Web-based mapping functions with their library geospatial information holdings. Such efforts combine the locational aspects of the metadata descriptions of the geospatial data in the catalog with the mapping capabilities of GIS software. In building Web-based functions into their geospatial collections, some libraries have termed their collection a "digital geolibrary." Different perspectives about geolibraries have been discussed in the scholarly geography and information science literature such as from Boxall (2002) and Goodchild (1998). Goodchild (1998) refers to a geolibrary as being comprised of georeferenced information that can be accessed by a geographic footprint. Likewise, in his discussion of geolibraries, Boxall (2002) notes that their scope extends beyond the traditional map library if the construct of a geolibrary is based upon the idea that information has a geographic footprint (p. 2). His discussion references earlier research (Goodchild, 1992) that explains georeferenced information as including such things as photographs, videos, music, and literature that can be given a locational variable that defines a footprint. The collection areas of geolibraries extend beyond the traditional scope of map libraries and archives to include almost all information contained within libraries. He later mentions that it can include information outside of libraries as well (Boxall, 2002). This is the theoretical basis for what we now view as geolibraries.

Boxall (2002) discusses that a significant focus of geolibraries is on digital information and metadata, as well as the distributed nature of the libraries and "collections." He adds that "…Distributed geolibraries provide a useful framework for discussion of the issues of dissemination associated with the National Spatial Data Infrastructure (NSDI)" (p.1). The vision is readily extendible to a global context (National Research Council Mapping Science Committee, 1999). Boxall (2002) feels any discussion of "Digital Earth" (DE), the "Global Spatial Data Infrastructure" (GSDI), and "Distributed Geolibraries" should be framed around the broadest definitions of information and infrastructures; namely to include and focus upon the people, technology and organizations which give rise to and sustain such infrastructures" (p. 3).

An early significant effort at establishing a digital spatial library was the Alexandria Digital Library at the University of California Santa Barbara. The Alexandria Digital Library was one of six federally funded library projects. The U.S. National Science Foundation founded it in early 1994. The Library's collection and services focus on georeferenced information: maps, images, data sets, and other information sources with links to geographic locations (Hill, Carver, Larsgaard, Dolin, Smith, & Frew,, 2000). Much of the information in the collection was primarily of the University's service area, or Southern California. A key aspect of the collection is the ability to perform data queries and retrieve results by geography location. The basic means of describing and finding information is with a geographic footprint. A footprint depicts the location on the surface of the earth associated with either an object in the collection such as a map, remote sensed image, or aerial photograph, or with a user's query. The footprint may be represented as a point or polygon, with latitude and longitude coordinates (Hill et al., 2000). As a user queries the collection through a user interface, the user creates a footprint or an interactive map to indicate the area of interest (the query area). The query area is matched with the object footprints in the metadata to retrieve relevant objects about the query area. This approach to query structure allows the user to choose arbitrary query areas and is not limited to geographic areas with place names. The objects in the collection that fall within a particular query area do not have to have the names associated with them that the user enters for a text based query (Hill et al., 2000). By translating a user's text-based query into a footprint query for a certain geographic area, the user can retrieve all types of information about a location such as remote sensed images, data sets, aerial photographs, and textual information. The catalog for the Alexandria Digital Library is configured for searches that will retrieve objects that are either in an online format or physical location as a map (Hill et al., 2000).

Using the Alexandria Digital Library as a model, the Idaho Geospatial Data Center was started in 1996 by a team of geographers, geologists, and librarians. The aim of the team was the establishment of digital library of public domain geographic data for the state of Idaho. As a theoretical and practical foundation for their digital spatial collection, which they termed a "geolibrary," the team used a set of parameters as defined by Goodchild (1998). Goodchild (1998) defined a geolibrary's components as including a browser or specialized software application running on the user's computer and providing access to the geolibrary through a computer network. A geolibrary also includes a "basemap" or geographic frame of reference for the browser's queries. A basemap would provide an image of an area corresponding to the geographical extent of geolibrary collection. The basemap would depend on the scale of the search being performed from a large geographic area, such as a state, or a smaller location, as a city block. A gazetteer or index that would link place names to a map would also be included. A large collection of collection catalogs would be maintained on distributed computer servers. The servers would be accessed over a network with the browser, using basic server-client architecture. A geolibrary would ideally provide open access to many types of information with geographic referenced queries regardless of the storage media (Jankowska & Jankowski, 2000)

Using a grant from the Idaho Board of Education's Technology Incentive Program, the team built a geographic digital data repository or the Idaho GeoSpatial Data Center (IGDC). The library contained a number of digital geospatial datasets that was searchable through a flexible browsing tool. The collection contained a number of public domain information such as Digital Line Graphs and Digital Raster Graphics from the United States Geological Survey and U.S. Bureau of the Census TIGER boundary files for the state of Idaho. The site provided an interactive visual analysis of selected demographic/economic data for Idaho counties. The site also contained interactive links to other Idaho and national spatial data repositories (Jankowska & Jankowski, 2000)

A key aspect of the IGDC's collection was the development of the GeoLibrary's browser, which was implemented using MicroSoft Visual Basic 5.0 and ESRI MapObjects technology. The interface of the browser consisted of three panels resembling the Microsoft Outlook user interface. A first panel, a map panel, would be used to explore the geographic coverage of the geolibrary and to select an area of interest. A second panel in the interface was where the query would be performed. The final panel is where the query results would be displayed for analysis and to download spatial data (Jankowska & Jankowski, 2000).

In many ways, librarians are well aware of the ideas outlined in the discussion on the concepts and components for digital data infrastructure. Library collections and services in electronic form, such as a digital library, integrate many aspects of the client-server architecture in providing access to materials in an online distributed environment. Digital libraries have in them collections of digital information objects of various formats in an ordered database defined by descriptive data standards. Organized in a regular manner, the database is searchable using query applications via a user interface, usually through an online catalog. Information is compartmentalized in short descriptions using metadata tags to ease search and retrieval applications. The overall structure is guided by administrative oversight that takes into account the needs of the various users involved in the community. Researchers, such as Goodchild, envision an intersection of the concepts of the digital library and that of the spatial data infrastructure in a sort of "geolibrary" (Goodchild, 1998). Goodchild envisions a geolibrary as a library filled with "georeferenced information that can have a geographic footprint." Including multimedia, images, and music that could be assigned a location attribute, geolibraries would thus extend beyond the scope of a traditional map library (Nogueras-Iso, Zarazaga-Soria, and Muro-Medrano. 2005, p. 6), and provide new services and resources for users to discover.

The survey of the characteristics of spatial data databases and related infrastructure indicate the involvement of many different factors in assembling a data infrastructure. However, in addition to data providers, databases, metadata, data networks, technology, and institutional arrangements, there is also a significant in organizational and human resource factors. Nogueras-Iso et al. (2005) offer six talking points that are essential to consider in building a spatial data infrastructure. These include technology, policies and standards, human resources, institutional arrangements, spatial data and metadata, and data networks. They suggest that spatial data infrastructure should not recreate the wheel in its technology development. It would be prudent to look at what information technology has accomplished to date, and then integrate that knowledge with GIS expertise, regardless of the difficulties involved. Integration of existing technology also applies to standards and policies. Standards enhance communication and development with a common language and concepts, leading to guidelines that affect architecture, processes, methods, or policies. There should be a common consensus of minimal guidelines that can accommodate those working with geospatial data, retrieval, and discovery. User-guided development is also critical when examining the human resources side of GIS information and technology. Further, the use of qualified researchers and developers is essential. To create institutional frameworks, agreements have to be certified to establish national, regional, and global spatial data infrastructure. These spatial data infrastructures should be created over the geographic data, stored in the spatial databases, and their description in rich, descriptive metadata. Finally, open systems and ease of access is of strategic importance to ensure quality and accuracy from remote systems (Nogueras-Iso et al., 2005). The need for data description standards for digital geospatial data is especially apparent in the proliferation of WebGis applications in the wake of spatial data distribution enabling initiatives, such as the NSDI. WebGIS (also known as Internet GIS) describes a type of geographic information system. WebGIS basically consists of client, server, and network model, wherein the client is integrated in a Web browser. While creating many opportunities for librarians in making available unique geospatial datasets, the tremendous growth in WebGIS has created many challenges for libraries in trying to incorporate digital geospatial data in their services and legal considerations in the use of distributed data.

Web GIS and Libraries

In recent years the number and scope of WebGIS applications has grown tremendously. In their survey of WebGIS applications, Yang et al. (Yang, Wong ,Yang, Kafatos, & Li 2005), discuss a wide number of interactive mapping sites available on the Internet. They note that many users of the Internet have already experienced using WebGIS software tools through interactive mapping tools available on the Internet. Such sites include: Yahoo Maps (http://maps.yahoo.com), National Map http://nationalmap.gov), and GoogleEarth (http://earth.google.com) Other WebGIS applications that are available on the Internet give access to a varying amount of specialized geospatial data. Many local government Web sites for city and county governments across the United States offer interactive mapping sites such as the City of Albuquerque (http://http://www.cabq.gov/gis/) and City of Durham (http://gisweb2.ci.durham.nc.us/sdx/) for users to access technical data about their municipalities. These mapping sites give access to socioeconomic, demographic, and planning data. Many government sites use ESRI software products such as ArcGIS and ArcWEB services to facilitate data searches for information.

Other WebGIS applicatiosn include environmental planning (Rao, Fan, Thomas, Cherian, Chudiwale, & Awawdeh, 2007) , agricultural planning (Zhang, Li, & Zhu, 2004) wetlands management (Mathiyalagan, Grunwald, Reddy, K& Bloom, 2005), archaeological research (Jahjah, Ulivieri, Invernizzi, & Parapetti, 2007), health planning research (Boulos, 2005), transportation planning (Lu, 2006; Sutton, 2005), citizen political participation (de Souza Baptista, Leite, Rodrigues da Silva, & Paiva, 2004), and education planning (Baker, 2005). Other developments in WebGIS include different data visualizations to offer unique perspectives on information for analytical purposes, such as three-dimensional modeling (Qunyong, Han, Qinmin, & Chenghu, 2005) and combinations of software such as multimedia and WebGIS to offer comprehensive views of geospatial data A net effect of the increasingly popularity of WebGIS.

The popularity of WebGIS applications among the public has led to unequal data distribution across the Internet, since bandwidth is constrained. In their study of WebGIS, Yang et al. (2005) identify two issues related to improving access to digital geospatial data on the Interent. The first issue concerns the sharing of and interoperability for heterogeneous data among different systems, different communities, and different users. The second issue is a quality of service issue, that of how to improve the system performance so data are delivered to the users within a reasonable time span (Yang et al., 2005). In regards to the first issue, the authors identify international organizational efforts at creating accepted conventions in data descriptive standards in regards to interoperability and data access, such as the OpenGIS Consortium (OGC) (http://www.opengis.org/) and the Technical Committee 211 of the International Organization of Standards http://www.isotc211.org/.

The strides being made by the aforementioned organizations and agreement between international geospatial data producers and providers is creating an environment wherein libraries have access to geospatial Web Services, such as ESRI's Geography Network (http://geographynetwork.com) and Geosptatial One Stop (http://www.geodata.gov). Web service Internet sites essentially provide access to GIS data, software, educational and mapping, or related services. For libraries, the services can give access to current and large amounts of geospatial data are usually available in a timely manner. In his study of how academic libraries can integrate digital geospatial data into their collections and services, Morris (2006) provides a good overview of the issues involved in trying to facilitate Web Service data in an academic library. Some drawbacks of attempting to integrate Web services into a library environment can include lingering technical issues like linking data resources, developing sustainable licensing models, and negotiating access rights to digital geospatial data.

Conclusion/Summary

The current environment of academic libraries are made complex with the addition of many technological factors such as: electronic bibliographic databases, numerous operating systems, Web browsers, HTML, and other markup languages (Drake, 2000). Increasing amounts of information like text, video, audio, and graphics are available in an online format. Many print materials are being digitized into electronic format. As a result of the changes in technology, librarians are required to learn about metadata and description for electronic data especially for GIS data. In digital collections, such as the FGDL, cataloging and data access issues are even more significant to librarians.

In comparing spatial data infrastructures and geolibraries Nogueras-Iso et al. (2005) offer some discussion about their characteristics that are valid here.

Spatial data infrastructures have an important political and social component. Government involvement in the creation and maintenance of spatial data has created a hierarchy of local, state, national and regional levels that can affect access to data.

Standardization processes in spatial data infrastructures involve not only the organization of data, but also issues related to the capture and integration of these data. These problems are inherent to digital libraries but are aggravated in the case of georeferenced data. The wide range of data users can have a significant effect on visualization and manipulation of data.

Necessity of interoperability with other digital library systems. As a consequence of the necessity for managing heterogeneous information, spatial data infrastructure services could be developed over the services provided by other digital library systems. (pp. 8-9)

Traditionally, geospatial data has been produced in limited formats, either as a paper map, atlas, or as reproductions of aerial photographs. With the application of data processing capabilities, advances in graphic presentation, and production, digital spatial data presents unique problems in description and classification. Amendments to Chapters III and IX of the AACR2 have enabled the description of digital spatial data elements within the standard form for traditional maps, thus enabling access by title, and subject. The high variability of the subject content of digital spatial data elements, though, may require additional metadata structure for better access by users.

Since the 1960s, digital spatial data has been produced in a variety of formats and in a number of carriers. Digital spatial data can include information digitized and recorded by a single researcher or be in the form of large-scale geographic coverages processed and packaged by firms like International Computer Works, Inc. Much digital spatial data is also generated by government agencies at the city, state, and federal level and is issued in a variety of formats like on CD-ROMs with computer files with accompanying attribute data in relational databases (Decker, 2001). Chapter IV discusses cataloging and description of geospatial data especially in an online environment of the digital library. Chapter V contains further elaboration about data standards and the interoperability of data in an Internet environment.

References

Aronoff, S. (1993). *Geographic information systems: A management perspective.* Ottawa, Canada: WDL Publications.

Ayres, R. U., & Williams, E. (2004). The digital economy: Where do we stand? *Technological Forecasting and Social Change, 71*(4), 315-339.

Baker, T. R. (2005). Internet-based GIS mapping in support of k-12 education. *The Professional Geographer, 57*(1), 44-50.

Baran, P. (1964). *On distributed communications* (RM No. 3420-PR). Santa Monica, CA: RAND.

Bobak, A. R. (1996). *Distributed and multi-database systems.* Boston, MA; London: Artech House.

Boulos, M. (2005). Web GIS in practice III: Creating a simple interactive map of England's Strategic Health Authorities using Google Maps API, Google Earth

KML, and MSN Virtual Earth Map Control. *International Journal of Health Geographics, 4*(1), 22.

Boxall, J. (2002). Geolibraries, the global spatial data infrastructure and digital Earth: Atime for map librarians to reflect upon the moonshot. *INSPEL, 36*(1), 1-21.

Brodeur, J., Bédard, Y., Edwards, G., & Moulin, B. (2003). Revisiting the concept of geospatial data interoperability within the scope of human communication processes. *Transactions in GIS, 7*(2), 243-265.

Decker, D. (2001). *GIS data sources*. New York, NY: John Wiley & Sons.

de Souza Baptista, C., Leite, F. L., Rodrigues da Silva, E., & Paiva, A. C. (2004). Using open source GIS in e-government applications. In R. Traunmüller (Ed.). *Electronic government: Third International Conference, EGOV 2004, Zaragoza, Spain, August 30-September 3, 2004. Proceedings* (*Lecture Notes in Computer Science, 3183* (pp. 418-421). Berlin, Germany: Springer.

Drake, M. A. (2000). Technological innovation and organizational change revisited. *Journal of Academic Librarianship, 26*(1), 53-59.

Federal Geographic Data Committee. (2006a). *2005 annual report*. Reston, VA: The Committee. Retrieved June 21, 2006, from http://www.fgdc.gov/fgdc-news/2005-annual-report

Federal Geographic Data Committee. (2006b). *National spatial data infrastructure*. Reston, VA: The Committee. Retrieved June 21, 2006, from http://www.fgdc.gov/nsdi/nsdi.html

Gabe, T. M., & Abel, J. R. (2002). The new rural economy: Deployment of advanced telecommunications infrastructure in rural America: Measuring the digital divide. *American Journal of Agricultural Economics, 84*(5), 1246-1252.

Goodchild, M. F. (1992). Geographical information science. *International Journal of Geographical Information Systems, 6*(1), 31-45.

Goodchild, M. F. (1998). The GeoLibrary. S. Carver *Innovations in GIS 5: Selected papers from the Fifth National Conference on GIS Research UK (GISRUK)* (pp. 59-68). London; Bristol, PA: Taylor & Francis.

Green, D., & Bossomaier, T. R. J. (2002). *Online GIS and spatial metadata*. London; New York: Taylor & Francis.

Hill, L. L., Carver, L., Larsgaard, M. L., Dolin, R., Smith, T. R., Frew, J., et al. (2000). Alexandria Digital Library: User evaluation studies and system design. *Journal of the American Society for Information Science, 51*(3), 246-259.

Hu, S. (2006). Design issues associated with discrete and distributed hypermedia GIS. In E. Stefanakis, M. P. Peterson, C. Armenakis, & V. Delis (Eds.) *Geographic hypermedia: concepts and systems* (Lecture Notes in Geoinformation and Cartography, pp. 37-51). Berlin, Germany: Springer-Verlag.

Jahjah, M., Ulivieri, C., Invernizzi, A., & Parapetti, R. (2007). Archaeological remote sensing application pre-post war situation of Babylon archaeological site—Iraq. *Acta Astronautica, 61*(1-6), 121-130 [Bringing Space Closer to

People: Selected Proceedings of the 57th IAF Congress, Valencia, Spain, 2-6 October, 2006].

Jankowska, M. A., & Jankowski, P. (2000). Is this a geolibrary? A case of the Idaho Geospatial Data Center. *Issues in Science & Technology Librarianship, 19*(1), 4-10.

Kainz, W. (2004). *Geographic information science* (Version 2.0 ed.). Vienna, Austria: Institut für Geographie und RegionalforschungUniversität Wien Universitätsstraße. Retrieved November 2006, from http://www.geografie.webzdarma.cz/GIS-skriptum.pdf

Kessler, G. C. (1990). *ISDN: Concepts, facilities, and services.* New York: Mc-Graw-Hill.

Lu, X. (2006). Develop Web GIS based intelligent transportation application systems with Web service technology. In G. Wen, S. Komaki, P. Fan, & G. Landrac (Eds.) *ITST 2006: 2006 6th International Conference on ITS Telecommunications Proceedings, June 21-23 2006, Chengdu, China* (pp. 159-162). Chengdu, China; Piscataway, NJ: University of Electronic Science and Technology of China; IEEE.

Lutz, M., Riedemann, C. R., & Probst, F. (2003). A classification framework for approaches to achieving semantic interoperability between GI Web services. In W. Kuhn, M. F. Worboys , & S. Timpf (Eds.), *Spatial information theory* (pp. 186-203). Berlin,Germany: Springer.

Malecki, E. J. (2003). Digital development in rural areas: Potentials and pitfalls. *Journal of Rural Studies, 19*(2), 201-214.

Maryland State Geographic Information Committee. (2002). *Maryland's Geospatial Data Implementation Team plan.* Baltimore, MD: MSGIC. Retrieved November 2006, from http://www.msgic.state.md.us/publicat/MDITeamPlan.pdf

Mathiyalagan,V., Grunwald, S., Reddy, K. R., & Bloom, S. A. (2005). A WebGIS and geodatabase for Florida's wetlands. *Computers and Electronics in Agriculture, 47*(1), 69-75. McGraw-Hill Communication Series, 1990

Morris, S. P. (2006). Geospatial Web services and geoarchiving: New opportunities and challenges in geographic information services. *Library Trends, 55*(2), 285-303.

National Research Council Mapping Science Committee. (1999). *Distributed geolibraries: Spatial information resources: Summary of a Workshop held June 15-16, 1998.* Washington, DC: National Academy Press. Retrieved from http://www.nap.edu/html/geolibraries

Nogueras-Iso, J., Muro-Medrano, P. R., & Zarazaga-Soria, F. J. (2005). *Geographic information metadata for spatial data infrastructures: Resources, interoperability, and information retrieval* (1st ed.). Berlin; New York: Springer.

Parsons, E. (2003). *The role of Web services for spatial data delivery.* Paper presented at the annual conference of the Geospatial Information & Technology Association (GITA): "Wide Open Spatial Frontiers: Adding Value to Your

Business" San Antonio, TX. Retrieved February 2006, from http://web-services.gov/web_services.pdf: The Association

Prieger, J. E. (2003). The supply side of the digital divide: Is there equal availability in the broadband Internet access market? *Economic Inquiry, 41*(2), 346-363.

Qunyong, W., Han, C., Qinmin, W., & Chenghu, Z. (2005). WebGIS-based surface 3D reconstruction of pipeline. In *Geoscience and Remote Sensing Symposium, 2005. IGARSS '05. Proceedings 2005 IEEE International, vol. 2* (pp. 25-29). Piscataway, NJ: IEEE.

Rao, M., Fan, G., Thomas, J., Cherian, G., Chudiwale, V., & Awawdeh, M. (2007). A Web-based GIS Decision Support System for managing and planning USDA's Conservation Reserve Program (CRP). *Environmental Modelling & Software, 22*(9), 1270-1280.

Reid, J. S., Higgins, C., & Medyckyi-Scott, D. (2004). Spatial data infrastructures and digital libraries: Paths to convergence. *D-Lib Magazine, 10*(4). Retrieved July 2006, from dlib.anu.edu.au/dlib/may04/reid/05reid.html

Rhind, D., & Connolly, T. (1993). *Understanding GIS: The ARC/INFO method: Self-study workbook: PC version.* Redlands, CA; Somerset, NJ: Environmental Systems Research Institute; J. Wiley.

Strover, S. (2001). Rural Internet connectivity. *Telecommunications Policy, 25*(5), 331-347.

Sutton, J. C. (2005). GIS applications in transit planning and operations: A review of current practice, effective applications and challenges in the USA. *Transportation Planning and Technology, 28*(4), 237-250.

Tsichritzis, D. C., & Klug, A. C. (1979). *The ANSI/X3/SPARC DBMS framework: Report of the Study Group on Database Management Systems.* Montvale, NJ: AFIPS Press.

Yang, C., Wong, D. W., Yang, R., Kafatos, M., & Li, Q. (2005). Performance-improving techniques in Web-based GIS. *International Journal of Geographical Information Science, 19*(3), 319-342.

Zhang, J., Li, J., & Zhu, Y. (2004). Integrating crop simulation models with WebGIS for remote crop production management. In *Geoscience and Remote Sensing Symposium, 2004. IGARSS '04. Proceedings of 2004 IEEE International, vol. 4* (pp. 2598-2600). Piscataway, NJ: IEEE.

Zhang, M., Yang, D., Deng, Z., Wu, S., Li, F., & Tang, S. (2004). WebGIS-RBDL: A rare book digital library supporting spatio-temporary retrieval. In Z. Chen, (Ed.) *Digital libraries: International collaboration and cross-fertilization* (*Lecture Notes in Computer Science, 3334*, pp. 255-265). Berlin, Germany: Springer Verlag.

Chapter IV

Describing Geospatial Information

Ardis Hanson, University of South Florida Libraries, USA

Susan Heron, University of South Florida Libraries, USA

Overview and Introduction

To be optimally useful, geospatial resources must be described. This description is referred to as **metadata.** Metadata tells "who, what, where, when, why, and how" about every facet of a piece of data or service. When properly done, metadata answers a wide range of questions about geospatial resources, such as what geospatial data is available, how to evaluate its quality and suitability for use, and how to access it, transfer it, and process it. To ensure consistency for access and retrieval, metadata can be standardized to provide a common set of terms, definitions, and organization.

In the desire to adequately and accurately describe geospatial resources, cataloging codes and practices have been established to accommodate these new resources (known in library parlance as "works"), to provide networked access to these resources, and to respond more effectively to an increasingly broad range of user expectations and information needs. Issues surrounding the quality and relevance of metadata (bibliographic access) become more critical in online venues, especially with geospatial data. What kind of bibliographic records or metadata will be required to meet the different uses of geospatial information and user needs? How should these bibliographic data or metadata be organized and structured for intellectual and physical access to the works?

This chapter will provide an overview of current academic cataloging principles, issues in handling evolving formats, and challenges for academic catalogs. It will also discuss the development of a MARC geospatial information record and the issues involved in adequately describing these works.

Descriptive Standards in Libraries

In libraries, the use of metadata, MARC, or any other knowledge organizational tool is based upon some form of Cutter's principles of organization. Cutter's *Objects* were to (1) enable a person to find a book for which either the author, title, or subject is known; (2) show what the library has by a given author, on a given subject, or in a given kind of literature, and (3) assist in the choice of a book, as to its edition (bibliographically) or to its character (literary or topical). His *Means*, or method of doing so, provides numerous access points, including author-entry with necessary references; title-entry or title-reference; subject-entry, cross-references, and classed subject-table; form-entry; edition; and imprint, with notes when necessary (Cutter, 1904). Today, increasing numbers of library patrons see libraries more as remote resources, rather than as walk-in facilities. To meet this need, libraries must have sustainable systems of access and databases and durable objects that fulfill the three R's: reliability, redundancy, and replication of results (Cline, 2000). These "three R's" can be seen in the development of the functional requirements for the bibliographic record (FRBR) entity relationship model for works, expressions, manifestations, and items.

An international analysis of user needs determined that there are four generic information tasks users perform: "*finding materials that correspond to the user's stated search criteria* (e.g., in the context of a search for all documents on a given subject, or a search for a recording issued under a particular title); *using the data retrieved to identify an entity* (e.g., to confirm that the document, described in a record corresponds to the document sought by the user, or to distinguish between two texts or recordings that have the same title); *using the data to select an entity that is appropriate to the user's needs* (e.g., to select a text in a language the user understands, or to choose a version of a computer program that is compatible with the hardware and operating system available to the user); [and] *using the data in order to acquire or obtain access to the entity described* (e.g., to place a purchase order for a publication, to submit a request for the loan of a copy of a book in a library's collection, or to access online an electronic document stored on a remote computer)." (IFLA Study Group on the Functional Requirements for Bibliographic Records, 1998).

These principles are still the foundation of best cataloging practice, including the notion of specificity, the consideration of the user as the principal basis for subject-heading decisions, the practice of standardizing terminology, the use of cross-refer-ences to show preferred terms and hierarchical relationships, and solving the problem of the order of elements (Heron & Hanson, 2003). They organize the information in such a way that allows the user to eliminate irrelevancies or false cognates and to focus on specifics, thereby reducing cognitive overload. To organize information, libraries use a variety of standards to describe data, including structural frameworks, cataloging rules and interpretations, class schedules, and subject access.

MARC

The structural framework most commonly used in libraries is MARC (Machine Readable Cataloging), a communications format developed in the 1960s by the Library of Congress (LC). This format was created to represent and communicate LC's bibliographic and related information in computer-stable form. A MARC re-cord is comprised of three elements: the record structure, the content designation, and the data content of the record. A set of codes and content designators define how the records will be encoded for the five types of data: bibliographic, holdings, authority, classification, and community information (Library of Congress, 1989). The MARC record has evolved with networking advances. Its uniform data structure was enhanced with the development of the Z39.50 standard for the electronic shar-ing of data and with its XML format, which will be discussed later in this chapter and in Chapter V.

Within the United States, an American Library Association (ALA) committee, the ALCTS/LITA/RUSA Machine-Readable Bibliographic Information Committee (MARBI), is responsible for developing official positions on standards for the representation in machine-readable form of bibliographic information. For the international library community, the International Federation of Library Associa-tions and Institutions (IFLA) works with national cataloging and standardization committees, multinational organizations, ISO (International Standards Organization) committees "to promote the development of an international cataloguing code for bibliographic description and access" (International Federation of Library Associa-tions and Institutions, Cataloguing Section, October 17, 2005).

In today's "metadata" terminology, MARC is a schema supplemented by syntax rules with a set of constraints. Schemas typically restrict element and attribute names and their allowable containment hierarchies. The constraints in a schema may also include data type assignments that affect how information is formatted and processed. For example, date elements would be formatted in a specific way, such as MM/DD/YYYY or MM/DD/YY. MARC is based upon ISO 2709:1996, Format for Information Exchange (INEX). USMARC is based on ANSI Z39.2, American National Standard for Bibliographic Information Interchange. These standards will be discussed in more detail in Chapter V. The MARC schema has three structural frameworks: bibliographic, holdings, and authority.

Bibliographic Structure

The MARC bibliographic structure uses a combination of fixed- and variable-length fields. Fixed fields contain excerpted information in predetermined length strings to allow ease in searching the datasets. Variable fields have no predetermined lengths, can contain extensive amounts of information, and are of variable length because the amount of information differs for each item. All information is entered into defined fields, designating the type of data that is then further subdivided into discrete pieces. For example, an author can be one of three types of data: a person, an entity, or a conference. Depending upon its type, the data are delineated into its three-character field. The field is then subdivided by additional information. National standards exist for minimal, core, and full coding. If materials are permanent parts

Figure 1. MARC bibliographic format

LDR *****nam##22*****#a#4500
001 <control number>
003 <control number identifier>
005 19920331092212.7
007 ta
008

| 820305 s | 1991#### n | yu#### | ###### | #001#0# e | ng## |

020 ## $a0845348116 :$c$29.95 (£19.50 U.K.)
020 ## $a0845348205 (pbk.)
040 ## $a<organization code>$c<organization code>
050 14 $aPN1992.8.S4$bT47 1991
082 04 $a791.45/75/0973$219
100 1# $aTerrace, Vincent,$d1948-
245 10 $aFifty years of television :$ba guide to series and pilots, 1937-1988 /$cVincent Terrace.
246 1# $a50 years of television
260 ## $aNew York :$bCornwall Books,$cc1991.
300 ## $a864 p. ;$c24 cm.
500 ## $aIncludes index.
650 #0 $aTelevision pilot programs$zUnited States$vCatalogs.
650 #0 $aTelevision serials$zUnited States$vCatalogs.

of a collection, full coding, which provides the maximum amount of information to the user of the online catalog, is preferable.

Holdings Structure

The holdings describe what parts of a work a library owns or can access. The holdings record defines the codes and conventions (tags, indicators, subfield codes, and coded values) that identify elements for single items, multipart sets, or serials. The holdings structure is based on the American National Standards for Holdings Statements. Holdings information is important for local and union catalogs, cooperative acquisitions and collection development, preservation programs, and as a link in fully automated interlibrary loan and document delivery systems. The holdings records are linked to the bibliographic records; holdings format information may also be "embedded" within bibliographic records as a "holdings cluster." The CONSER Publication Patterns and Holdings Project is using the MARC holdings record as the vehicle to develop a "universal" holdings record for serial titles (CONSER Task Force on Publication Patterns and Holdings, 2006). To develop a "universal" holdings record, a pattern (i.e., a publication pattern) describes the captions and pattern of issuance (publication) of a serial or multipart item. Data elements include

Figure 2. MARC holdings format and OPAC display

LEVEL 1 <Location Identifier> Main
LDR *****nx###22*****1##4500
001 <control number>
852 ## $a**<location identifier>**$bMain
LEVEL 2 <Location Identifier> Main 19870414 (0,ta,4,2,8)
LDR *****nx###22*****2##4500
001 <control number>
008 8902202p####8###4001aa###0870414
852 ## $a**<location identifier>**$bMain

OPAC display

Field guide and reference manual series /

[Toronto, Ont.] : Geological Association of Canada, c1985-

v. : ill., maps ; 28 cm.

Description

Location: LIBRARY -- Circulating Collection -- QE376 .F54

Holdings: no.1 (1985)–no. 12 (1997).

frequency for each bibliographic unit; its secondary or lower bibliographic units and the relationship of the lower numbering system to the higher (whether it restarts or is continuous); the calendar change point; and variations in the intervals of issuance (CONSER Task Force on Publication Patterns and Holdings, 2006). For maps and other geospatial information that are issued in sets or successive editions, holdings information indicates what the library owns, and in a more enhanced record, what the ideal run is.

Authority Structure

Authority control is the semantic interoperability of controlled subject terminology and classification data. Authority control allows the user to find all items on a particular topic or by an author, regardless of the many term variations, synonymous terms, or languages that might be available in the catalog. The AUTHOR and SUBJECT entries in a library catalog are controlled, recognized forms of name and subject headings. They are established in authority records and used as access points in the bibliographic record. Unlike bibliographic records, which represent items owned by a library, authority records are tools that librarians use to achieve consistency among bibliographic records, and to provide a linking framework for related names and subjects in a catalog. An authority record includes three basic components: headings, cross references, and notes. The heading is the standardized "authoritative" form of a name, subject, or title that is used on bibliographic records. There are two types of cross references. The first type, a *see* reference, directs a user from a variant form of a name or subject to the authoritative form. The second type is a *see also*

Figure 3. MARC authority format

LDR *****nz###22*****n##4500
001 <control number>
003 <control number identifier>
005 19870121083133.6
008

870121 #	n#ac a	naa	bn###	####	###n#	aaa## #	##u

010 ## $aex#86114834#
040 ## $a<organization code>$c<organization code>
100 1# $aCameron, Simon,$d1799-1889
670 ## $aNUCMC data from NJ Hist. Soc. for Bradley, J.P. Papers, 1836-1937$b(Simon Cameron)
670 ## $aLC data base, 1-21-87$b(hdg.: Cameron, Simon, 1799-1889)
670 ## $aDAB$b(Cameron, Simon, 1799-1889; Sen. from Pa. (Republican boss); financier; Sec. War under Lincoln; Min. to Russia; s. Charles & Martha (Pfoutz) C.; newspaper editor; owner Harrisburg Republican; commis. to settle claims of Winnebago Indians; m. Margaret Brua; father of: J.D. Cameron (1833-1918)).

reference. The *see also* reference directs the user from one authoritative form to another authoritative form because they are related to one another. The references are carried or "traced" on the record for the authoritative heading. Notes contain general information about standardized headings or more specialized information, such as citations for a consulted source in which the librarian verified a form of name or a definition.

There have also been suggested changes to fields within the authority record to accommodate for and enhance resource discovery of geographically centered items. In library catalog records for maps and other cartographic materials, field 034 (Coded Cartographic Mathematical Data) and field 255 (Cartographic Mathematical Data) unambiguously identify the location of a coverage area with geographic coordinates. However, these fields are generally used for print items only when coordinates actually appear on the cartographic item being cataloged. Other formats of materials may not have this information easily available or apparent, reducing the richness of the record-describing data that could be used for GIS layering.

Modifying field 034 in authority records for geographic coordinates associated with places would eventually form the basis for coordinates-based retrieval of all cataloged records containing geographic terms (George A. Smathers Libraries & ALA/MAGERT Cataloging and Classification Committee, 2005). Since different geographic areas may exist during different periods of time or coordinates may be recorded from different sources and possibly in different formats, repeatability of the field is essential. Two other areas for consideration included coordinates for the surface of planets and for the position of stars, and further exploration of the recording source of information in the field (Smathers & ALA/MAGERT, 2005). The first indicator and subfields $a, $b, and $c were determined not to be applicable and needed no modification (Smathers & ALA/MAGERT, 2005). Note fields have also been a repository for geographic coordinates, such as field 670, that includes cites to authoritative sources to document the place name. Repeating geographic coordinates in field 670 in a specific data field for machine retrieval would aid in user discovery of relevant materials. Further, since authority records establish the authoritativeness of the information or the issuing body, such as the USGS Geographic Names Information System (GNIS) record for the place or the U.S. Board on Geographic Names, which often gives lists of coordinates for places, it makes sense to place coordinates in authority records. Records for place names, geographic features, and some subject headings, such as battles, buildings, and so forth, would also be affected (Smathers & ALA/MAGERT, 2005).

Several concerns remain regarding the inclusion of geographic coordinates in field 034. Should the decimal format of data in the 034 field remain in the decimal format or be converted to degrees, minutes, and seconds? (Smathers & ALA/MAGERT, 2005). This may require catalogers use additional tools to convert coordinates, adding another layer in the cataloging process. However, to increase resource discovery, this will be a crucial component. The Board of Geographic Names uses point coordinates (with one latitude and one longitude), not the bounding box coordinates typically used in bibliographic records, another problem to be resolved (Smathers & ALA/MAGERT, 2005). Should bibliographic record field 034 be added *en toto* to the authority record or only those fields directly applicable to the authority record? (Smathers & ALA/MAGERT, 2005). Repeatability is also an issue. Since Board of Geographic Names place records often provide a list of coordinates that are centers of quadrangles that together cover the area for a place, multiple coordinate fields with an additional subfield describing the specific area may be necessary (Smathers & ALA/MAGERT, 2005). How will the date be handled? A subfield for date might be subfield y to be consistent with subject date subfields or, perhaps the 045 "Time period of heading" field already defined for authority records (which includes both BCE and AD dates) may work. However, to link the 045 to the 034 would require the use of multiple 034 fields subfield |8 (Smathers & ALA/MAGERT, 2005). A subfield to document source would also be helpful, since the Board of Geographic Names does not establish all forms of name and place (Smathers & ALA/MAGERT, 2005).

MARC XML

With the emergence of extensible mark-up languages (XML), the Library of Congress' Network Development and MARC Standards Office developed a MARC XML framework to represent a complete MARC record in XML. There are XML description schemas that correspond to MARC bibliographic and authority formats. The Metadata Object Description Schema (MODS) corresponds to MARC 21 bibliographic (Library of Congress, 2006, April 52006, April 5 #145). The Metadata Authority Description Schema (MADS) is an authority element set that provides metadata about agents (personal authors and corporate bodies), events, and terms (e.g., subject headings, geographic headings, and genre headings) (Library of Congress, December 14, 2005). The Metadata Encoding and Transmission Standard (METS) schema is a standard for encoding descriptive, administrative, and structural metadata regarding objects within a digital library, expressed using the XML schema language (Library of Congress, 2003).

The framework has additional schemas, stylesheets, and software tools. MARC XML can also be used to represent metadata for OAI harvesting, for original resource description, and for electronic resource metadata packaged with the item. Toolkits are available to convert to and from XML and MARC using Java. XML is an important new language and tool since much of the geospatial metadata is created using XML generators.

AACR2r

Cataloging rules and rule interpretations allow standardization or, from another perspective, the intellectual collocation of works and items, which aids in retrieval. This collocation facilitates "cognitive miserliness"(Fiske & Taylor, 1991), which is the concept that the human mind is the most efficient in acquiring information when it has to do the least work. One way that librarians have facilitated researchers is by creating a standard record format that accepts data into predefined fields and is governed by a set of rules that determine what information goes where, thereby allowing the user to predict where relevant data will be located and to ignore the irrelevant. Further, the standardization of data allows the user to concentrate on the decision-making process to assure relevance and precision in his or her search results. The current guide used by librarians is the *Anglo American Cataloging Rules, 2nd edition, revised* (*AACR2r*). Simply, *AACR2r* provides the standard for structuring catalogs (or datasets in today's larger perspective) with headings and references to provide links between items with similar or related characteristics.

The American Library Association's Committee on Cataloging: Description and Access serves (CC:DA) is the body within the United States that facilitates the continued development of those standards and formulates the official ALA policy on descriptive cataloging. A representative from the CC:DA is an ex officio member of the international committee responsible for updating and revising the *AACR2r*, the Joint Steering Committee for the Revision of Anglo-American Cataloguing Rules, which implements updates to the *AACR2r*. In the international community, IFLA also promotes the Functional Requirements for Bibliographic Records (FRBR) model and develops new descriptive standards and standards for access points (International Federation of Library Associations and Institutions, Cataloguing Section, October 17, 2005).

As an international standard, *AACR2r* facilitates data sharing among disparate systems and frees the cataloger from having to reinvent the rules for each dataset. Another advantage is that it attempts to identify the most relevant features of the item it is describing, ensuring the capture of an accepted set of information. *AACR2r* prescribes what type of data should be captured, but allows the cataloger to determine the extent to which it is recorded.

Resource Description and Access (RDA)

An exciting initiative of the Joint Steering Committee for Revision of AACR (JSC) is their work on a new standard, *RDA: Resource Description and Access*, scheduled for release in early 2009. The *RDA* "provides a set of guidelines and instructions

Figure 8. RDA to FRBR mapping (http://www.collectionscanada.ca/jsc/docs/5rda-frbrmapping.pdf)

RDA Element	Corresponding FRBR entity	Corresponding FRBR attribute/relationship
Scale	expression	~
Scale of still image or three-dimensional form expression	expression	n/a
Scale of cartographic content	expression	scale (cartographic image/object)
Additional scale information [cartographic content]	expression	scale (cartographic image/object)
Variations in scale [cartographic content]	expression	scale (cartographic image/object)
Non-linear scale [cartographic content]	expression	scale (cartographic image/object)
Vertical scale [cartographic content]	expression	scale (cartographic image/object)
Projection of cartographic content	expression	coordinates (cartographic image/object)
Coordinates of cartographic content	work	coordinates (cartographic work)
Longitude and latitude	work	coordinates (cartographic work)
Strings of coordinate pairs	work	
Ascension and declination work	work	coordinates (cartographic work) / equinox (cartographic work)
Equinox work equinox	work	n/a
Epoch	work	n/a
Magnitude of cartographic content	expression n/a	~
Other details of cartographic content ~	expression	~
Other mathematical data	expression	scale / projection / coordinates / geodetic, and vertical measurements (cartographic image/object)
Other features of cartographic	expression	special characteristic (remote sensing image) / +

(Joint Steering Committee, June 14, 2007b. Available from http://www.collectionscanada.ca/jsc/docs/5rda-frbrmapping.pdf

on formulating descriptive data and access point control data to support resource discovery" (Joint Steering Committee, June 14, 2007a, p. 1). The purpose of the RDA is to ensure metadata standards for record elements, controlled vocabularies, and the overall structures of bibliographic and authority records. The point of the RDA, as with its predecessor *AACRs*, is to create a formal model, with rules and frameworks to ensure consistency in data across disparate content providers (cataloguers). The structure of *RDA* aligns more directly with the *FRBR* and *FRAR* (*Functional Requirements for Authority Records*, now *Functional Requirements for Authority Data*) models. One of the changes we will see in the RDA is a change in how map elements are named, described, and structured. Figure 8 provides an example of the RDA to FRBR mapping for the cartographic term, scale.

The most significant area the conversion from AACR2r to RDA will have is in the ability to better describe and embed FGDC data elements and description into the bibliographic and authority records. However, that does mean more of a challenge for professional and paraprofessional staff in cataloguing and technical services areas to fully describe the work before them. It also presents a challenge for public services staff to communicate user questions and suggestions regarding how works are named and described. This is essential, since one of the foci of the RDA is on user needs: "The descriptive data provided for in the guidelines and instructions should enable the user to: identify the resource described (i.e., to confirm that the resource described corresponds to the resource sought, or to distinguish between two or more resources with similar characteristics); select a resource that is appropriate to the user's requirements with respect to content, format, etc. The access points provided for in the guidelines and instructions should enable the user to locate: all resources described in the catalogue that embody a particular work or a particular expression of that work; all resources described in the catalogue that embody works and expressions of works associated with a particular person, family, or corporate body; a specific resource described in the catalogue that is searched under a title appearing in that resource; works, expressions of works, and manifestations represented in the catalogue that are related to those retrieved in response to the user's search" (Joint Steering Committee, 2005, p. 3). As the *RDA* is finalized, the impact of increased access and description will better serve the geographic and GIS communities.

Classification Schedules

Arranged by subject in a logical, hierarchical manner, a classification scheme divides a field of knowledge into main classes (each class covering a particular discipline, knowledge domain, or subject area). The main classes are divided into subclasses, representing branches of the main discipline, domain, or subject area. Within each subclass, further subdivisions are made to specify form, place, time, and subject (or topical) aspects. These subdivisions, from the most general to the most specific, create a hierarchical display. The more hierarchical levels, the more specificity in the classing of the item. In the United States, the USMARC Format for Classification Data defines the codes and conventions (tags, indicators, subfield codes, and coded values) that identify the data elements in USMARC classification records.

Classification schedules contain a series of numbers, captions, instructions, and notes. Most academic libraries in the United States use the Library of Congress classification schedules. There are 21 categories, alphanumeric, from A-Z (missing I, O, W, X, and Y) with one, two, or three letters and a set of numbers denoting the class or subclass within the schedule. The Library of Congress Classification System organizes material in libraries according to 21 branches of knowledge. The original organization of the classification was based on the "academic world of knowledge" at the beginning of the 20th century. Each LC class schedule was developed separately, following a logical order based upon that discipline's understanding of its domain; therefore, each schedule has unique features, making it difficult to generalize about the schedules as a whole.

In MARC, field 084 is designed to be the carrier for information about the classification scheme used, the classification number, captions (describing what the classification number range covers), and edition of the scheme. Authoritative classification schemes, such as the LC class schedules, Dewey Decimal class schedules, or the National Library of Medicine classification scheme, are the most commonly used. Since the classification schedules are considered a "living" work, with new sections of classes being added or obsolete sections being deleted, the edition used in creating a class number is an important historical element.

Classification numbers are used in both MARC bibliographic and authority records. Classification numbers form the basis of an item's call number. The traditional purpose of the class number is to place an item in its respective part of a knowledge domain, that is, on the library shelf among like items. Libraries with closed stacks needed a method of uniquely identifying titles so that a page could retrieve the desired material. Libraries developed various schemes to satisfy this need. The successful scheme resulted in "call number," a unique extension for each item. Even libraries with open stacks understood the desirability of a unique identifying call number and adopted the practice. A patron browsing the shelves would find similar, or "like," materials grouped around that specific item. In an online library catalog, the class number provides the same service, allowing the user to "virtually" browse the library's collection, whether the item is located in a physical collection or accessed off-site. The need to "call" for electronic titles is unnecessary, but the class number is still useful for information discovery and evaluation of the collection..

For earth sciences libraries, the U.S. Geological Survey (USGS) Library Classification System (Sasscer, 2000). It is also seen as a retrieval system to access materials through the subject and geographic numbers. Unlike the LC class system, the USGS class system contains seven schedules: a subject schedule, a geological survey schedule, an earth science periodical schedule, a government document periodical schedule, a general science periodical schedule, an earth science map schedule, and a geographic schedule (Sasscer, 2000) These schedules mirror the seven collections contained within most earth sciences libraries.

The general subject collection is comprised of the earth sciences disciplines, geology, petrology, mineralogy, paleontology, and biology. Pure sciences, physics, chemistry, engineering, mathematics, and computer sciences, are included only as complements or augmentations to the earth sciences. The general subject collection consists chiefly of monographs, but may also include periodicals that are narrow in scope, international in scope, or issued by an international agency (Sasscer, 2000).

Also unlike the LC schedules, the USGS system segments four collections by issuance. The geological survey collection contains the monographs, periodicals,

Figure 6. Example of a USGS and LC call number

USGS	LC
Title: Multivariate geostatistics / Authour: H. Wackernagel, 1995.	
Class no.: 208.2	QE33.2.S82
Shelf list no.: W323	Cutter: W32
Title mark: m	
Date: 1995	Date: 1995
Call number: 208.2 W323m 1995	Call number: QE33.2.S82 W32 1995

and monographic series issued by the geological surveys of the world. The earth science periodical collection contains publications issued by earth science societies, associations, and earth science departments of universities. The government documents collection contains periodicals and monographic series issued by federal, state, provincial, and local governments, nationally and internationally. The general science periodical collection contains science periodicals issued by national and international societies, associations, and universities (Sasscer, 2000).

The next collection, the earth science map collection, is segmented by format. It contains earth science maps issued by national and international governments, societies, associations, and departments of universities (Sasscer, 2000). The last schedule, the geographic schedule, consists of numbers enclosed in parentheses that can be combined with notation from the other schedules. For example, the geographic number for the United States is (200). This parenthetical enclosure is immediately recognizable when scanning call numbers.

Once the class number has been established, to create a call number in the USGS system, the first element, an uppercase letter, is taken from the first letter of the first word of the main entry. The second element, a three-digit number, is taken from Library of Congress shelf listing tables. The final element is the title mark taken from the first letter in the title.

For the USGS, the general class is "208 Geological technique." The next subsection, 208.2, is for "Mathematical geology, statistic in geology, and geostatistics." A unique identifier is then added, created by using the first letter of the author's surname, a number from a Cutter table, and a workmark from the first significant word in the title – W323m.

In the LC system, QE places the item in Geology, 33.2 A-Z is "special topics" and, within that, S82 is "statistical methods." This classification number is "cuttered" by taking the first letter from the author's last name and appending a numeric value that will represent the second letter so that the names will fall in alphabetical order on the shelf. Finally, the year of publication is added. The LC subject is "Geology—Statistical methods."

Subject Access

Researchers who are serious about retrieving the most targeted materials on a topic use subject access as opposed to keyword searching. The cataloger can use a universal thesaurus, such as the *Library of Congress Subject Headings*, one that covers a broad subject area, such as the Medical Subject Headings (MeSH), a discipline-based controlled vocabulary, such as the *Thesaurus of Psychological Index Terms*, depending upon the needs and focus of the library and its patrons.. Typically, thesauri come in three display formats: "explicitly hierarchical" displays; alphabetical displays, showing references between related terms or references from unused synonym terms to the "postable" or preferred term; or rotated displays, in which each word in every natural language term is used as an index term to indicate the several phrases in which it appears.

A well-developed thesaurus uses a hierarchical structure (or a set of structures) in which clusters of concepts that share common characteristics are organized in facets, and represented by natural language terms useful in the context in which the thesaurus will be used. For example, Alabama, Georgia, New York, Rhode Island, and Oregon all share the common characteristic of being STATES, in addition to whatever other characteristics some of them may share, such as History. The terms constitute the beginnings of a STATES facet. Sometimes the terms in a facet can be divided into subfacets by secondary characteristics: Within the STATES facet, Alabama and Georgia are SOUTHEASTERN STATES; New York and Rhode Island are NORTHEASTERN STATES. Georgia is also part of a larger regional facet, APPALACHIAN REGION, SOUTHERN. Facets and subfacets are then arranged as simple hierarchies of terms, from general to specific.

There are several benefits to a faceted approach. By reviewing a small, related group, the user can see naming consistency, order, hierarchical relationships, relationships to other groups (broader or narrower concepts), and current or superseded terms. Since a faceted approach is extensible, there is greater flexibility in adding new terms or establishing new relationships among existing terms without disturbing the rest of the thesaurus. From a psychological perspective, faceted approaches also embrace the concept of cognitive miserliness. Clearly, it is much easier for an indexer or user to understand a set of hierarchically organized facets as a conceptual map, which shows the precise level and set of associations of a term, than to negotiate a long list of alphabetized terms.

To provide the most effective subject approaches to traditional and networked resources, librarians must account for the different patron strategies and models currently used in information retrieval. The Boolean model, which uses exact match across inclusive and exclusive groupings (AND, OR, NOT), tends to provide more precision in one's search. Ranking algorithms, vector, and probabilistic models compensate for their loss of precision to a certain degree by methods of statistical ranking and computational linguistics, based on term occurrences, term frequency, word proximity, and term weighting.

Controlled vocabulary offers the benefits of consistency and accuracy with better recall through synonym control and term relationships and greater precision through homograph control. Although controlled vocabulary will not replace the apparent "ease" of keyword searching, it supplements and complements keyword searching to enhance retrieval results. The use of controlled vocabulary in the description of metadata places the burden on the indexer rather than the user (Svenonius, 2000) and, as further observed, "There is a burden of effort in information storage and retrieval that may be shifted from shoulder to shoulder, from author, to indexer, to index language designer, to searcher, to user. It may even be shared in different proportions. But it will not go away" (Batty, 1998).

Experiments conducted on subject access systems in WebPACs and metadata-processed systems demonstrate the potential benefit of structured approaches to the description and organization of Web resources (Chan, 2001). This would involve the use of established subject heading schemes and thesauri at a general level, recognizing that more local or specific schemes may also be necessary to provide more detailed indexing. However, the success of this endeavor will depend on trained catalogers for their proper application according to current, and often complex, policies and procedures, the cost of maintenance, and their incompatibility with most tools now used on the Web (Chan, 2001).

Finding Geographic Information in Libraries

Even though maps have been used throughout the centuries as essential tools for analyzing current and historical conditions, the history of map librarianship in the United States really begins at the end of the Second World War. Prior to 1945, only 30 or so libraries employed full-time map librarians. Initially, there was little research concerning the management of this material.

In 1945, the *Classification and Cataloging of Maps and Atlases* (Boggs & Lewis, 1945) was published and referenced the 1941 ALA cataloging rules. The authors stressed the order of importance for entry in this order: first, area, second, subject, third, date of situation of the map (not date of publication or reprinting), fourth, author, and, lastly, title. Further, the authors stressed the importance of previous descriptive elements and requested notes that are more detailed.

From 1945 to the 1960s, cartographic collections become part of mainstream library collections and services for academic and larger public libraries. With the advent of geographic information systems (GIS) and the digitization of data, maps and other geospatial data formats have become critical elements in libraries. Librarians are generally not cartographic experts. Nevertheless, to be effective in cataloging maps, it is useful to know some key considerations about maps and in map making.

What is the Purpose of the Map?

Maps are used to perform three distinct functions: *reference, analysis*, and *persuasion*. As reference tools, maps display "geographically referenced" information, such as cadastral maps or topographic maps. They also provide information on the location and the spatial relationships of the features displayed on the map. Reference maps typically incorporate a lot of information about the features they describe, for example, most roadmaps show cities, counties, county seats, interstate highways, local roads, state routes, and historical or geographic landmarks.

As analytical tools, maps explore the spatial dimensions of, and interrelationships between, phenomena and activities located in space and time. Unlike reference maps, an analytic map is a working document and is used to examine different data sets and explore spatial relationships. For example, the Annie E. Casey Foundation's mapmaker (http://www.aecf.org/kidscount/sld/databook.jsp) on social indicators is used to generate the maps that cover time, space or place, and specific datasets on economic, social, or health issues, using data at the county level, city level, metropolitan statistical area, or congressional district. Geographic information systems (GIS) are also used as an analytical tool.

Persuasive, or thematic maps, are often used dramatically to illustrate a specific point of view or an argument in print and media publications or in public presentations. Thematic maps are created to make a specific point and are often used to persuade or influence a person's decision. For this reason, the ability of the mapmaker to design an effective image is critical to a thematic map's success. Examples of thematic maps are those that chart earnings, projections of an activity, or demographics. For a more in-depth discussion on the display of quantitative visual information, we recommend *The Visual Display of Quantitative Information* (Tufte, 1983).

Issues Regarding Scale

A mapmaker may take a section of a map and magnify or "zoom in" to look at this section at a larger scale. This does not change the accuracy of the map; it simply changes the scale of the map, that is, the relationship between the distances shown on the map and distances on the Earth's surface. The importance of scale issues has led researchers to propose a "science of scale" that would study which measures or properties are invariant with respect to scale, methods to change scale, measures of the impact of scale change, scale as a parameter in process models, and implementation of multiscale approaches (Goodchild & Quattrochi, 1997).

Two terms that are critical to cartographers (and users of maps) when reviewing scale are *spatial accuracy* and *spatial precision*. Spatial accuracy measures how close a recorded location is to its true position on the Earth. It is determined by two factors: (1) the care used in conducting the initial land survey and (2) establishing the original map scale. Spatial precision measures how exactly a location is depicted on a map; therefore, the precision of a map is limited by the width of the smallest line that can be displayed at a given scale.

If the finest line on a 1:24,000 quad sheet is 0.5 mm wide, the smallest distance (or "minimum map unit") that can be recorded true to scale is 39 feet. At that scale, 1 inch represents 2,000 feet. Zooming in by a factor of 10 changes the scale to 1 inch representing 200 feet, making it theoretically possible to locate features with a precision of 4 feet. However, this precision is misleading, because features cannot be located any more accurately than they were recorded on the original map, that is, with an accuracy of 39 feet (Robinson, Morrison, Muehrcke, Kimerling, & Guptill, 1995).

Making the Map

The process of compiling statistical or thematic data should be largely the same as for preparing a base map. The mapmaker consults existing maps and data sources, assesses their reliability and applicability, selects the most useful and dependable information, and presents it on the completed map. Unfortunately, the process of map compilation is not always as rigorous as it should be. Remember, a map is no better than the spatial and attribute data it contains. Librarians may not have the time or expertise to evaluate the reliability and accuracy of the information incorporated in maps. Fortunately, the federal government has established standards requiring that data distributors provide metadata or "data about data" describing the content, quality, source, and other characteristics of their spatial and non-spatial data (Federal Geographic Data Committee, 1999, 1998; U.S. Geological Survey, 1998) (The FGDC record format and metadata requirements will be discussed in more detail in Chapter V). Librarians should become familiar with these efforts, support them, and do their part by preparing metadata for the data they distribute. In addition, they should be sure that the sources for all cartographic and statistical data are clearly identified on their maps.

Bibliographic Issues for Maps

Map cataloging requires knowledge of maps. Maps combine characteristics of both books and pictures, making their bibliographic description more difficult and time-consuming to create. Since the utility of each map is directly affected by the quality of the descriptive cataloging, the consequences of sloppy cataloging practices can be significant.

The principles of current descriptive cataloging codes, developed during the first half of the 1900s, were authored with the book format in mind. The few librarians who championed map cataloging called for the development of specialized map bibliographic records. The map cataloging practices they preferred were those of Boggs and Lewis, published in 1945 by the American Geographical Society. Neither the *Rules for Descriptive Cataloging* (1947) nor the *Anglo-American Cataloging Rules* (1967) adequately met the needs of librarians for maps or for standard map reference resources, such as gazetteers or atlases.

Since the most important consideration is that all maps should be able to stand alone (e.g., their basic message and purpose should be evident to the map viewer without requiring any supplemental textual information), a map should always contain five basic elements:

1. A short, concisely worded title that indicates the map's subject, geographical location, and time frame (when appropriate).
2. A legend or key delineating the meaning of all point symbols, line symbols, and/or area fills used on the map.
3. A map scale to relate distances on the map to distances on the Earth's surface.
4. A locational identifier that places the mapped area in its appropriate spatial context (e.g., a small inset map that places the mapped area within a larger or better known geographic entity).
5. A source citation identifying the source for the spatial and thematic data portrayed on the map, the organization that produced the map, and the date on which it was produced.

If the mapped area is not a closed shape, it should also include a border that provides a frame for the map image.

Bibliographic Description and Access Points:
Title, Author, and Mathematical Data

As we look at a MARC record, the first point of access for descriptive information about geographic information appears in the fixed field area, in the 008 field. The 008 field is a fixed field data element that contains 40 character positions (00-39) (Network Development and MARC Standards Office, 2005). Useful for retrieval and data management purposes, these character positions provide coded information about (1) the record as a whole and (2) special bibliographic aspects of the item. Character positions 18-21 are used to describe relief maps, 22-23 describe projections, 25 describes the type of cartographic material (e.g., map series, map serial, a separate map supplement to another work, or a map bound as part of another work), 28 describes government publications (multilocal, federal, state, national, international, etc.), 29 describes form of item (e.g., microfiche, large print, electronic, etc.), 31 indicates if the item or accompanying material has a location index or gazetteer, and 33-34 indicates if there is any special format characteristics of the map. Character positions 24, 26-27, 30, and 32 are undefined (Network Development and MARC Standards Office, 2005).

In bibliographic records, geographic coordinates also appear in field 034 and field 255. The data in the 034 field is formatted and may be given as decimal degrees, or as degrees, minutes, and seconds, with the presence of the decimal indicating which is used. Four subfields define a bounding box (subfields |d, |e, |f, and |g). These are generally adequate for expressing the westernmost, easternmost, northernmost, and southernmost extent of coverage of a map. The west, east, north, and south hemispheres are indicated either with the first letter of the direction or by a plus or minus symbol. Subfields |s and |t describe G-ring latitude and longitude. The rest of the information included in field 034 applies to the manifestation (i.e., the resource described in the bibliographic record).The data in the 255 field is in a textual form in subfield |c. A cataloger transcribes the coordinates as they are written on the map being catalogued into field 255 subfield |c. He or she then codes the same data into the 034 field of the same record. In addition, in the 255 field, there are symbols for degrees, minutes, and seconds that do not appear in the 034 field.

Field 084 defines the classification system used. Most academic libraries use the Library of Congress classification scheme, although there are map libraries that retain the USGS classification system.

Generally, the most significant access point for a map is the *geographic area* of coverage. However, the cataloging codes consider the geography of a map to fall under SUBJECT. Therefore, like books, the most significant access points for maps in most library catalogs tend to be authors (1xx) and titles (130, 24x).

Figure 7. 034 Coded cartographic mathematical data (R)

Indicators
First - Type of scale
0 - Scale indeterminable/No scale recorded
1 - Single scale
3 - Range of scales
Second - Type of ring
- Not applicable
0 - Outer ring
1 - Exclusion ring
Subfield Codes
$a - Category of scale (NR)
a - Linear scale
b - Angular scale
z - Other type of scale
$b - Constant ratio linear horizontal scale (R)
$c - Constant ratio linear vertical scale (R)
$d - Coordinates--westernmost longitude (NR)
$e - Coordinates--easternmost longitude (NR)
$f - Coordinates--northernmost latitude (NR)
$g - Coordinates--southernmost latitude (NR)
$h - Angular scale (R)
$j - Declination--northern limit (NR)
$k - Declination--southern limit (NR)
$m - Right ascension--eastern limit (NR)
$n - Right ascension--western limit (NR)
$p - Equinox (NR)
$s - G-ring latitude (R)
$t - G-ring longitude (R)
$6 - Linkage (NR)
$8 - Field link and sequence number (R)

Author as Main Entry

For maps, author vs. area as the principle access point or "main entry" was a major discussion point in the *AACR*. There were, and are, arguments against and for the use of author main entry for maps. Supporting area main entry is the contention that "recognition of the differences between maps and books and the need for separate rules for cataloging of maps is long overdue" (Woods, 1959). Arguments supporting author main entry for maps were based on historicity: "For an historical library, it is important to establish responsibility for the map. Thus the identification of the 'author,' or cartographer, the individual or the institution, government agency or publisher responsible for the map, is an important responsibility of the cataloger" (White, 1962). Early map cataloging codes either considered the author and publisher to be separate entities or grouped them into one category. Prior to the 1988 revision of *AACR2*, cataloging rules stated explicitly "cartographers are the authors of their maps." *AACR2r* added a section on corporate authorship. This shift to allow corporate authorship of maps was an important step in the recognition of the format-specific needs of map users and catalogers. The cartographer, publisher, or government agency responsible for producing the map could end up as the "author" of a map. The cartographer's name, publisher's name, and government agency are all valid access points in the catalog record.

Title

Map titles are often nondistinctive, such as "Map of the Known Universe." The map title, which might be prominent on the map's face, may be different on the map panel (the panel is the portion of the map that is displayed when the map is folded). All of these various titles are given access points in the catalog record, so that the map can be retrieved regardless of which title is used. If no title is available, the cataloger is directed on how to establish a title for the map.

Mathematical Data Area

Immediately after the title and edition statements is the *mathematical data area*, which includes *scale, projection*, and *coordinates*. Scale, defined as the ratio of distances on a map, globe, relief model, or section to the actual distances on the ground they represent, is generally represented with a bar scale, a written scale (1 inch = 100 miles), or a fractional scale (1: 10,000). Scale is extremely important in reference service, because it relates to the level of detail and area covered by a specific map. A large-scale map will cover a smaller geographic area with more detail than a small-scale map. The scale may be stated textually (e.g., 1 mile to the inch), but must also be converted to a representative fraction (e.g. 1:63,360) in the catalog record. If no scale is given, it must be computed. In addition to its analytical functions, scale also affects map display, which may be a critical point for the user. The smaller a map's scale, the more generalized the map's features are. A road that is very curvy in physical space may be represented by a fairly straight line on a map, or if on a very small scale-map, be dropped altogether.

Projection is an equally important but more complex concept. The projection is the type of distortion that occurs when a curved surface is forced onto a flat surface. Converting information from the curved surface of a globe (latitude and longitude) to a flat one (x,y coordinates) of a map involves a mathematical formula called a map projection, which uses a projected coordinate system. However, this flattening process causes distortions in one or more of the following spatial properties: distance, area, shape, and direction. No projection can preserve all these properties, and as a result, all flat maps are distorted to some degree. However, the type of projection can assist in the choice of the map. *Equal Area* projections preserve area and are used in the making of many thematic maps, for example, maps of the United States commonly use the Albers Equal Area Conic projection. *Conformal* projections preserve shape and area. Although shape is preserved for small areas, large area shapes, such as a continent, will be significantly distorted. Conformal projections are often used for navigational charts, weather maps, and Mercator maps. *Equidistant* projections preserve distance true from one point (or a few points) to all other points or along all meridians or parallels. Equidistant map projections are often used to find geographic or man-made features that are within a certain distance of other features. *Azimuthal* projections preserve direction from one point to all other points. Azimuthal projections can be combined with equal area, conformal, and equidistant projections. When available, information about coordinates is also given in the mathematical data area.

Bibliographic Notes

Bibliographic notes for maps serve a dual purpose: to explain elements of the bibliographic record and to provide information about the map's contents. Although map catalogers are instructed to favor bibliographic over content notes, limit the total number of notes, and simplify contents notes, these restrictions can have a negative impact on OPAC searching by limiting what can be retrieved using a keyword search.

The bibliographic notes area of a map record follows a set order, outlined in AACR2r. Notes contain information on any accompanying material (typically text or charts), the map's publishing history, its overall scope, additional geographic area covered, cartographic details (such as the type of map), or other helpful information (e.g., insets, indexes, and legends). All of these notes greatly enrich the catalog user's knowledge of the content of each map and allow patrons to decide if the map is potentially useful to them or not.

Dates

Librarians need to be aware of the meaning and complexity of map dates. The date of publication is the printing date. This might not be the same as the date of situation, defined as the date of the information displayed on the map. Furthermore, a map can have many dates of situation; there might be a different date for each type of information displayed, for example, the date of field check, date of survey, and so forth. For research purposes, the date of situation is generally the most important. However, maps can lack a publication date.

Descriptive cataloging rules emphasize the publication date, placing the date of situation to the notes area. AACR2r does not specifically mention the date of situation in the notes section of chapter three on cartographic materials, though an example of the date of situation is included. The reader is invited to review additional manuals developed to assist to catalogers in understanding the intricacies of map information (Andrew & Larsgaard, 1999; Larsgaard, 1998; Mangan, 2003).

Primary Access Points: Geographic Area, Subject Analysis, and Classification

The primary access points for maps are geographic area, subject analysis, and classification. How do you ask for a map? What terms do you use to get at the subject or content of a map? Geographic area headings are a challenge to the cataloger, particularly in determining the name of a geographic area.

Subject Headings: Geographic Area and Geographic Place Names

Jurisdictional names are based upon political jurisdictions, for example, a country, state, or city, such as France, Louisiana, or Apopka. Maps with jurisdictional boundaries are important in the location of events and individuals. Many governmental records, including genealogical records, are created and stored by jurisdictional entities that may not be the actual place where the event occurred or the individual resided. Noting jurisdictional boundaries also allows the user to locate a historical location that is now divided between two or more modern locations. Jurisdictional names (country, state, or city) are constructed following rules in *AACR2r*. *Nonjurisdictional names* are based on geographic features. These include named rivers, lakes, or mountains, such as the Seine (France), Okeechobee, Lake (Florida), and McKinley, Mount (Alaska), or named places (e.g., battlefields, national parks, or forests, such as Gettysburg National Military Park, Pa., Ocala State Forest (Florida), and Yellowstone National Park. Inverted forms of the name are used when the natural feature begins with a generic term, such as Mount or Lake, except when it is an integral part of the name, for example, Mount Baker National Forest (Wash.). Nonjurisdictional names are constructed using the *Library of Congress Subject Cataloging Manual*. State names are abbreviated when used as a qualifier to a place, but are not abbreviated when used as a subject subdivision. Therefore, both the abbreviated version of the state name (as specified in *AACR2r*) and full name must be used to find all appropriate geographic items. For example, when searching for maps of Ocala, Florida, a general map will be found under the subject heading OCALA (FLA.)—MAPS, while a thematic map may receive a subject heading formulated as PARKS—FLORIDA—OCALA—MAPS.
The most difficult factor in determining map subject headings is determining the

area covered by the map. Geographic areas are not tidy plots of land clearly defined by straight lines. Often a map of one area will include a nearby area. Sometimes these nearby areas are mentioned in the title or notes. Catalogers use their judgment as to when these nearby places should receive subject headings. There are similar problems with boundary maps that do not typically incorporate the whole of a political or geographic area.

Name changes of the geographic area shown on the map are also problematic. Currently, catalogers assign subject headings under the latest name of political jurisdiction, regardless of when the map was made. This has implications not just for bibliographic control but also for authority control, both of which affect searchers. For example, with the fall of the Soviet Union, new subject headings were created for each distinct geographical entity. The old heading, Soviet Union, was no longer an accurate subject heading. Another example is Siam, which is modern day Thailand.

Subject Headings: Analytical/Topical

For libraries using Library of Congress classification and subject headings, the first heading will be the one that most nearly represents the predominant topic of the work. Thematic maps use the topical heading to which the geographic location is attached as the first subject heading (e.g., SPIRIT CENTERS—UNITED STATES—MAPS). A single map can contain a variety of information requiring several subject headings. The range of topics in a map collection is very wide as well. Two examples of Library of Congress subject headings are ELECTION DISTRICTS—UNITED STATES—MAPS, and MIDDLE EARTH (IMAGINARY PLACE)—MAPS, which refer to the titles *The Almanac of State Legislatures: Changing Patterns 1990-1997* and *A Map of Middle-Earth*, respectively. Another example is topographic quadrangles. At a minimum, these maps receive a general heading, STATE—MAPS, TOPOGRAPHIC. In large collections, individually cataloged quadrangles would be augmented with more specific subject headings for better retrieval. Specificity, again, is critical in the cataloging of geographic material.

Classification

The classification of a map is seemingly straightforward. Contrasted to books, which are classified by topic and then by place, the map is classified *first* by place and then by topic. Although classification reflects the geographical area depicted, complications arise when the map contains two or more areas. With two areas, catalogers are instructed to class the map under the first one named in the title. For maps that contain three or more areas, catalogers are instructed to use the next higher administrative or regional class number that includes them all.

Set or Separate Decision

Most cartographic collections include a category of multipart items traditionally stored together and controlled by one catalog record. Maps are often distributed as part of a set or a *series*. The map series or set is printed on a number of sheets. Some are published at one time, some over many years, with no predictable seriality. The series has repeated and connective information on each sheet enables the sheets to be used singly or in conjunction with all or part of the remaining sheets. Among the common map series are the topographic maps from the U.S. Geological Survey. These are not the same as a *multisheet single map*, which is usually published all at one time and contains fewer sheets, all intended to be thought of as a single map and used together. Some map sets include thousands of sheets, each with unique sheet level data elements.

Libraries can choose to catalog each component on a separate record or put them together on one record, depending on their local needs. This decision affects access on many levels. For instance, a map set that is cataloged as a set will receive one bibliographic record and contain general subject headings for the whole. The titles of each individual sheet will be accessed only through a contents note, and the item will be filed together. When each part is cataloged on its own record, the same map set will receive more specific subject headings and title access for the individual sheet. The set title will remain accessible as a series title, and each unit may possibly be filed in different areas. Sheet level data elements may include some or all of the following types of elements: sheet titles, sheet number, coordinates, edition or version, issuing body, ISBN, stock or publisher number, date, supplements/separate indexes, or grid structure/number. Caveats in cataloguing sheet maps include multiple languages or scripts used in the sheet title, number, supplement, or index; sheets may or may not be textually enumerated or be an alphanumeric system relating to the grid structure of the set; edition or version of a map set may include updates during the span of the set edition; and sheets in the same set may be issued by differing entities. Another area of concern is how to address holdings.

MARC Discussion Paper No. 2006-DP07 (ALA/MAGERT Holdings Task Force, 2006) proposes several possible ways of recording information for multipart cartographic materials data using MARC 21 Bibliographic and/or Holdings Formats. However, the discussion paper does not provide a clear solution that can satisfy the needs for enhanced records or address the realization that there is increasingly a diminished amount of time spent on cataloging full records, even by large academic/research institutions. The Task Force recommendations ranged from creating a full level record for each sheet, giving all data and access points, even if they are already on the record for the set, to creating a record for each set plus brief records for each sheet, with links to the set bibliographic record through 773 fields (ALA/MAGERT, 2006). However, both of these approaches present difficulty in the display of collective and individual data in an understandable format to the user. Another suggested approach was the use of the 505 notes field, again problematic in the sense that one cannot parse out discrete subfields for searching and for displaying results. The use of the 774 fields (Constituent unit entry) in the bibliographic record for the set as a whole was also suggested, allowing the parsing of data elements into separate subfields (ALA/MAGERT, 2006). The use of holdings fields 844, 853, and 863 was another alternative. Field 034 (Coded Cartographic Mathematical Data) could be added to the holdings format, since coordinates will be needed at the sheet map level. However, that presented several problems, such as separate holdings records for each sheet map, linking multiple 844s to the appropriate 85X/86X pair, and the change of the 844 field $a (Name of Unit) to a repeatable subfield to handle parallel or variant titles (ALA/MAGERT, 2006). Disadvantages to the use of this approach included the inability to generate accurate institutional holdings based on the data in multiple separate holdings records, resistance to adding bibliographic information to the holdings format, and search issues at a systems level (ALA/MAGERT Holdings Task Force, 2006).

Similar to print maps, geospatial data also comes in single-type collections or multiple-type collections (Federal Geographic Data Committee, 1998). A single-type collection is an aggregate collection of multiple data units collected in similar conditions with equivalent semantic content. Each unit can stand on its own as a geographic information item and does not require additional processing to generate it. These would be classed as a spatial collection (spatial division), a temporal collection (time-series or periodicity), or a spatio-temporal collection that has both spatial distribution and periodicity. A multiple-type collection would have data layers or components from two or more different datasets that require compilation to generate a product based on user queries. Creating nested collections, or sets/series, would aid in resource discovery and guide users to layered, hierarchical items more easily.

To date, there has been little exploitation of MARC to create "intellectually related" collections. A minimal record will not provide the richness or completeness of information to guide the user to a specific item with any ease. Aggregate collections require both the aggregate information, as well as the single item information, to be updated whenever there is an addition or removal of an item to the collection or when an item is modified. An advantage is that much of the base information, or description, for the item and the collection level record is similar, or "inheritable." This creates the possibility of generating new content from existing records more easily, a feature that catalog developers should consider when creating new library management systems. Another important component of spatial-temporal data to consider is that spatial-temporal values of the descriptive elements may themselves be aggregated or averaged over the values of the items within the collection. Collection-level display in a catalog should be different from the item level display, providing the user the possibility of reviewing an aggregated view of search results, grouped by scenes/data/time series (Longley, Goodchild, Maguire, & Rhind, 2001).

Conclusion/Summary

In 1967, it was estimated that the annual cartographic output ranged from 60 to 100 thousand sheets, including new and revised editions, with American university and public libraries adding an estimated 15 thousand sheets per year (Gerlach, 1956). Over the past 50 years, an exponential amount of geographic information was created, produced, and distributed in digital form, particularly by the United States government. There are map-based graphical interfaces to perform geographical queries. The Web-based catalogs, hyperlinked with the MARC 856 field, allow patrons merely to click on the index to the map set in order to determine which map a library actually houses. In many libraries, maps are being digitized for online use. In the midst of this abundance of raw data, "skills that assure consistency, predictability, and repeatability of access are as needed as ever" (p. 50) (Zyroff, 1996) in order to ensure that users can find relevant materials. However, there are still a number of challenges that must be dealt with, such as legacy data, regularizing OPAC displays of bibliographic information for geographic data, and the possibility of incorporating thesauri data. Nogueras-Iso, Muro-Medrano, and Zarazaga-Soria (2005) suggest there are additional problems that also hinder the correct use of metadata for geographic and geospatial resources. The sheer amount of geographic resources that has been created without associated documentation makes it difficult to identify groups of related resources (Nogueras-Iso et al., 2005). Hierarchical identification of collections and subcollections by the use of the series field in MARC, a mechanism frequently used in print collections, should be used in the cataloging of these related resources, facilitating resource discovery, and creation of metadata. The diversity and heterogeneity of metadata standards is also problematic (Nogueras-Iso et al., 2005). Legacy data and different software applications require interoperability and crosswalks to ensure transferability and, again, resource discovery. The use of a summary view of the data using MARC and specific geographic metadata, such as ISO 19115, ensures access by the general public and discovery agents. Finally, heterogeneity of metadata content makes it difficult to identify the values given to metadata element in two different records that mean the same, conceptually, but use different terms (Nogueras-Iso et al., 2005). Authority records, a common feature of library catalogs, can address this problem. However, it will require a commitment on the part of library administration to see the value of enhancing both bibliographic and authority records.

At the 1997 International Conference on the Principles and Future Development of AACR, the charge was to review the underlying principles of AACR, and take into account present and future trends in information resources and information management. Among the recommendations was to advance the discussion on the primacy of intellectual content over physical format (The current rules call for cataloging each item based on the physical form of the item in hand). IFLA's FRBR requirements emphasize the importance of ensuring access based on user behaviors while not forgetting the primary function of the librarian is to best organize and make accessible the items held within its collections, physical and digital (IFLA Study Group on the Functional Requirements for Bibliographic Records, 1998). In 1999, the ALCTS Subcommittee on Metadata and Subject Analysis recommended the use of a combination of keywords and controlled vocabulary in metadata records for Web resources (ALCTS/CCS/SAC/Subcommittee on Metadata and Classification, 1999).

The focus on integration for the user has led the library community to standardize the data exchange format (MARC) and the content description (*AACR2, LCSH,* and other subject authority lists, internationally accepted classification schemes). The adaptation of *LCSH*'s rich vocabulary, syntax, and application rules to a Web-based environment would create a system "relatively easy to apply and maintain–and easier and more effective on the searching end as well" (Chan, Mai, & Hodges, 2000), particularly using a faceted, postcoordinate approach. User query analysis could be used as a means to develop or link existing terms to user-centered terms in authority files. This would move online catalogs away from data information retrieval (user-worded query statements) to information retrieval strategies.

Another area that will be equally important to consider is the capability to use seamless languages. If English is assumed to be the primary language, then titles and subject areas will need to be enhanced with translations, particularly for transliteration for non-Latinate languages (Hanson & Heron, 2003). Variant or translated titles/abstracts, and so forth, for non-English materials will need to be created to provide access to those items. Finally, a link to a translation engine to create some sort of translation of the item (if in HTML, Word, etc.) and vice-versa may also be required (Hanson & Heron, 2003). Differing formats and hardware/software necessary to view content will require notes to inform the user that to view this data one would need X software/plug-in application(s), Y amount of space on their drive to install and run said plug-in, and so forth (Hanson, 2006). Plug-ins may be critically important to visually display or load updated geographic or spatial data.

Search languages, that is, the language of the catalogue, will need to ensure consistency, accuracy, precision, and negotiation power between the remote parties as well as to accommodate whatever communication languages will be needed for disadvantaged users (Hanson & Heron, 2006). As the reader will see in Chapters V and VI, there is a need to establish naming conventions for geographic/GIS access points to enhance precision and relevance in one's retrieval when searching. In fact, the literature attests to user frustration when trying to find a relevant something and then having to sift through hits that are contextually irrelevant, although their term(s) might be somewhere in the record. Thesauri and ontologies that can create the hierarchical, attribute, entity, and bibliographical relationships among content and context of items are critical. Catalogers attempt to create listings of various depths and degrees of detail to record the existence of research materials. Researchers then search for answers to their questions and to make the best possible use of recorded knowledge. As Smiraglia (2002) states "That is, they [researchers] seek to exploit what is already known, so as to create new knowledge."

The content richness of the MARC database makes it desirable to maintain and add MARC records as the digital environment evolves. Increasing the interoperability of MARC with XML to integrate with other types of metadata records is a considerable return on the enormous investments that have gone into and go into preparing MARC records. This will allow libraries to improve the interoperable capabilities of existing bibliographic data assets and to advance integration of bibliographic systems in a manner that is sensible and best practice in the cataloging of geographic and spatial data.

In 1994, it was suggested that "[t]he twenty-first century will see geographic information transported from remote nodes using computer networks to support decision making throughout the nation. ... Timely use of these data would be difficult due to ill-defined format, quality, and accuracy. National or regional decision making would be severely impaired because most data sets are not adequately characterized." (National Research Council, Mapping Science Committee, 1994). We posit that the use of library best practices can successfully address the description, querying, and discovery of geospatial information. All libraries containing digital collections of geospatial data would become geolibraries, allowing searches by geographic location, returning any item in any format, from maps, images, books, reports, photographs, music, art, archaeology, and so forth, identified with a particular location.

References

ALA/MAGERT Holdings Task Force. (2006). *Recording set information for multipart cartographic materials* (MARC discussion paper No. 2006-DP07). Washington, DC: Library of Congress. Retrieved November 2006, from http://www.loc.gov/marc/marbi/2006/2006-dp07.html

ALCTS/CCS/SAC/Subcommittee on Metadata and Classification. (1999). *Subject data in the metadata record: Recommendations and rationale: A report from the ALCTS/CCS/SAC/Subcommittee on Metadata and Subject Analysis*. Retrieved from http://www.ala.org/alcts/organization/ccs/sac/metarept2.html

Andrew, P. G., & Larsgaard, M. L. (1999). *Maps and related cartographic materials: Cataloging, classification, and bibliographic control*. Binghamton, NY: Haworth Information Press.

Batty, D. (1998). WWW: Wealth, weariness or waste: Controlled vocabulary and thesauri in support of online information access. *D-Lib Magazine, November*. Retrieved from http://www.dlib.org/dlib/november98-11batty.html

Boggs, S. W., & Lewis, D. C. (1945). *The classification and cataloging of maps and atlases*. New York, NY: Special Libraries Association.

Chan, L., Mai, & Hodges, T. (2000). Entering the millennium: A new century for LCSH. *Cataloging & Classification Quarterly, 29*(1/2), 225-234.

Chan, L. M. (2001). Exploiting LCSH, LCC, and DDC to retrieve networked resources: Issues and challenges. In A. M. Sandberg-Fox (Ed.), *Proceedings of the Bicentennial Conference on Bibliographic Control for the New Millennium: Confronting the challenges of networked resources and the Web. Washington, D. C. November 15-17, 2000* (p. 159-). Washington, DC: Library of Congress, Cataloging Distribution Service. Retrieved from http://www.loc. gov/catdir/bibcontrol/chan_paper.html

Cline, N. M. (2000). Virtual continuity: The challenge for research libraries today. *EDUCAUSE Review, 35*(5/6), 22-28. Retrieved 09/23/2001, from, http://www. educause.edu/ir/library/pdf/ ERM0032.pdf

CONSER Task Force on Publication Patterns and Holdings. (2006). *CONSER Publication Pattern Initiative.* Retrieved from http://lcweb.loc.gov/acq/conser/patthold.html

Cutter, C. A. (1904). *Rules for a dictionary catalog* (4th ed.). Washington, DC: Government Printing Office.

Federal Geographic Data Committee. (1998). *Content standard for digital geospatial metadata.* Reston, VA: The Committee.

Fiske, S. T., & Taylor, S. E. (1991). *Social cognition* (2nd ed.). New York, NY: McGraw-Hill.

George A. Smathers Libraries, & ALA/MAGERT Cataloging and Classification Committee. (2005). *Recording geographic coordinates in the MARC 21 Authority Format* (MARC discussion paper No. 2006-DP01). Washington, DC: Library of Congress. Retrieved November 2006, from http://www.loc. gov/marc/marbi/2006/2006-dp01.html

Gerlach, A. C. (1956). *An adaptation of the Library of Congress classification for use in geography and map libraries; annex to report of the Commission on Library Classification of Geographical Books and Maps.* Washington, DC: Reproduced by the National Academy of Sciences of the United States of America.

Goodchild, M. F., & Quattrochi, D. A. (1997). Introduction: Scale, multiscaling, remote sensing, and GIS. In D. A. Quattrochi & M. F. Goodchild (Eds.), *Scale in remote sensing and GIS* (pp. 1-11). Boca Raton, FL: Lewis Publishers.

Hanson, A. (2006). Organization and access to electronic resources. In V. L. Gregory, *Selecting and managing electronic resources* (Rev. ed.), (pp. 45-64). New York, NY: Neal-Schuman.

Hanson, A., & Heron, S. J. (2006). *Impact of globalisation on information seeking: The role of cultural lenses and indigenous knowledge.* Paper presented at the annual meeting of ACURIL XXXVI: Information and Human Rights: The Social, Cultural, and Ethical Aspects of the Information Society, Oranjestad, Aruba, 29 May 2006.

Heron, S. J., & Hanson, A. (2003). From subject gateways to portals: The role of metadata in accessing international research. In N. Callaos (Ed.), *Conference Proceedings of the SCI 2003: The 7ᵗʰ World Multiconference on Systemics, Cybernetics, and Informatics* (pp. 529-533). Orlando, Florida: International Institute of Informatics and Systemics.

IFLA Study Group on the Functional Requirements for Bibliographic Records. (1998). *Functional requirements for bibliographic records: Final Report: IFLA Study Group on the Functional Requirements for Bibliographic Records.* Frankfurt, Germany: Deutsche Bibliothek: Frankfurt am Main. Retrieved 09/23/2001, from http://www.ifla.org/VII/s13/frbr/frbr.pdf

International Federation of Library Associations and Institutions, Cataloguing Section. (2005). *Cataloguing section strategic plan 2005-2007*: International Federation of Library Associations and Institutions. Retrieved July 2006, from http://www.ifla.org/VII/s13/annual/sp13.htm

Joint Steering Committee (2005). *Draft statement of objectives and principles for RDA.* Retrieved June 2007, from http://www.collectionscanada.ca/jsc/docs/5rda-objectives.pdf

Joint Steering Committee (2007a,) *RDA: Resource description and access: Scope and structure.* Retrieved June 2007, from http://www.collectionscanada.ca/jsc/docs/5rda-objectives.pdf

Joint Steering Committee (2007b) *RDA-FRBR Mapping.* Retrieved June 2007, from http://www.collectionscanada.ca/jsc/docs/5rda-scoperev.pdf

Larsgaard, M. L. (1998). *Map librarianship: An introduction* (3rd ed.). Englewood, CO: Libraries Unlimited.

Library of Congress. (1989). *The USMARC formats: Background and principles.* Washington, DC: Library of Congress.

Library of Congress. (2003). *Metadata encoding and transmission standard (METS).* Washington, DC: Library of Congress. Retrieved May 2006, from http://www.loc.gov/standards/mets/

Library of Congress. (2005). *Metadata Authority Description Schema (MADS).* Washington, DC: Library of Congress, Network Development and MARC Standards Office. Retrieved May 2006, from http://www.loc.gov/standards/mads/

Library of Congress. (2006). *Metadata object description schema (MODS).* Washington, DC: Library of Congress, Network Development and MARC Standards Office. Retrieved May 2006, from http://www.loc.gov/standards/mods/

Longley, P. A., Goodchild, M. F., Maguire, D. J., & Rhind, D. W. (2001). *Geographic information systems and science.* Chichester, England: John Wiley & Sons Ltd.

Mangan, E. U. (2003). *Cartographic materials: A manual of interpretation for AACR2, 2002 revision* (2nd ed.). Chicago, IL: American Library Association, Anglo-American Cataloguing Committee for Cartographic Materials.

National Research Council, Mapping Science Committee. (1994). *Promoting the national spatial data infrastructure through partnerships.* Washington, DC: National Academy Press.

Network Development and MARC Standards Office. (2005). *MARC 21 concise formats.* Washington, DC: Cataloging Distribution Service, Library of Congress.

Nogueras-Iso, J., Muro-Medrano, P. R., & Zarazaga-Soria, F. J. (2005). *Geographic information metadata for spatial data infrastructures: Resources, interoperability, and information retrieval* (1st ed.). Berlin; New York: Springer.

Robinson, A. H., Morrison, J. L., Muehrcke, P. C., Kimerling, A. J., & Guptill, S. C. (1995). *Elements of cartography* (6th ed.). New York, NY: New York: John Wiley and Sons.

Sasscer, R. S. (2000). *U.S. Geological Survey library classification system* (U.S. Geological Survey Bulletin No. 2010). Reston, VA: U.S. Geological Survey. Retrieved May 2006, from http://pubs.usgs.gov/bul/b2010/b2010.pdf

Smiraglia, R. (2002). The progress on theory in knowledge organization. *Library Trends, Winter,* 20. Retrieved from http://www.finarticles.com/articles/mi_m1387/is_3_50/ai_88582618/print

Svenonius, E. (2000). *The intellectual foundation of information organization.* Cambridge, MA: MIT Press.

Tufte, E. R. (1983). *The visual display of quantitative information.* Cheshire, CT: Graphics Press.

U.S. Geological Survey, (1998). *SDTS: Spatial data transfer standard.* Reston, VA: Geological Survey.

White, J.B. (1962). Further comment on map cataloging. Library Resources and Technical Services, 6 (Winter), p. 78. Retrieved from http://www.sla.org/speciallibraries/ISSN00386723V58N6.pdf

Woods, B. (1959). Map cataloging: Inventory and prospect. *Library Resources and Technical Services, 3*(Fall), 257-273.

Zyroff, E. (1996). Cataloging is a prime number. *American Libraries, 27*(5), 47-50.

Chapter V

From Print Formats to Digital:
Describing GIS Data Standards

Ardis Hanson, University of South Florida Libraries, USA

Susan Heron, University of South Florida Libraries, USA

Introduction

The preceding chapter discussed how geographic and cartographic materials are traditionally described in libraries. With the growth of geospatial data, new methods of description needed to be developed to allow users, often with very different information needs, to find and retrieve relevant resources across different platforms and software systems. Geographic information systems are designed to allow the management of large quantities of spatially referenced information about natural and man-made environments, covering areas such as public health, urban and regional planning, disaster response and recovery, environmental assessments, wetlands delineation, renewable resource management, automated mapping/facilities management, and national defense. Powerful computers, advanced network capacities, and enhanced, distributed GIS software allowed the growth of the National Spatial Data Infrastructure (NSDI). Established by Executive Order 12906 in April 1994, the NSDI assembles "technology, policies, standards, and human resources to acquire, process, store, distribute, and improve utilization of geospatial data for a variety of users nationwide" (Federal Geographic Data Committee, 2006a). The goal of the NSDI is to "reduce duplication of effort among agencies, improve quality and reduce costs related to geographic information, to make geographic data more accessible to the public, to increase the benefits of using available data, and to establish key partnerships with states, counties, cities, tribal nations, academia and the private sector to increase data availability" (Federal Geographic Data Committee, 2006b). However, the success of a national spatial data infrastructure depends on the development of a series of standards for that infrastructure. Infrastructure components encompass a variety of elements. Hardware and physical facilities store, process, and transmit information; software applications and software allow access, structure, and manipulation of information; and network standards and transmission codes facilitate interorganizational and cross-system communication (Hanson, 2006). When reviewing standards for geospatial data, one must look at standards for cartography, hardware and software, telecommunications, and information technology standards at national and international levels. Several thousand standards apply to computers, and this can be multiplied geometrically, if not exponentially, with the advent of network standards and integrated data formats. This chapter will address standards in geospatial data, interoperability and transferability, mark-up languages, and the development of the federal metadata standard for geospatial information.

What is Spatial Information?

By understanding how spatial information is defined and described, users can better access and retrieve the specific items they want. At the simplest level, spatial data is comprised of coordinates. A *coordinate* is a number that denotes either a position along an axis relative to an origin, given a unit of length or a direction relative to a base line or plane, given a unit of angular measure, such as latitude or longitude. The definitions of coordinates, points, lines, planes of reference, units of measure, and other associated parameters are referred to collectively as a *coordinate system*. Each coordinate system has its own distinct parameters and definitions. Two types of coordinate systems are geographic and projected coordinate systems. A geographic coordinate system is a reference system that uses a three-dimensional spherical surface to determine locations on the earth. Any location on earth can be referenced by a point with latitude and longitude coordinates based on angular units of measure. A projected coordinate system is a flat, two-dimensional representation of the earth. Using Cartesian (rectilinear) coordinates based on linear units of measure, a projected coordinate system is based on a spherical (or spheroidal) earth model, and its coordinates are related to geographic coordinates by a projection transformation. Geodetic data is spatial data expressed in latitude and longitude coordinates, in a coordinate system that describes a round, continuous, closed surface. One of the

Figure 1. Sample of GIS layers that create a multipurpose and functional view of dimensional spatial data

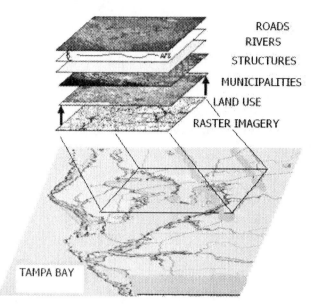

overarching concerns for digital geoscience cartography is to establish standards for the generation of geologic elements (base map elements, thematic elements, mappable geologic units, symbology, positional accuracy, stratigraphic nomenclature, and colors and patterns) in addition to those of other disciplines (Solem, Chalmers, Dibiase, Donert, & Hardwick, 2006).

Creating a National Spatial Digital Infrastructure

Geographic and cartographic standards are not new. In 1971, criteria were developed for a land-use and land-cover classification system that could effectively employ orbital and high-altitude remote sensor data (Anderson, 1971). Almost 30 years later, in 1998, with the continued evolution of paper to data, a proposed standard for digital cartographic data for base maps was published. A parallel standard for geologic maps was also developed. During the debate surrounding a single spatial data transfer standard, both the base map and the geologic map standards were discussed. Both standards have four major components that include definitions and references, spatial data transfer specification, digital cartographic quality, and cartographic features. However, there are differences in how each uses geospatial data.

The first component, definitions and references, addresses how each discipline describes its universe of knowledge. Points, lines, curves, and areas can geometrically define cartographic objects in a zero-, one-, or two-dimensional space. Geologic maps also address items in zero-, one- or two-dimensional space, but have the added dimension of volume. Geologic definitions must also provide age attributes to geologic features as well as geologic time. Entity types, entity instances, and features address both base map and geologic map requirements. An *entity type* is a general spatial phenomenon. When an entity type is digitally translated, it is referred to as an "entity object class." An example is a geologic layer in a group of map layers. An *entity instance* is a specific spatial phenomenon that, after digital translation, becomes an entity object having a fixed location in time and space. A *feature* is the combination of the spatial phenomenon and its digital representation.

The second component, spatial data transfer specification, addresses interoperability and portability. It allows the transfer of generic map data between spatial data handling systems that may have different file formats. The specification, ISO 8211, will be addressed more fully later in the chapter.

The third component, digital cartographic quality, contains quantitative or qualitative information that describes the lineage, the positional accuracy, the attribute accuracy, the logical consistency, and completeness of the map. This component is critical for user evaluation to determine the appropriateness of the map for a specific use. In addition, the use of global positioning system (GPS) has required digital standards be adapted to handle the quantitative statements of positional accuracy relative to geodetic positions.

The final component is cartographic features. This should list the entity types and attributes required to create the map. Features are defined based on a number of earth science disciplines, such as geology, and formats, such as general topographic maps and nautical charts. Earth science feature symbols have a symbol code, a graphic representation of the symbol, an explanation of the symbol, specifications for publication (font, symbol size, height, etc.), and notes on usage.

Table 1. Base map standards and geologic cartographic standards

	Base Map Standards	Geologic Cartographic Standards
Definitions and references	Zero- one-, and two-dimensional data	Zero- one-, and two dimensional, and volume data
Spatial data transfer specification	ISO 8211	ISO8211
Digital cartographic quality	Lineage, positional accuracy, attribute accuracy, logical consistency, completeness	Lineage, positional accuracy, attribute accuracy, logical consistency, completeness
Cartographic features	Entity types and attributes (General topographic maps and nautical charts)	Entity object class specific to geologic maps

The spatial data transfer standard (SDTS) is comprised of three parts. Part 1 addresses the logical specifications for conformance, the conceptual model, quality specifications, the data structure model, and the transfer format. Part 2 provides a standards list and definitions of spatial features. Spatial features are classed and defined as entity types and entity instances. Part 3 deals with the implementation of the SDTS *vis-à-vis* the ISO 8211 standard to ensure information interchange, including the syntax and semantics necessary to transport files, records, fields, and subfields, by their data description, into machine-readable form across platforms in a compatible format. The early standard defined zero, one-, and two-dimensional objects only and supported three major types of spatial data operations: geometry only, geometry and topology, and topology only.

Geodetic control is "accepted as being survey control at the highest accuracy levels, connected to the national spatial reference framework, and used for as reference to lower accuracy surveys" (Federal Geodetic Control Subcommittee, s.d.). Geodetic control provides a common reference system for establishing the coordinate positions of all geographic data to support land information compatibility. By tying all geographic features to common, nationally used horizontal and vertical coordinate systems, geodetic control information plays a crucial role in developing framework data and user applications data. It also allows data quality assessment, collection, and conversion, as well as retrofitting new areas of data into existing covered areas. Geodetic control is typically separated into a horizontal (latitude/longitude) component and a vertical (elevation) component. Each component is based on different concepts and measurement methods. While there are geodetic-quality values for both horizontal and vertical components, the methods of determining these values will be different. Extremely high-accuracy results can be provided by today's global positioning systems; however, a more traditional method is required to establish vertical control. The National Spatial Reference System (NSRS), administered by the National Geodetic Survey (NGS), is separated into horizontal and vertical sections. Each control point is classified based on accuracy, which can be affected by the purpose of the control point, the type and condition of equipment used, field procedures adopted, and the experience and capabilities of personnel employed.

Within geodetic control, there are four standards covering collection, content, transfer, and metadata. Collection standards establish the submission, processing, and database format standards for data sources. *Input Formats and Specifications of the National Geodetic Survey Data Base* (Dewhurst, 1985), commonly referred to as the "Blue Book," describes the formats and procedures for submitting data into the National Geographic Survey (NGS) database. There are separate volumes for horizontal control data, vertical control data, and gravity control data. Content standards for geodetic control data contain code types and database elements to handle holdings data for storage and management. The primary source of reference for content standards is the NGS database data dictionary. Transfer standards manage the importing and exporting of digital data exchange standards necessary for data transfer. The spatial data transfer standard (SDTS) has all of the characteristics of an "open systems" standard, namely, it is modular, growth-oriented, extensible, and flexible. The SDTS specification is organized into the base specification (Parts 1-3) and multiple profiles (Parts 4-6). Parts 1 to 3 cover logical specifications, spatial features, and ISO 8211 encoding, respectively. Parts 4 to 6 each define specific rules and formats for applying SDTS for the exchange of particular types of data in SDTS: topological vector profile, raster profile, and point profile, respectively (U.S. Geological Survey, 1999). Part 7, Computer Aided Design and Drafting Profile (CADD), contains specifications for an SDTS profile for use with two- and three-dimensional vector-based geographic data as represented in CADD software. This allows the translation of vector-based geodata between CADD packages and between CADD and mainstream GIS packages. Approved as Federal Information Processing Standard (FIPS) Publication 173 (FIPSPUB 173-1, 1994), the current version, known as ANSI NCITS 320-1998, superseded FIPSPUB 173-1 in 1998. (For more information about the SDTS, the reader is referred to Morrison & Wortman, 1992.) The Federal Geographic Data Committee (FGDC) geodetic metadata standards provide cataloging and accessibility standards for data retrieval. Executive Order 12906 requires all Federal agencies to comply with the provisions of the metadata standard.

Standards Developers

Standards promote maximum reusability, interchangeability, and mergeability. In the United States, the Federal Information Processing Standards (FIPS) are government standards for federal agencies and organizations. Established in the 1960s, FIPS is administered by the National Institute of Standards and Technology (NIST). FIPS may evolve into national and international standards. In 1980, the U.S. Geological Survey (USGS) was designated the lead agency in developing earth science data standards for the federal government. Another U.S. standards organization is the American National Standards Institute (ANSI), whose InterNational Committee for Information Technology Standards (INCITS-L1) is its technical advisory group to the International Standardization Organization's (ISO) geospatial standards technical committee. Other standards organizations include the Institute of Electrical and Electronic Engineers (IEEE, the Open Geospatial Consortium, Inc. (OGC), and OASIS. International standards are determined by the International Standards Organization (ISO), the International Telecommunication Union (ITU), and International Electrotechnical Commission (IEC). Within the ISO, the Technical Committee 211 (ISO/TC 211) Geographic information/Geomatics is responsible for the ISO geographic information series of standards.

American National Standards Institute (ANSI)

Since 1918, the American National Standards Institute (ANSI) has administered and coordinated the private sector voluntary standardization system in the United States. Originally founded in 1918 by five engineering societies and three government agencies, ANSI is a private, nonprofit membership organization committed to assisting the development of standards based on openness, balance, consensus, and due process. ANSI itself does not develop American National Standards (ANSs). It provides a neutral venue by where interested parties may meet and create standards. ANSI accredits the procedures of standards developing organizations who work cooperatively to develop voluntary national consensus standards. ANSI accreditation ensures that the procedures to develop ANSs meet ANSI's essential requirements for openness, balance, consensus, and due process (American National Standards Institute, 2006). ANSs are "open" standards, a collaborative and consensus-based approval process used by a recognized body for developing and approving a standard. ANS standards may relate to products, processes, services, systems, or personnel. In its role as the only accreditor of U.S. voluntary consensus standards developing organizations, ANSI helps to ensure the integrity of the standards developers (American National Standards Institute, 2006).

The American National Standards Institute's Technical Committee for Geographic Information Systems (InterNational Committee for Information Technology Standards, INCITS-L1) is ANSI's technical advisory group to the ISO's Geospatial Standards Technical Committee (TC211).

Parallel to standards for geographic information are a number of library standards, such as ANSI Z39.2, American National Standard for Bibliographic Information Interchange, computer-to-computer communications protocols, such as Z39.50, the American National Standard Information Retrieval Application service definition and protocol specification for open systems interconnection, and ANSI/NISO Z39.7-2004, information services and use: Metrics and statistics for libraries and information providers—Data Dictionary.

In 2000, the first-ever U.S. National Standards Strategy (NSS) was approved. A roadmap for the development of reliable, market-driven standards in the public and private sectors, it is a standardization framework built upon consensus, openness, and transparency. The framework includes due process and flexibility allowing different methodologies and technologies to meet industry needs. It also stresses coherence and timeliness, avoiding overlap or conflict in development, and ensuring a streamlined administrative and accrediting review process. The NSS also emphasises that:

- "Standards are *relevant*, meeting agreed criteria and satisfying real needs by providing added value.

- "Standards are *responsive* to the real world; they use available, current technology and do not unnecessarily invalidate existing products or processes.
- "Standards are *performance-based*, specifying essential characteristics rather than detailed designs" (American National Standards Institute, 2000, p. 4).

ANSI organizational and company members, as well as ASC Z80 and INCITS (ANSI Accredited Standards Developers), and JTC 1, the U.S. Technical Advisory Group endorse these standards. Canada, Germany, the United Kingdom, and China have developed their own national standards strategies.

The International Organization for Standardization (ISO)

The International Organization for Standardization (ISO) was established as a nongovernmental organization in Geneva, Switzerland in 1947. It is an umbrella organization for national standardization activities. Members are countries represented by their respective national standards groups or external liaison organizations comprised of professional associations, such as the International Cartographic Association, and consortia, such as the OpenGIS Consortium. The ISO has a significant influence on the standardization of geographic information.

ISO publishes the ISO Standard, the ISO/PAS (Publicly Available Specification), the ISO/TS (Technical Specification), the ISO/TR (Technical Report), and the IWA (International Workshop Agreement). There are differences among these products. Although most ISO standards are highly specific, that is, "documented agreements containing technical specifications or other precise criteria to be used consistently as rules, guidelines, or definitions of characteristics to ensure that materials, products, processes and services are fit for their purpose" (International Organization for Standardization, [s.d.]), the ISO 9000 and ISO 14000 series, for example, are generic management system standards. An ISO Standard is reviewed at least every

Table 2. Selected ANSI geographic standards

ANSI INCITS 61-1986 (R2002)	Geographic Point Locations for Information Interchange, Representation of (formerly ANSI X3.61-1986 (R1997))
ANSI INCITS 353 2004	Information Technology - Geographical Information Systems - Spatial Data Standard for Facilities, Infrastructure, and Environment (SDSFIE)
AS 3736 1990	Geographical information systems - Bibliographical elements on maps (foreign standard)
ASTM D5714	Standard Specification for Content of Digital Geospatial Metadata
ASTM E2468	Standard Practice for Metadata to Support Archived Data Management Systems
ANSI T1.253-1999	Information Interchange - Code Description and Codes for the Identification of Location Entities for the North American Telecommunications System (Revision and Consolidation of ANSI T1.201-1987 and ANSI T1.205-1988)
ANSI/ISO 19101-2002	Geographic information -- Reference model

5 years for currency, technological evolution, methods and materials, and quality and safety requirements. A Publicly Available Specification (PAS) is reviewed every 3 years; however, after 6 years the PAS must become an international standard or be withdrawn. If a proposed standard did not receive enough support by ISO members, it can be submitted as a Technical Specification for a review by a wider audience. Similar to the ISO/PAS review, the ISO Technical Specification must advance to an international standard within 6 years or be withdrawn. Of the three types of ISO Technical Reports, the first two types are either standards that did not achieve consensus of the members or are reports of emerging/experimental standards. These will no longer be published as technical reports but as technical specifications. A technical report will only cover documentation or information on de facto standards. The International Workshop Agreement is one of ISO's strategies to create responses to the demands of a fast moving, open marketplace. These agreements may eventually be formalized into international standards. From a library perspective, understanding the differences among the types or versions of publications or documents promulgated by an agency or organization is important for collection development and reference.

Within the ISO, the Technical Committee 211 (ISO/TC 211) Geographic information/ Geomatics is responsible for the ISO geographic information series of standards. Its base standards include the reference model, feature definition, spatial and temporal schema, coordinate reference system, portrayal, encoding, quality, and metadata, to name just a few of the standards it has developed since 1994. Recently ISO/TC 211 reorganized itself into five working groups: geospatial services, imagery, information communities, location-based services, and information management.(International Organization for Standardization, 2004).

There are approximately 28 standards *specific* to geographic information in the ISO 19100 family (Kresse & Fadaie, 2004)(see table 3). These include standards for the representation of latitude, longitude, and altitude for geographic point locations; a reference model; conceptual schema language; spatial, temporal, and application schemas; methodologies for feature cataloging; spatial referencing; profiles; conformance and testing; quality principles and evaluation procedures; metadata; services, personnel; positioning services; encoding; location-based services; and imagery and gridded data; codes and parameters, and Web server interfaces.

ISO/IEC 10746, the reference model for open distributed processing in information technology, is the basis for the ISO 19100 family. Two of 10746's viewpoints, information and computational, are particularly relevant to the 19100 standards (Kresse & Fadaie, 2004). The information viewpoint is the most important since it deals with semantics of information and information processing within a GIS; the computational viewpoint deals with the patterns of interaction among services within a larger system (Kresse & Fadaie, 2004). These two viewpoints drive how large, distributed-GIS databases are decomposed in the design process, ensuring the standardization of geomatics.

The ISO 19100 family of standards is not a rigid or hierarchical set of standards. Instead, it is best viewed as guidelines describing the origin and quality of geographic information. For example, one "side" of the family deals with data capture. ISO 19113 describes quality principles, 19114 describes quality evaluation procedures, and 19115 describes metadata, with 19115:2 addressing metadata for imagery. Another "side" handles data storage. ISO 19109 covers the rules for application schema for objects, or features, in datasets. (As mentioned earlier, features may have attributes and operations. An attribute may be a point, a curve, or a surface. An operation may be the change in the feature due to an external influence). To assist in the description of the dataset, ISO 19110 provides a methodology for feature cataloging using general rules and a catalog template to create a complete listing of feature attributes and operations, while 19107 provides the geometry classes and rules for relationships. Since all spatial datasets use a coordinate reference system,

Table 3. ISO geographic information standards

ISO 6709:1983	Standard representation of latitude, longitude and altitude for geographic point locations
ISO 19101:2002	Geographic information -- Reference model
ISO/TS 19103:2005	Geographic information -- Conceptual schema language
ISO 19104	Geographic information --Terminology
ISO 19105:2000	Geographic information -- Conformance and testing
ISO 19106:2004	Geographic information -- Profiles
ISO 19107:2003	Geographic information -- Spatial schema
ISO 19108:2002	Geographic information -- Temporal schema
ISO 19109:2005	Geographic information -- Rules for application schema
ISO 19110:2005	Geographic information -- Methodology for feature cataloguing
ISO 19111:2003	Geographic information -- Spatial referencing by coordinates
ISO 19112:2003	Geographic information -- Spatial referencing by geographic identifiers
ISO 19113:2002	Geographic information -- Quality principles
ISO 19114:2003/ Cor 1:2005	Geographic information -- Quality evaluation procedures
ISO 19115:2003/ Cor 1:2006	Geographic information -- Metadata
ISO 19116:2004	Geographic information -- Positioning services
ISO 19117:2005	Geographic information -- Portrayal
ISO 19118:2005	Geographic information -- Encoding
ISO 19119:2005	Geographic information -- Services
ISO/TR 19120:2001	Geographic information -- Functional standards
ISO/TR 19121:2000	Geographic information -- Imagery and gridded data
ISO/TR 19122:2004	Geographic information/Geomatics -- Qualification and certification of personnel
ISO 19123:2005	Geographic information -- Schema for coverage geometry and functions
ISO 19125-1:2004	Geographic information -- Simple feature access -- Part 1: Common architecture
ISO 19125-2:2004	Geographic information -- Simple feature access -- Part 2: SQL option
ISO/TS 19127:2005	Geographic information -- Geodetic codes and parameters
ISO 19128:2005	Geographic information -- Web map server interface
ISO 19133:2005	Geographic information -- Location-based services -- Tracking and navigation
ISO 19135:2005	Geographic information -- Procedures for item registration

ISO 19111 provides the guidelines for defining the geographic positions of the data. Yet another "side," ISO 19117, handles how data is displayed while ISO 19118 addresses how data will be exchanged between datasets. The infrastructure "side" is comprised of ISO 19103 and 19104, which address the conceptual schema language and terminology, respectively, while ISO 19105 handles conformance and testing, and 19106 handles the profile (Kresse & Fadaie, 2004).

Each individual element of a standard may be addressed by another standard, much like building blocks. Although ISO 19115 describes the general content of the metadata and relationships between metadata elements, it does not state how metadata records should be built and formatted. ISO 19139 creates an XML (extensible markup language) schema that prescribes the format of the metadata record. The ISO 19139 standard incorporates undefined metadata elements referenced in ISO 19115, such as entity and attribute descriptions addressed by ISO 19109 geospatial data standard. ISO 19139 provides an encoding schema for describing, validating, and exchanging metadata about geographic datasets, dataset series, individual geographic features, feature attributes, feature types, feature properties, and more.

In addition to these standards, there are additional ISO standards that apply to representation, transmission, interchange, processing, storage, input, and presentation. For example, the Universal Multiple-Octet Coded Character Set (UCS), defined jointly by the Unicode Standard [Unicode] and ISO/IEC 10646, allows Web documents authored in the world's scripts (and on different platforms) to be exchanged, read, and searched by Web users globally (ISO/IEC, 2003). Another example is MARC (MAchine-Readable Cataloging) and its suite of related standards (US-MARC, Can/MARC, InterMARC, UKMARC, CCF, etc.), used for bibliographic control within the library science and digital libraries communities. USMARC, now known as MARC 21, is based upon ISO 2709:1996, Format for Information Exchange (INEX) and ANSI Z39.2, American National Standard for Bibliographic Information Interchange.

Open Geospatial Consortium, Inc.

The Open Geospatial Consortium, Inc. (OGC) is an international industry consortium of **over 300** private sector companies, public sector agencies, and universities. Much like ASNI and ISO, members of the OGC participate in an open, consensual process to develop publicly available interface specifications. These specifications, known as **OpenGIS®** Specifications, support interoperable ("plug and play" solutions) for Web, wireless, and location-based services. OpenGIS® Specifications come in two types: abstract and implementation. The Abstract Specification provides the conceptual foundation for specification development and the reference model to ensure interoperability as open interfaces and protocols are built and tested. The

Table 4. OpenGIS implementation specifications

04-021r3	Catalogue Service
01-009	Coordinate Transformation
04-095	Filter Encoding
05-047r3	GML in JPEG 2000
03-064r10	Geographic Objects
03-105r1	Geography Markup Language
01-004	Grid Coverage Service
05-016	Location Services (OpenLS)
05-016	Simple Feature Access 1
05-134	Simple Feature Access 2
99-054	Simple Features (SF) CORBA
99-050	Simple Features (SF) OLE/COM
02-070	Styled Layer Descriptor
03-065r6	Web Coverage Service
04-094	Web Feature Service
05-005	Web Map Context
03-109r1	Web Mapping Service
05-008c1	Web Service Common

Binary Extensible Markup Language (BXML) Encoding Specification	03-002r9	Binary encoding format for scientific data characterized by arrays of numbers in XML documents
Definition identifier URNs in OGC namespace	06-023r1	Formats used by these URNs, plus a set of specific URNs for specific definitions. changes in change requests OGC 05-091r2 and 05-060
Gazetteer Service - Application Profile of the WFIS Specification	05-035r2	Search and retrieval of georeferenced vocabulary of well-known place-names.
ISO19115/ISO19119 Application Profile for CSW 2.0 (CAT2 AP ISO19115/19)	04-038r2	Profile for data metadata, services, metadata, and application metadata organization, retrieval, and management
Catalogue Services - Best Practices for Earth Observation Products	05-057r4	Identification and acquisition of earth observation data from identified collections
Geography Markup Language (GML) Encoding Specification	03-105r1	XML encoding for the transport and storage of geographic information, including both the geometry and properties of geographic features
Sensor Model Language (SensorML)	05-086	General models and XML encodings for sensors
Web services architecture description	05-042r2	Description of a service-oriented architecture, with all components providing one or more services to other services or to clients
Recommended XML/GML 3.1.1 encoding of common CRS definitions	05-011	Recommended standard XML encodings of data defining some commonly-used coordinate reference systems, including geographic, projected, and vertical CRSs
Recommended XML/GML 3.1.1 encoding of image CRS definitions	05-027r1	Recommended standard XML encodings of data defining monoscopic image coordinate reference systems
Units of Measure Recommendation	02-007r4	Common semantic for units of measurement to be used across all OGC specifications
Web Coverage Processing Service (WCPS)	06-035r1	Retrieval and processing of geo-spatial coverage data, currently constrained to equally spaced grids

Implementation Specifications detail the interface structure between software components and use specific schemas found in the OGC Schema Repository.

The OGC has created a number of specifications that cover a range of language, protocol, and application concerns, including binary extensible markup language (BXML), geographic markup language (GML), universal resource names (URNs), a gazetteer service profile, identification and ordering catalog service. The binary extensible markup language (BXML) encoding specification (Bruce, 2006) addresses a binary encoding format for scientific data that is characterized by arrays of numbers in XML formatted documents. The OGC's Universal resource names (URNs) specifications (Whiteside, 2006) address definition identifier URNs in OGC namespace, such as "authority" and "objectType" values, URNs for specific data types and OGC implementation specifications. Just as paper gazetteers provided access by place names, the OGC Gazetteer Service Application Profile (Fitzke & Atkinson, 2006) allows a client to search and retrieve elements of a georeferenced vocabulary of well-known place-names. Catalog services define how geospatial information is to be organized and implemented for the discovery, retrieval, and management of data metadata, services metadata and application metadata. The OGC has one standard based on the ISO19115/ISO19119 Application Profile (Voges, Senkler, & Müller, 2004) and another that describes the minimum interface necessary to identify earth observation data products from previously identified data collections, such as satellite operators and data distributors (Martin, 2006). Model languages include geography markup language (GML) (Cox, Daisey, Lake, Portele, & Whiteside, 2004) and sensor model language (Botts, 2005).

Organization for the Advancement of Structured Information Standards

The Organization for the Advancement of Structured Information Standards (OASIS) produces standards for Web services, security, and e-business for public sector and application-specific markets. OASIS has more than 5,000 participants representing over 600 organizations and individual members in 100 countries. Originally founded in 1993 as SGML Open, OASIS' early work focused on interoperability guidelines for products using standard generalized markup language (SGML). By 1998, its expansion into extensible markup language (XML) and other related standards resulted in a name change that described its wider area of standards development. A new OASIS technical subcommittee created for published subjects for geography and languages will define sets of published subjects "for language, country, and region subjects, in accordance with the guidelines for published subjects to be laid down by the OASIS Published Subjects TC." Published subjects are a form of controlled vocabulary allowing "unambiguous indication of the identity of a subject." They are defined in the ISO 13250 Topic Maps standard and further refined in the XML Topic

Table 5. OASIS standards

Application Vulnerability Description Language (AVDL) v1.0 Common Alerting Protocol v1.0
Common Alerting Protocol (CAP) v1.1 Darwin Information Typing Architecture (DITA) v1.0 Directory Services Markup Language (DSML) v2.0 DocBook v4.1 ebXML Collaborative Partner Profile Agreement (CPPA) v2 ebXML Message Service Specification v2.0 ebXML Registry Information Model (RIM) v2. ebXML Registry Information Model (RIM) v3.0
ebXML Registry Services Specification (RS) v2.0 ebXML Registry Services Specification (RS) v3.0
Emergency Data Exchange Language (EDXL) Distribution Element v1.0
Election Markup Language (EML) v4.0
Extensible Access Control Markup Language (XACML) v1.0
eXtensible Access Control Markup Language TC v2.0 (XACML)
OpenDocument Format for Office Applications (OpenDocument) v1.0
Security Assertion Markup Language (SAML) v1.0
Security Assertion Markup Language (SAML) v1.1
Security Assertion Markup Language (SAML) V2.0
Service Provisioning Markup Language (SPML) v1.0
Service Provisioning Markup Language (SPML) v2.0
Universal Description, Discovery and Integration (UDDI) v2.0
Universal Description, Discovery and Integration (UDDI) v3.0.2 Universal Business Language (UBL) v1.0
Universal Business Language Naming & Design Rules v1.0 (UBL NDR) WS-Reliability (WS-R) v1.1 Web Services Resource Framework (WSRF) v1.2
Web Services for Remote Portlets (WSRP) v1.0 Web Services Security v1.0 (WS-Security 2004) Web Services Security v1.1 Web Services Security SAML Token Profile v1.0 and REL Token Profile v1.0 WSDM Management Using Web Services v1.0 (WSDM-MUWS) WSDM Management Using Web Services v1.0 (WSDM-MOWS) XML Catalogs v1.1 XML Common Biometric Format (XCBF) v1.1

Maps (XTM) 1.0 Specification. The committee will review the existing published subjects sets for topic maps using existing ISO, MARC 21[1], and UNSD standards. Two "Published Subjects – Languages" will be based on USMARC and ISO 639, respectively. Two "Published Subjects - Countries and Regions" will be based on MARC 21 and ISO 3166, respectively. A "Published Subjects -Geographic areas" will be based on MARC 21 and a "Published Subjects – Regions" will be based on UNSD Standard Country or Area Codes.

Another standard of interest is the Darwin information typing architecture (DITA) OASIS Standard, which is a document creation and management specification that builds content reuse into the authoring process. The DITA TC subcommittee is developing a general, top-level design for structured, intent-based authoring of learning content with good learning architecture. This standard has implications for educational use and distribution.

Table 6. Selected W3C standards

Architecture of the World Wide Web
Authoring Tool Accessibility Guidelines 1.0 (2000)
Cascading Style Sheets Specifications
Character Model for the World Wide Web
Composite Capability/Preference Profiles (CC/PP)
Document Object Model (DOM)
Extensible HyperText Markup Language (XHTML)
Extensible Markup Language (XML)
Extensible Stylesheet Language (XSL)
Hyper Text Markup Language (HTML)
Mathematical Markup Language (MathML)
Namespaces in XML
OWL Web Ontology Language
PICS (Platform for Internet Content Selection)
Platform for Privacy Preferences Specification
Portable Network Graphics (PNG) Specification
Scalable Vector Graphics (SVG) Specification
Mobile SVG Profiles: SVG Tiny and SVG Basic
QA Framework: Specification Guidelines
Resource Description Framework (RDF)
SOAP Message Transmission Optimization Mechanism
Speech Synthesis Markup Language (SSML) Version 1.0 (2004)
Synchronized Multimedia Integration Language (SMIL)
User Agent Accessibility Guidelines
Voice Extensible Markup Language (VoiceXML) Version 2.0 (2004)
Web Content Accessibility Guidelines
WebCGM 1.0 Second Release
Web Services Addressing 1.0 - Core; SOAP Binding
XForms
xml:id
XPointer element() Scheme
XSL Transformations (XSLT) Version

Figure 2. Matrix of Web services, applications, languages, and protocols

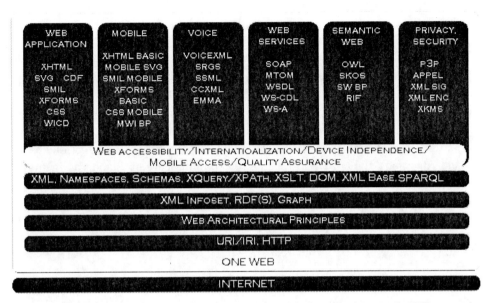

The World Wide Web Consortium (W3C)

The World Wide Web Consortium (W3C) was created by in 1994 by Tim Berners-Lee, the inventor of the World Wide Web. W3C is an industry consortium dedicated to building consensus around Web technologies. Since 1994, it has published more than 90 standards and guidelines, called W3C Recommendations. Based on Director Berners-Lee's philosophy of open standards for languages and protocols, it is dedicated to improving "Web interoperability." Although W3C does not concentrate on geographic standards, the display and transmission of geographic information is certainly indebted to W3C for its focus on interoperability and open standards development.

W3C has produced more than 90 W3C Recommendations. A W3C Recommendation is the equivalent of a Web standard, that is, that the specification is stable, is Web interoperable, and has been reviewed by the W3C Membership, who favor its adoption by the industry. Development of Web infrastructure is the focus of most of the work at WC3 with foci on accessibility, internationalization, device independence, mobile access, and quality assurance. URIs, HTTP, XML, and RDF supports pursuits in those areas.

CEN

CEN, the European Committee for Standardization (Comité Européen de Normalisation, Europäisches Komitee für Normung), was founded in 1961 by the national standards bodies in the European Economic Community and EFTA countries. With its sister organizations, CENELEC (European Committee for Electrotechnical Standardization) and ETSI (European Telecommunications Standards Institute), CEN posits that its principal critical success factors are "quality and efficiency of the standardization process; relevance of the work for the market; communication about the importance of ENs [European Standards]; links with international standardization; balanced participation of interested parties; availability of up-to-date

Figure 3. Number of European standards at end of 2005 developed by CEN by industry

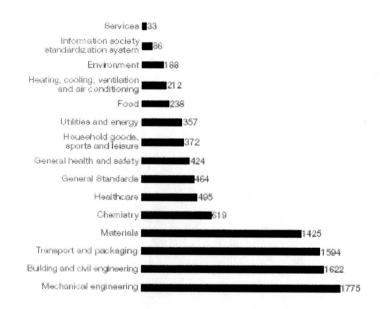

Reprinted with permission of publisher. Source: CEN (European Committee for Standardization). (2005). Annual report: European Committee for Standardization. Brussels, Belgium: CEN Management Centre, p. 73. Available http://www.cenorm.be/cenorm/aboutus/information/annual+report/ar05. pdf [Rapport annuel: Comité Européen de Normalisation; Jahresbericht: Europäisches Komitee für Normung].

communication platforms; support to the development of international standards which are implemented directly as European and national standards" (CEN European Committee for Standardization, s.d., p. 6).

Although CEN is deeply rooted in Europe through its 30 National Members, its Technical Assistance programmes and Partner Standardization Bodies broaden its scope to other continents and developing nations. All CEN Standards reflect national, European, and international standards. The CEN/CENELEC Ad Hoc Group on ICT defines a common approach to strategic issues in ICT standardization. The newly reconstituted Technical Committee 287 for Geographic Information has translated the first nine standards in the EN ISO 191XX series of standards to international standards (CEN European Committee for Standardization, 2005). A new CEN/ISSS Workshop is working on Web accessibility and certification of Web sites' standards that comply with the W3C Web Accessibility Initiative guidelines.

CEN publishes several principal products: its European Standards (designated "EN"); technical specifications (CEN TS), which are normative documents when the state-of-the-art is still in flux; technical reports (CEN TR), which deal with information and the transfer of information; and CEN workshop agreements, which are consensual agreements developed in open workshops. CEN developed 86 standards related to the information society, however, there are other standards hidden within its other categories, such as general standards and engineering standards.

Reprinted with permission of publisher. Source: CEN (European Committee for Standardization). (2005). *Annual report: European Committee for Standardization*. Brussels, Belgium: CEN Management Centre, p. 73. Available http://www.cenorm. be/cenorm/aboutus/information/annual+report/ar05.pdf [Rapport annuel: Comité Européen de Normalisation; Jahresbericht: Europäisches Komitee für Normung].

Languages and Protocols

One of the critical issues facing standards developers is the wide array of languages and protocols necessary to accommodate platforms, applications, services, and users. This is exacerbated by the numerous still-operational legacy systems and unofficial "standards" that abound in the networked world. This section discusses several of the more common languages and protocols used by the major standards developers, or those open source standards incorporated into network applications and services that meet the quality of service standards of various industries and countries.

Unified Modeling Language (UML)

The unified modeling language (UML) initiative was developed by Booch, Jacobson, and Rumbaugh (Rumbaugh, Jacobson, & Booch, 1999). In 1996, unified modeling language (UML) became a non-proprietary industry-standard language used to model objects and their behavior after its adoption by the Object Management Group. An ISO standard (ISO/IEC 19501), UML is "a language for specifying, visualizing, constructing, and documenting the artifacts of software systems, as well as for business modeling and other non-software systems" (OMG, 2001, Section 1.1). A third-generation method for specifying, visualizing, and documenting the artifacts of an object-oriented system under development, UML provides a standard way to write a system's framework conceptually (business processes and system functions) and concretely (programming language statements, database schemas, and reusable software components). Since UML is a language and not a methodology, UML easily fits into most modeling methodologies. Why is this important? Modeling is the designing of software applications before coding. A model plays the same role in software development as blueprints play in the architecture of a building. More importantly, UML can be easily converted into another language, such as XML, using translator tools.

eXtensible Markup Language (XML)

XML is viewed as a data exchange format language. Created by the World Wide Web Consortium (W3C), XML (extensible markup language) is a simpler, more concise dialect of SGML (Standard Generalized Markup Language; ISO 8879:1985). XML is often defined as a metalanguage, that is, a language that describes other languages. XML is simultaneously a human- and a machine-readable format in that it uses language-based tags rather than numeric tags (MARC) but it can be mapped for both. It supports Unicode, an industry standard designed to allow text and symbols from all languages to be consistently represented and manipulated by computers. This allows Web documents authored in the world's scripts across different platforms to be exchanged, read, and searched by Web users globally.

Figure 4. The UML 2.0 semantics framework

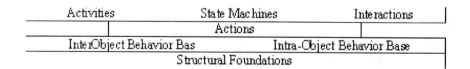

XML uses syntax tags to identify various types of data in a file. In the record framework (data model), a document type definition (DTD) is used to define the structure, or "tree" of an XML document. The fundamental unit in XML is the character, as defined by the Universal Character Set (ISO/IEC 10646). Characters are combined in certain allowable combinations to form an XML document. The document consists of one or more entities, each of which is typically some portion of the document's characters, encoded as a series of bits and stored in a text file. Languages based on XML, for example, geography markup language (GML), are formally defined, and allow programs to modify and validate documents in these languages without prior knowledge of their form.

How does XML differ from HTML (hypertext markup language)? HTML is primarily a display language, that is, it tells the browser *how* to display the information encoded on a Web page. For example, HTML tells the browser that this block of information is to be displayed in a table format but does not describe or identify what the information is. HTML also uses cascading style sheets (CSS) to tell the browser how to display very specific elements of a Web page, such as Times Roman 12 point text with a Times Roman 14 point header, across a Web site. The data (information) within an HTML page cannot be reused or manipulated for other purposes.

A Web page written in XML contains data that can be extracted, recycled, and manipulated by other database systems. Similar to HTML and the use of CSS, XML has its own stylesheet language, **XSL** (extensible stylesheet language). Unlike CSS, XSL serves two functions. The first is to handle the graphical display of information; the second function, XSLT (XSL Transformations), is to contain instructions on transforming the data into other formats, such as e-commerce. XSLT generates a file different in structure from the original, allowing data to be "pushed forward," moving data successfully between networks and programs for processing purposes.

In summary, XML has three primary uses: "as a representation language that enables the transport of bibliographic data in a way that is technologically independent and universally understood across systems and domains; as a language that enables the specification of complex validation rules according to a particular data format such as MARC; and, finally, as a language that enables the description of services through which such data can be exploited in alternative modes that overcome the limitations of the classical client-server database services" (de Carvalho & Cordeiro, 2002).

Geography Markup Language (GML)

The OpenGIS® geography markup language (GML) encoding specification 3.1.1 03- (Cox et al., 2004) uses XML encoding for the transport and storage of geographic information, including both the geometry and properties of geographic features, such as feature, geometry, coordinate reference system time, dynamic feature, coverage (including geographic images), unit of measure, and map presentation style.

Figure 5. ISO-GML relationship (simplified)

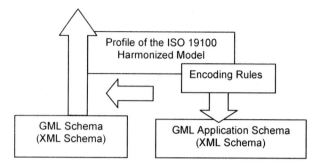

GML enables Internet-based tools, such as Google Maps®, to access geographical information, such as merchant locations and traffic conditions. GML is also in the United States National Information Exchange Model, a federal, state, local, and tribal interagency initiative between the U.S. Department of Justice and the U.S. Department of Homeland Security. Originally based on the W3C's Resource Description Framework (RDF), GML also serves as a modeling language for geographic systems. GML implements concepts found within the ISO 19100 (19103, 19107, 19108, 19109, 19111, 19117, 19118, 19123, and 19136) series to support spatial and non-spatial properties of objects.

There are several components, profiles, applications, geometries, features, to GML that are of interest. Part of GML namespaces, *GML profiles* are logical restrictions (restricted subsets) and may be expressed by either a document or an XML schema or both (Lake, Burggraf, Trninic, & Rae, 2004). GML-specified profiles include a Point Profile for applications that do not need full GML grammar and a GML Simple Features profile, which is a more complete profile of GML than the Point Profile and supports a wide range of vector feature objects, vector feature requests, and transactions (Lake et al., 2004). There is even a GML profile for RSS (a family of Web feed formats used to publish frequently updated digital content). Unlike other GIS models, GML defines *features* as different from geometry objects[2]. In GML, a feature has a set of geometric properties that describe geometric aspects or characteristics of the feature or share a geometry property with one another (Lake et al., 2004). Features with similar characteristics are grouped to feature types. This structure is specified in an *Application schema* that builds either a specific GML profile or the full GML schema set. The schema describes the object types in the data, such as roads, viewpoints, churches, oceans, and so forth. Those object types, in turn, reference the primitive object types defined in the GML standard (Lake et al., 2004). GML encodes the *GML geometries*, or geometric characteristics (point, linestring, and polygon), of geographic objects as elements within GML documents (Lake et al., 2004). Coordinates in GML must be specified with a Coordinate Reference System (Lake et al., 2004).

Structured Query Language (SQL)

Structured query language (SQL) is a standard interactive and programming language used to query and update information in databases. Although SQL is both an ANSI and an ISO standard, many commercial database products add proprietary extensions to standard SQL. Queries take the form of a command language that lets you select, insert, update, find out the location of data, and so forth. There is also a programming interface. The query language SQL is extended to manipulate spatial data as well as descriptive data. New spatial types (point, line, region) are handled as base alphanumeric types.

There are a number of standards that use SQL, ranging from frameworks, call-level interfaces (SQL/CLI), storage, spatial multimedia and applications, spatial schema, and GIS services. ISO 19125:2004 specifies an SQL schema that supports storage, retrieval, query, and update of simple geospatial feature collections using the SQL Call Level Interface (SQL/CLI), and establishes architecture for the implementation of feature tables. ISO 19125:2004 defines terms to use within the architecture of geographic information, defines a simple feature profile (ISO 19107), and describes a set of SQL Geometry Types. It also standardizes the names and geometric definitions of the SQL Types for Geometry and the names, signatures, and geometric definitions of the SQL Functions for Geometry. ISO 13249-3 describes the profiles for Geometry Types and Functions.

Simple Object Access Protocol (SOAP)

The simple object access protocol (SOAP) defines the use of XML and HTTP to access services, objects, and servers in a platform independent manner. SOAP bridges technologies and facilitates interoperability with a three-part protocol: an "envelope" defining a framework description and processing rules, a set of encoding rules for application defined data types, and a convention for remote procedure calls and responses (Hanson, 2006). When SOAP is attached with hypertext transfer protocol (HTTP), organizational firewalls become virtually transparent for the defined SOAP services. This leads to unforeseen possibilities in cross-organizational interoperability. Other SOAP-related technologies include the universal description, discovery, and integration (UDDI) and Web services description language (WSDL). The UDDI protocol is a key member of the group of interrelated standards that comprise the Web services stack. It defines a standard method for publishing and discovering the network-based software components of a service-oriented architecture.

As Figure 6 shows, each new online application or service requires integration with existing standards or the creation of new standards for applications and services still to be built. Clearly, the importance of standards cannot be ignored. With so many organizations working together to establish standards on national and international

Figure 6. Schematic of Web services

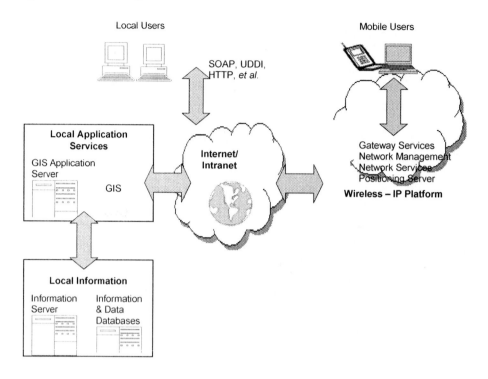

levels, it is critical that librarians working in GIS centers, with GIS data, or with researchers, keep current with the evolution of Web, library, and GIS standards. With the evolution of the current protocols and new applications, an imminent change is occurring in the way we think about Internet and cross-organizational interoperability.

Content Standard for Digital Geospatial Metadata

The content standard for digital geospatial metadata (CSDGM), Version 2 (FGDC-STD-001-1998) is the U. S. federal metadata standard (Federal Geographic Data Committee, 1998; Federal Geographic Data Committee , 2000). Originally adopted by the Federal Geographic Data Committee, revised in 1998, all federal agencies are ordered to use this standard to document geospatial data created as of January, 1995.

The creation of a content standard for geographic information is very similar to the MARC record format in Chapter IV. Both standards have a formal structure; provide identifying information, such as title and author, attribute and entity (description of various parts of the work) that provide subject areas (themes), currency (date of content) and/or publication, use restrictions, publisher and place of publication, and other pertinent information. However, for geospatial data, there are other considerations that need to be included in the descriptive information. These are much like the fields for geographic data in the MARC format, including spatial data organization, spatial reference (coordinate system), encoding systems, and person responsible for describing the data. As with any cataloging or classification system, the more descriptive information about an item provides the user with enough information to make an informed and judicious choice of spatial data. Therefore, additional information on positional and attribute accuracy, completeness of dataset, data consistency, statistical processes used to model the data, spatial data models, and number of spatial objects, to name a few, are critical in the querying and selection of a geospatial dataset.

Similar to MARC and AACR2r, the CSDGM provides guidelines to develop geospatial profiles, allows the use of user-defined metadata entities and elements, creates short names/tags for all entities/elements, includes spatial reference, allows the use of free-text, and includes a glossary. Similar to the minimum standards for a MARC record, the CSDGM established mandatory (minimum) elements, "mandatory if applicable (must be provided if the data set exhibits the defined characteristic), and optional (provided at the discretion of the producer of the data set)" (Federal Geographic Data Committee, 2000, p. 7).

Similar to Cutter's rules to ensure that the user finds and acquires the specific item for which he or she is searching, the CSDGM defines that a prospective user can determine (1) "the availability of a set of geospatial data, (2) … the fitness of a set of geospatial data for an intended use, (3) … the means to access the set of geospatial data, and (4) [how] to transfer the set of geospatial data successfully"(Federal Geographic Data Committee, 2000, p. 6).

Like the AACR2r, the CSDGM specifies the information required to describe a set of digital geospatial data, including a common set of terminology and definitions for concepts related to metadata. These concepts include "the names of data elements and compound elements (groups of data elements) to be used, the definitions of these compound and data elements, and information about the values that are to be provided for the data elements" (Federal Geographic Data Committee, 2000, p. 7).

The CSDGM is organized into sections, component elements, and data elements. Numbered chapters, called "sections," have a specific name and a definition. Each section is organized into three parts: the section definition, the production rules, and a list of component elements. "The *section definition* includes the name and definition of the section. The *production rules* describe the section in terms of lower-level component elements. Each production rule has an identifier (left side) and an expression (right side) connected by the symbol "=," meaning that the term on the left side is replaced by or produces the term on the right side. (This is similar to an authority file or the "used for" note to indicate obsolete or changes in terminology in a MARC authority file or in a thesaurus, such as the Library of Congress Subject Headings). The production rules allow matching terms, mandatory elements, and repeatable elements ... The *list of component elements* provides the name and definition of each component element in the section, and information about values to be provided for data elements" (Federal Geographic Data Committee, 2000, p. 12).

The CSDGM contains seven sections: Identification Information, Data Quality Information, Spatial Data Organization Information, Spatial Reference Information, Entity and Attribute Information, Distribution Information, and Metadata Reference Information. Only 1 and 7, identification information and metadata reference information, are generally required. Sections 2-6 are mandatory only if applicable.

Table 7. CSDGM sections and metadata section attributes

	CSDGM Sections
1.	Identification Information
2.	Data Quality Information
3.	Spatial Data Organization Information
4.	Spatial Reference Information
5.	Entity and Attribute Information
6.	Distribution Information

7.	Metadata Reference Information
	Metadata_Reference_Information:
	  Metadata_Date:
	  Metadata_Review_Date:
	  Metadata_Future_Review_Date:
	  Metadata_Contact:
	    Contact_Information:
	      Contact_Organization_Primary:
	      Contact_Person:
	      Contact_Organization:
	     Contact_Address:
	      Address_Type:
	      Address:
	      City:
	      State_or_Province:
	      Postal_Code:
	      Country:
	     Contact_Voice_Telephone:
	   Contact_Facsimile_Telephone:
	     Contact_Electronic_Mail_Address:
	  Metadata_Standard_Name:
	  Metadata_Standard_Version:
	  Metadata_Time_Convention:

In Table 7, only the Metadata Reference Information is shown. Readers are referred to the appendix at the end of this chapter to see the remaining six sections.

In addition to the CSDGM, there have been extensions created to accommodate the needs of other disciplines. These include the *CSDGM - Extension for Remote Sensing Data*, the *CSDGM - Biological Data Profile*, and the *CSDGM - Metadata Profile for Shoreline Data*. These extensions allow the documentation of geospatial data that is pertinent to individuals working in or accessing information about these areas. For example, the extension for remote sensing includes elements that describe the remote sensing platform and its sensors, and to document data collected directly from the sensor, not derived data. The biological data profile addresses items such as laboratory results, field notes, specimen collections, research reports, and requires the use of a specified taxonomical vocabulary. The shoreline data profile provides a controlled vocabulary of terms and data elements specific to shoreline and coastal datasets.

Conclusion/Summary

With the increased use of digital computation, data, information, and networks to replace and extend traditional research, description for digital data, applications, and services becomes increasingly more complex. Classic scientific research, often defined as "theoretical/analytical" and "experimental/observational," incorporates temporal and spatial factors. The new geospatial cyberinfrastructure can be described as a "layer of enabling hardware, algorithms, software, communications, institutions, and personnel" that lies between a layer of "base technologies . . . the integrated electro-optical components of computation, storage, and communication" and a layer of "software programs, services, instruments, data, information, knowledge, and social practices applicable to specific projects, disciplines, and communities of practice" (Atkins, Droegemeier, Feldman, Garcia-Molina, Klein, et al., 2003). Further, it is clear that considering the amount of private and public sector involvement, a geospatial infrastructure is required for a knowledge economy. With the extension into national and international arenas, there are a number of players and concerns that need to be addressed.

National and International Considerations

Since the United States is also a member of ISO, it must revise the CSDGM in accord with ISO 19115. Currently, ISO 19115 allows a national profile to be developed centering around the ISO 13 core elements. The FGDC is working on the U.S. profile.

Table 8. ISO 19115 core elements

Mandatory Elements (7)	Conditional Elements (6)
Dataset title	Dataset responsible party
Dataset reference date	Geographic location by coordinates
Dataset language	Dataset character set
Dataset topic category	Spatial resolution
Abstract	Distribution format
Metadata point of contact	Spatial representation type
Metadata date stamp	Reference system
	Lineage statement
	On-line Resource
	Metadata file identifier
	Metadata standard name
	Metadata standard version
	Metadata language
	Metadata character set

Other new elements in the standards address clarifying roles, and adding new areas. For example, in the CSDGM, the element "Originator" is problematic, since many organizations and agencies have datasets created for them by external contractors or acquired from other agencies. Organizations trying to ensure "authorship" of the data were encouraged to list themselves (owners) as the "Originator," or as one of multiple "Originator" elements, and reference the contractor under "Data Set Credit" field. ISO 19115 provides a field for "Responsible Party" that is further subdivided by a number of "Role" codes (for example, originator, custodian, publisher, owner) to better establish the relationship of each organization to the dataset.

International concerns, such as *data set language* and *metadata language,* are included. Element applications are also broadened, for example, "Extent" now includes *geographic extent, temporal extent,* and *vertical extent.* Since there is no standardized theme keyword thesaurus, there is a "Topic category" that should allow a standardized thesaurus to aid in resource discovery. Suggested theme terms include boundaries, oceans, health, transportation, economy, sociology, biota, structure, environment, utilities, and so forth. ISO 19115 is also moving to more fixed domains and code lists in lieu of free text.

As substantiated by library and information science research, controlled vocabulary helps eliminate false cognates, directs users to the appropriate terms, and improves resource discovery, especially in the assurance of data quality. Data quality, which is how well the characteristics of the data meet the requirements of the user, should have a high information value for the user and low uncertainty (as in accuracy of descriptive information) to ensure the fitness of a dataset for its user.

The responsibility for the following themes are based on the National Academy of Public Administration recommendations (National Academy of Public Administration, 1998).

Table 9. ISO 19115 topic categories

farming	Intelligence - military
biota	inland waters
boundaries	location
climatology, meteorology, atmosphere	oceans
economy	planning - Cadastre
elevation	sociology
environment	structure
geoscientific information	transportation
health	utilities - Communication
imagery, base maps, earth cover	

Table 10. Potential responsibilities for data layers in a spatial information infra-structure

Theme	Federal	State	Local
geodetic control	primary	supplementary	supplementary
cadastral data	supplementary	supplementary	primary
political boundaries	primary for states and international	primary for counties and state reserves	primary for municipalities and local areas
base cartographic and elevation	primary for scales smaller than 1:24,000	supplementary for road building and state projects	supplementary for local projects
bathymetric	primary for offshore areas, int'l waters	supplementary for lakes and reservoirs	supplementary for ponds
geologic	primary	supplementary	supplementary
hydrography	primary	supplementary (water rights)	supplementary
transportation & utilities	supplementary	primary for highways	primary for some utilities
soils	primary for coordination	supplementary	primary for survey
vegetation	primary for federal lands	primary for state lands	primary for local lands
wetlands and wildlife habitat	primary	supplementary	supplementary
cultural and demographic	primary	supplementary	supplementary
digital orthoimagery (scale dependent)	primary at coarse resolutions	**supplementary**	**primary at fine resolutions**
statistical base maps & address files	supplementary	supplementary	primary
land cover & land use (added to napa list)	primary for land cover	supplementary for both	primary for land use
NOTE: Bold text boxes include the seven NSDI Framework themes.			

The format of the standards is also changing. Formerly published as a document, the newer ISO standards, such as ISO 19115, are published as a UML diagram and a supporting data dictionary. ISO 19139 will be published as an XML schema with supporting documentation

Further, although the new standards formats will better support digital geospatial technologies, such as GIS, remote sensing, automated mapping, and GPS, most geospatial data developers will need to learn UML and XML or an interface tool to work directly with the standard. This is a reminder that all changes to new technologies and formats require changes in workflow processes and adding new tools to developers' kits. On a positive note, these changes will allow users to create more consistent, robust metadata in multiple formats (txt, doc, html, etc.) with the least robust .txt format serving as the primary metadata transfer format, to internally harvest metadata from the data, and to better manage data. Further, ISO-to-CSDGM and ISO-to-DC crosswalks have been drafted, conversion software is in development, and a new version of the CSDGM Workbook ("Green book") is also planned. OSI-poised software includes ESRI ArcCatalog, Intergraph/SMMS, and MetaD.

FGDC Metadata and MARC

Using a postcoordinate approach to subject data in metadata records offers several advantages, (Chan, Mai, & Hodges, 2000). A postcoordinate approach is more adaptable to rapid changes in the online environment, and a faceted thesaurus is easier for non-catalogers to understand and use. Cognitive miserliness is also enhanced, since a postcoordinate subject vocabulary shares structural and syntactical elements with other controlled vocabularies, maximizing knowledge transfer. A postcoordinate approach is easier to map to other controlled vocabularies, to other languages, or to classification schemes (Chan et al., 2000). Most importantly, a postcoordinated controlled vocabulary, based on or compatible with the Library of Congress subject headings, would increase interoperability between MARC and other resource description models (Chan et al., 2000). As Chan and Hodges (2000) suggest: "The content richness of the MARC database makes it desirable to maintain MARC records even in the digital environment and to make them interoperable with other types of metadata records; even more desirable would be to provide the potential for integrating various types of metadata records and MARC records into a single system. Being able to do so would extend the returns on the enormous investments that have gone into preparing MARC records in the past" (pp. 232-233).

References

American National Standards Institute. (2000). *National standards strategy for the United States.* New York, NY: ANSI. Retrieved November 2006, from http://publicaa.ansi.org/sites/apdl/Documents/News%20and%20Publications/Brochures/national_strategy.pdf

American National Standards Institute. (2006). *ANSI essential requirements: Due process requirements for American National Standards.* New York, NY: ANSI. Retrieved November 2006, from http://publicaa.ansi.org/sites/apdl/Documents/Standards%20Activities/American%20National%20Standards/Procedures,%20Guides,%20and%20Forms/ER0106.doc

Anderson, J. R. (1971). Land use classification schemes used in selected recent geographic applications of remote sensing. *Photogrammetric Engineering and Remote Sensing, 37*(4), 379-387.

Atkins, D. E., Droegemeier, K. K., Feldman, S. I., Garcia-Molina, H., Klein, M. L., Messerschmitt, D. G. *et al.* (2003). *Revolutionizing science and engineering through cyberinfrastructure: Report of the National Science Foundation Blue-Ribbon Advisory Panel on Cyberinfrastructure.* Washington, DC: National Science Foundation. Retrieved from http://www.nsf.gov/cise/sci/reports/atkins. pdf [Also known as the Atkins Report].

Botts, M. (2005). *OpenGIS® Sensor Model Language (SensorML) 1.0.0 05-086.* Wayland, MA: Open Geospatial Consortium Inc. Retrieved from http://portal. opengeospatial.org/files/?artifact_id=4700

Bruce, C. (2006). *Binary extensible markup language (BXML) encoding specification 0.0.8 03-002r9.* Wayland, MA: Open Geospatial Consortium Inc. Retrieved from http://portal.opengeospatial.org/files/?artifact_id=13636

CEN (European Committee for Standardization). (2005). *Annual report: European Committee for Standardization.* Brussels, Belgium: CEN Management Centre. Retrieved from http://www.cenorm.be/cenorm/aboutus/information/ annual+report/ar05.pdf [Rapport annuel: Comité Européen de Normalisation; Jahresbericht: Europäisches Komitee für Normung].

CEN (European Committee for Standardization). (s.d.). *CEN Strategy: 2010.* Brussels, Belgium: CEN Management Centre. Retrieved from http://www.cenorm. be/cenorm/aboutus/generalities/strategy/strategy.pdf

Chan, L., Mai, & Hodges, T. (2000). Entering the millennium: A new century for LCSH. *Cataloging & Classification Quarterly, 29*(1/2), 225-234.

Cox, S., Daisey, P., Lake, R., Portele, C., & Whiteside, A. (2004). *OpenGIS® geography markup language (GML) encoding specification 3.1.1 03-105r1 .* Wayland, MA: Open Geospatial Consortium Inc. Retrieved from http://portal. opengeospatial.org/files/?artifact_id=4700

de Carvalho, J., & Cordeiro, M. I. (2002). XML and bibliographic data: The TVS (Transport, Validation and Services) model. In International Federation of Library Associations (Ed.), *68th IFLA General Conference and Council: Libraries for Life: Democracy, Diversity, Delivery* (pp. [1-13]). Glasgow, Scotland: IFLA. Retrieved May 2006, from http://www.ifla.org/IV/ifla68/papers/075-095e.pdf

Dewhurst, W. T. (1985-). *Input formats and specifications of the National Geodetic Survey data base.* Rockville, MD: Federal Geodetic Control Committee: U.S. Dept. of Commerce, National Oceanic and Atmospheric Administration.

Federal Geodetic Control Subcommittee. (s.d.). *Part B-9: Lead agency/bureau and/or subcommittee/working group report:* Retrieved July 16, 2006, from http://www. fgdc.gov/library/whitepapers-reports/annual%20reports/2003%20reports/Geodetic_Control_2003.pdf

Federal Geographic Data Committee. (1998). *Content standard for digital geospatial metadata*. Washington, DC: Federal Geographic Data Committee.

Federal Geographic Data Committee. (2000). *Content standard for digital geospatial metadata workbook (For use with FGDC-STD-001-1998)* (2.0 ed.). Reston, VA: Federal Geographic Data Committee. Retrieved from http://www.fgdc. gov/metadata/documents/workbook_0501_bmk.pdf

Federal Geographic Data Committee. (2006a). *2005 annual report* . Reston, VA: The Committee. Retrieved June 21, 2006, from http://www.fgdc.gov/fgdc-news/2005-annual-report

Federal Geographic Data Committee. (2006b). *National Spatial Data Infrastructure*. Reston, VA: The Committee. Retrieved June 21, 2006, from http://www.fgdc. gov/nsdi/nsdi.html

Fitzke, J., & Atkinson, R. (2006). *Gazetteer service: Application profile of the Web feature service implementation specification 0.9.3 05-035r2*. Wayland, MA: Open Geospatial Consortium Inc. Available http://portal.opengeospatial.org/ files/?artifact_id=15529.

Hanson, A. (2006). Organization and access to electronic resources. In V. L. Gregory, *Selecting and managing electronic resources* rev. ed., (pp. 45-64). New York, NY: Neal-Schuman.

International Organization for Standardization. (2004). 19139 Geographic information - Metadata - XML schema implementation. *ISO Project Information Fact Sheet, 19139*, 1. Retrieved November 2006, from http://www.isotc211. org/Outreach/Overview/Factsheet_19139.pdf

International Organization for Standardization. ([s.d.]). *In the beginning.* Retrieved November 2006, from http://www.iso.org/iso/en/iso9000-14000/understand/ basics/general/basics_2.html

ISO/IEC. (2003). *Information technology: Universal multiple-octet coded character set (UCS)*. Geneva, Switzerland: International Organization for Standardization; International Electrotechnical Commission.

Kresse, W., & Fadaie, K. (2004). *ISO standards for geographic information*. New York: Springer-Verlag.

Lake, R., Burggraf, D., Trninic, M., & Rae, L. (2004). *Geography mark-up language: Foundation for the geo-web*. Chichester, England: John Wiley & Sons, Ltd.

Martin, J. (2006). *Best practices for Earth observation products 0.3 05-057r4*. Wayland, MA: Open Geospatial Consortium Inc. Retrieved from https://portal. opengeospatial.org/files/?artifact_id=6495

Morrison, J. L., & Wortman, K. (Eds.). (1992). *Cartography and geographic information systems* (vol. 19). Bethesda, MD: American Congress on Surveying and Mapping.

National Academy of Public Administration. (1998). *Geographic information for the 21st century: Building a strategy for the nation: A report*. Washington, DC: The Academy.

National Research Council, Mapping Science Committee. (2001). *National spatial data infrastructure partnership programs: Rethinking the focus.* Washington, DC: National Academy Press. Retrieved from http://books.nap.edu/catalog/10241.html

OMG. (2001). *Unified modeling language specification. Version 1.4.* Retrieved November 2006, from http://www.omg.org/uml

Rumbaugh, J., Jacobson, I., & Booch, G. (1999). *The unified modeling language reference manual.* Boston, MA: Addison Wesley Professional.

Solem, M., Chalmers, L., Dibiase, D., Donert, K., & Hardwick, S. (2006). Internationalizing professional development in geography through distance education. *Journal of Geography in Higher Education, 30*(1), 147-160.

U.S. Geological Survey. (1999). *Spatial data transfer standard (SDTS).* Reston, VA: U.S. Dept. of the Interior, U.S. Geological Survey.

Voges, U., Senkler, K., & Müller, M. (2004). *OpenGIS Catalogue Services Specification 2.0: ISO 19115/ISO 19119 Application profile for CSW 2.0.* Wayland, MA: Open Geospatial Consortium Inc. Retrieved from https://portal.opengeospatial.org/files/?artifact_id=6495

Whiteside, A. (2006). *Definition identifier URNs in OGC namespace 1.1.0 06-023r1.* Wayland, MA: Open Geospatial Consortium Inc. Retrieved from http://portal.opengeospatial.org/files/?artifact_id=16339

Endnotes

[1]   The U.S. MARC (MARC 21) language codes formed the basis of the ANSI/NISO Standard for Representation of Languages for Information Interchange, ANSI/NISO Z39.53-200X. The specification states "[A] standardized 3-character code to indicate language in the exchange of information is defined. Codes are given for languages, contemporary and historical. The purpose of this standard is to provide libraries, information services, and publishers a standardized code to indicate language in the exchange of information. This standard for language codes is not a prescriptive device for the definition of language and dialects but rather a list reflecting the need to distinguish recorded information by language."

[2]   A feature is an application object that represents a physical entity but may or may not have geometric aspects. A geometry object defines a location or region instead of a physical entity, and therefore is different from a feature.

APPENDIX 1. CSDGM

Identification_Information:
 Citation:
  Citation_Information:
   Originator:
   Publication_Date:
   Title:
   Geospatial_Data_Presentation_Form:
   Publication_Information:
    Publication_Place:
    Publisher:
   Online_Linkage:
 Description:
  Abstract:
  Purpose:
  Supplemental_Information:
 Time_Period_of_Content:
  Time_Period_Information:
   Single_Date/Time:
    Calendar_Date:
    Time_of_Day:
   Multiple_Dates/Times:
    Calendar_Date:
    Time_of_Day:
    Calendar_Date:
    Time_of_Day:
   Range_of_Dates/Times:
    Beginning_Date:
    Beginning_Time:
    Ending_Date:
    Ending_Time:
  Currentness_Reference:
 Status:
  Progress:
  Maintenance_and_Update_Frequency:
 Spatial_Domain:
  Bounding_Coordinates:
   West_Bounding_Coordinate:
   East_Bounding_Coordinate:
   North_Bounding_Coordinate:
   South_Bounding_Coordinate:
 Keywords:
  Theme:
   Theme_Keyword_Thesaurus:
   Theme_Keyword:
  Place:
   Place_Keyword_Thesaurus:
   Place_Keyword:
 Access_Constraints:
 Use_Constraints:
 Point_of_Contact:
  Contact_Information:
   Contact_Organization_Primary:

```
     Contact_Person:
     Contact_Organization:
    Contact_Address:
     Address_Type:
     Address:
     City:
     State_or_Province:
     Postal_Code:
     Country:
     Contact_Voice_Telephone:
     Contact_Facsimile_Telephone:
     Contact_Electronic_Mail_Address:
 Data_Set_Credit:
 Native_Data_Set_Environment:
 Cross_Reference:
  Citation_Information:
   Originator:
   Publication_Date:
   Title:
   Geospatial_Data_Presentation_Form:
   Publication_Information:
    Publication_Place:
    Publisher:
   Online_Linkage:


Data_Quality_Information:
 Attribute_Accuracy:
  Attribute_Accuracy_Report:
 Logical_Consistency_Report:
 Completeness_Report:
 Positional_Accuracy:
  Horizontal_Positional_Accuracy:
   Horizontal_Positional_Accuracy_Report:
  Vertical_Positional_Accuracy:
   Vertical_Positional_Accuracy_Report:
 Lineage:
  Source_Information:
   Source_Contribution:
  Process_Step:
   Process_Description:
   Process_Date:
   Process_Contact:
    Contact_Information:
     Contact_Organization_Primary:
      Contact_Person:
      Contact_Organization:
     Contact_Address:
      Address_Type:
      Address:
      City:
      State_or_Province:
      Postal_Code:
      Country:
     Contact_Voice_Telephone:
```

```
        Contact_Facsimile_Telephone:
        Contact_Electronic_Mail_Address:
    Cloud_Cover:

Spatial_Data_Organization_Information:
 Indirect_Spatial_Reference:
 Direct_Spatial_Reference_Method:
 Raster_Object_Information:
  Raster_Object_Type:
  Row_Count:
  Column_Count:
  Vertical_Count:
 Point_and_Vector_Object_Information:
  SDTS_Terms_Description:
   SDTS_Point_and_Vector_Object_Type:
   Point_and_Vector_Object_Count:
  VPF_Terms_Description:
   VPF_Topology_Level:
   VPF_Point_and_Vector_Object_Type:
   Point_and_Vector_Object_Count:


Spatial_Reference_Information:
 Horizontal_Coordinate_System_Definition:
  Geographic:
   Latitude_Resolution:
   Longitude_Resolution:
   Geographic_Coordinate_Units:
  Planar:
   Planar_Coordinate_Information:
    Planar_Coordinate_Encoding_Method:
    Planar_Distance_Units:
  Local:
   Local_Description:
   Local_Georeference_Information:
  Geodetic_Model:
   Horizontal_Datum_Name:
   Ellipsoid_Name:
   Semi-major_Axis:
   Denominator_of_Flattening_Ratio:
 Vertical_Coordinate_System_Definition:
  Altitude_System_Definition:
   Altitude_Datum_Name:
   Altitude_Distance_Units:
   Altitude_Resolution:
   Altitude_Encoding_Method:
  Depth_System_Definition:
   Depth_Datum_Name:
   Depth_Distance_Units:
   Depth_Resolution:
   Depth_Encoding_Method:


Entity_and_Attribute_Information:
 Detailed_Description:
```

Entity_Type:
  Entity_Type_Label:
  Entity_Definition:
Attribute:
  Attribute_Label:
  Attribute_Definition:
  Beginning_Date_of_Attribute_Values:
  Ending_Date_of_Attribute_Values:
  Attribute_Domain_Values:
   Enumerated_Domain:
    Enumerated_Domain_Value:
    Enumerated_Domain_Value_Definition:
   Codeset_Domain:
    Codeset_Name:
    Codeset_Source:
   Range_Domain:
    Range_Domain_Minimum:
    Range_Domain_Maximum:
    Attribute_Units_of_Measure:
    Attribute_Measurement_Resolution:
   Unrepresentable_Domain:

Distribution_Information:
 Distributor:
  Contact_Information:
   Contact_Organization_Primary:
    Contact_Person:
    Contact_Organization:
   Contact_Address:
    Address_Type:
    Address:
    City:
    State_or_Province:
    Postal_Code:
    Country:
   Contact_Voice_Telephone:
   Contact_Facsimile_Telephone:
   Contact_Electronic_Mail_Address:
  Distribution_Liability:
  Custom_Order_Process:
  Resource_Description:
  Technical_Prerequisites:
  Standard_Order_Process:
   Non-digital_Form:
   Ordering_Instructions:
   Fees:
   Turnaround:
   Digital_Form:
    Digital_Transfer_Information:
     Format_Version:
     Format_Name:
     Format_Specification:
     File_Decompression_Technique:
     Transfer_Size:

```
        Digital_Transfer_Option:
         Online_Option:
          Access_Instructions:
          Online_Computer_and_Operator_System:
          Computer_Contact_Information:
           Network_Address:
            Network_Resource_Name:
         Offline_Option:
          Offline_Media:
          Recording_Density:
          Recording_Density_Units:
          Recording_Format:
          Compatibility_Information:
    Available_Time_Period:
     Time_Period_Information:
      Single_Date/Time:
       Calendar_Date:
       Time_of_Day:
      Multiple_Dates/Times:
       Calendar_Date:
       Time_of_Day:
       Calendar_Date:
       Time_of_Day:
      Range_of_Dates/Times:
       Beginning_Date:
       Beginning_Time:
       Ending_Date:
       Ending_Time:


Metadata_Reference_Information:
 Metadata_Date:
 Metadata_Review_Date:
 Metadata_Future_Review_Date:
 Metadata_Contact:
  Contact_Information:
   Contact_Organization_Primary:
    Contact_Person:
    Contact_Organization:
   Contact_Address:
    Address_Type:
    Address:
    City:
    State_or_Province:
    Postal_Code:
    Country:
   Contact_Voice_Telephone:
   Contact_Facsimile_Telephone:
   Contact_Electronic_Mail_Address:
 Metadata_Standard_Name:
 Metadata_Standard_Version:
 Metadata_Time_Convention:
```

Chapter VI

Accessibility:
Critical GIS, Ontologies, and Semantics

Ardis Hanson, University of South Florida Libraries, USA

Introduction

With the creation of the Internet and the continued evolution of technologies in GIS, networking, and knowledge management, access to geospatial information is a critical component of research and practice. Interoperability is the "new paradigm for joining heterogeneous computer systems into synergistic units that facilitate a more efficient use of geographic information resources" (Harvey, Kuhn, Pundt, Bishr, & Riedemann, 1999, p. 213). As geographers reassess the description of geographic methodologies and techniques across different platforms in the online environment, so have researchers in other disciplines assessed the use of applied geographic techniques for a wide variety of analysis. Such efforts have led some researchers to use new descriptive classifications to identify functionalities in the new scholarship, such as in creating new ontologies for GIS (Fonseca, Davis, & Cmara, 2003; Goodchild, 2004; Goodchild & Haining, 2004; Mark, Skupin, & Smith, 2001). This chapter examines the impact of these new ontologies, reviews the impact standards have on access and issues for end-users in accessing geospatial information.

Interoperability and Accessibility

As discussed in Chapter V, geographic information standards apply to the definition, description, and management of geographic information and geospatial services. Although there are numerous reasons why standards are good, we will concentrate on three: to increase the "understanding and usage of geographic information," to increase the "availability, access, integration, and sharing of geographic information," and the "efficient, effective, and economic use of digital geographic information and associated hardware and software systems" (Albrecht, 1999, p. 151). All three are affected by interoperability.

Interoperability allows computers and users to share and access data and operations through information networks. It has been described as a voluntary, "bottom-up" approach where independently deployed heterogeneous systems, data sources, and data models exchange data, process queries/requests, and have a common understanding of the resource and user/system requests (Sondheim, Gardels, & Buehler, 1999).

Portability, a component of interoperability, implies the ability to transport application source code between computer platforms and operating systems, and data between databases. Standard specifications for data and for operations directed to data are necessary to communicate with one another and to exchange and use information, including content, format, and semantics. The U. S. National Institute of Standards and Technology (NIST, 1995) established the Open Systems Environment (OSE) to ensure that differing performance characteristics and capabilities between systems do not prevent portability. There are three fundamental entities in NIST's OSE: application software, application platform, and platform external environment. Interfaces are shared boundaries between entities, defined by functional characteristics. Services are capabilities provided by entities, falling into specific categories, such as operating systems services, human/computer interface services, data management services, data interface services, graphics services, and network services (National Institute of Standards and Technology, 1996, p. 11). Since networks must have a certain degree of structure and stability to be effective, the design of a network is strongly connected to the character of knowledge it is able to transmit (Batten, Karlsson, & Andersson, 1989).

Data interoperability is defined as the ability to access multiple, heterogeneous geoprocessing environments, either local or remote, by means of a single unchanging software interface (Buehler & McKee, 1996). ISO/TC-211 defines two types of interoperability: "Syntactical interoperability assures that there is a technical connection, that is, that the data can be transferred between systems. Semantic interoperability assures that the content is understood in the same way in both systems, including by those humans interacting with the systems in a given context." Research in geospatial interoperability must take into account not only data or structural issues but also semantics.

The Emergence of Critical GIS

Semantics are enmeshed in philosophy and perception. Maps or geospatial data are sites of critical inquiry. If "[m]aps are a technology of power, ... the key to this internal power is cartographic process ... the way maps are compiled and the categories of information selected; the way they are generalized, a set of rules for the abstraction of the landscape" (Harley, 1992, p. 245). Therefore, "[m]aps and GIS are important sources for the production of geographic knowledge. What are the power-knowledge relations of mapping as they occur against the historical horizon of possibilities and how can that horizon be enlarged?" (Crampton, 2003b, p. 53). Two definitions seem to cover the continuum of thought on what is critical GIS. It is a "part of a contemporary network of knowledge, ideology, and practice that defines, inscribes, and represents environmental and social patterns within a broader economy of signification that calls forth new ways of thinking, acting, and writing" (Pickles, 1995, p. 4). It is also "concerned with limitations in the ways that populations, locational conflict and natural resources are represented within current GISs, and the extent to which these limits can be overcome by extending the possibilities of geographic information technologies" (Crampton, 2003a, ¶4).

With the sophistication of present-day GIS applications and the emergence of critical geographic information science, GIS researchers and academic practitioners are more reflexive (Schuurman, 2000), that is, reflecting upon the relationships between events, actions, and observers. Kwan (2004) addresses the social-theory/spatial analysis split in geography. She suggests social-cultural and spatial-analytical geographies can "enrich each other in meaningful ways and to consider various possibilities to reconnect them" (Kwan, 2004, p. 757). This is echoed by others, who believe that the science can positively affect society (Elwood, 2006; Hannah & Strohmayer, 2001). Schuurman (2000) describes three waves of philosophical and epistemological debate between GIS practitioners and their critics in human geography. The first wave focused on the uses of GIS, with an emphasis on positivism. The second wave focused more on the social effects of GIS. In the third wave, debates about the technology shifted to the subtlety and granularity available with GIS technology (Schuurman, 2000; Schuurman, 2006).

As noted, the cartographic epistemology of maps is influenced by positivism and social constructionism. Positivists tend to focus their critiques on the value-neutral aspect of GIS technology, or, "retreating from knowledge to information" (Taylor, 1990, p. 212). Taylor (1990) suggests that "[k]nowledge is about ideas, about putting ideas together into integrated systems of thought we call disciplines" and information is about facts, about separating out a particular feature of a situation and recording it as an autonomous observation (p. 212). Therefore, "disciplines are defined by the knowledge they produce and not by facts: a 'geographical fact' that is not linked to geographical knowledge" (p. 212). Social constructionists see maps as "practices of power-knowledge" and geographic visualization as providing "multiple, contingent and exploratory perspectives of data" (Crampton, 2001, p. 235). They posit that GIS practices shape and are shaped by institutional contexts. How a discipline conceptualizes itself plays a large role in how it then perceives, defines, and extends its reach vis-à-vis ontological and epistemological research. With a new discipline, such as GIScience, the current emphasis of its ontology and spatial reasoning are "primarily concerned with implementation of complex philosophical and cognitive concepts in a computational environment" (Schuurman, 2006). Or, simply, "To catalogue the world is to appropriate it, so that all these technical processes represent acts of control over its image which extend beyond the professed uses of cartography" (Harley, 1992, p. 245).

Ontologies and Semantics

As the focus in GIS changes from format integration to semantic interoperability, new descriptive classifications identify functionalities, creating new ontologies for GIS (Fonseca *et al.*, 2003; Goodchild, 2004; Goodchild & Haining, 2004; Mark *et al.*, 2001; Schuurman, 2003). An ontology is a specification of a conceptualization of a knowledge domain. It can also be described as a controlled vocabulary or a faceted taxonomy with richer semantic relationships among terms and attributes that formally describes objects and the relations between objects. This vocabulary (or taxonomy) has its own grammar so the user can create meaningful expressions within the specified domain of interest. The vocabulary and grammar are used to make queries and assertions. Individuals or groups make ontological commitments, or agreements, to use the vocabulary in a consistent way for knowledge or data sharing. Often thought of as knowledge representation, ontologies play an important role in "supporting query disambiguation and query term expansion of the required query, relevance ranking of the retrieved search results, the creation of the spatial indexes to support the search and the annotation of Web resources, Web documents and geographic data sets" (Smart, Abdelmoty, & Jones, 2004, p. 175). Further, ontologies play a key role in enabling semantic interoperability. Since an ontology describes a specific reality with a specific vocabulary, using a set of assumptions regarding the intended meaning of the vocabulary words, it can be seen as an explicit specification of a conceptualization (Gruber, 1992), which is a formal structure of reality as perceived and organized by an agent, independent of the vocabulary used or the actual occurrence of a specific situation (Guarino, 1998). Defining geographic space requires the definition and study of geographic objects, their attributes, and relationships.

From the Physical Universe to Ontologies

A properly constructed ontology can integrate different ontological approaches in a unified system (Frank, 2001; Frank, 2003). The result is a formal framework that explains a mapping between a spatial ontology and a geographic conceptual schema. Three different levels of abstraction are used in the mapping of ontologies to conceptual schemas: the formal level, which uses highly abstract concepts to express schema and ontologies; the domain level, where the schema is one instance of a generic data model; and the application level, which addresses the particular case of necessary geographic applications (Fonseca et al., 2003).

Building on Frank's five tiers (human-independent reality, observation of physical world, objects with properties, social reality, and subjective knowledge), Fonseca *et al.* introduce the five-universes paradigm that attempts to provide the perspective from the geographic world (Fonseca *et al.*, 2003; Fonseca, Egenhofer, Davis, & Cmara, 2002). Each of the five levels in the model, the physical universe, the cognitive universe, the logical universe, the representation universe, and the implementation universe, deals with conceptual characteristics of the geographic phenomena of the real world. A geographic phenomenon in the physical world is first perceived by an individual. He or she then classes the phenomena according to his or her cognitive framework, providing explicit and formal structures (ontologies) based on the vocabulary of his or her logical universe. GIS reference systems, such as fields and objects, are part of the representation universe. The implementation universe occurs when components of the representation universe (fields and objects) are translated into data structures and computer language constructs (Fonseca et al., 2002).

To be used effectively, the computer must know a number of things, such as what application is being used, what language the "text" is in, encryption and encoding protocols, transmission protocols, and platforms. When one moves out of the monolingual state to a multilingual world, the computer's basic knowledge extends to meta-languages (e.g., SGML), mark-up languages (e.g., HTML) and applications, and software operating systems. Both language and script (writing system) become important information for the computer (and information provider and user) to know. Imagine, if you will, the sheer amount of information that must be considered for "intelligent" processing by a word processing program: spell-check, grammar-check, word wrapping, hyphenation, automated word correction, use of symbols for non-Latinate languages, use of accented characters for non-English words, *inter alia*. If numeric data, there is a similar host of commands that ensue in the creation and checking of data. Then, move to transmission over a network, across platforms, and receipt by the end user's computer. To effectively handle just the linguistic properties of text, standardized language codes must support document longevity and interoperability of computing and network solutions. The same applies to the creation of network and platform protocols. Standards, whether data, semantic, or syntactic, apply equally to querying, searching, and accessing information from both vendor and the end-user perspectives.

Since the 1990s, geospatial data interoperability has been the target of standardization bodies, developers, and the research community: "interoperability has been seen as a solution for sharing and integrating geospatial data, more specifically to solve the syntactic, schematic, and semantic as well as the spatial and temporal heterogeneities between various real world phenomena" (Brodeur, Bédard, Edwards, & Moulin, 2003, p. 243). The transition to a distributed, online environment has eased some of the issues based on previous monolithic architectures for GISs (Lutz, Riedemann, & Probst, 2003; Reid, Higgins, & Medyckyi-Scott, 2004). These new architectures, the Internet, and data mining technologies have led to a renewed emphasis on discovery, dissemination, and exploitation of geospatial data. Encoding and decoding functions are crucial components. They respectively generate and recognize geospatial conceptual representations.

Quality Assurance

The W3C has published several formalized quality assurance (QA) resources for various languages and protocols developed by them. Although some of the features of the QA specification are unique to W3C's technical process and tools, other features have broad applicability to formal specification development by other standards organizations. For example, there are guidelines on how to write unambiguous and clear specifications, how to define and specify conformance, and how a specification might allow variation among conforming implementations. W3C suggests that developers consider conformance models, normative language usage, test suites, extensibility, profiles, levels, validation services, and conformance claims during the development and documentation process. Design decisions of a specification's conformance model may affect its implementation and the interoperability of its implementations.

Variability

Specifications allow some sort of variation between conforming implementations. The concept of variability addresses how much these implementations may vary among themselves (Hazaël-Massieux & Rosenthal, 2005). There are seven dimensions of variability. These range from the most independence to the least independence from other design factors with a variety of possible associations, dependencies, and interrelationships among the dimensions (Hazaël-Massieux & Rosenthal, 2005).

The dimensions are classes of product, profiles, levels, modules, discretionary items, deprecation, and extensibility (Hazaël-Massieux & Rosenthal, 2005, ¶7). A product is a generic name for any group of items (such as application, platform, etc.) that would implement a specification for the same purpose. In technology subsets, a profile is tailored to meet specific functional requirements of a particular application community and defines how a set of technologies are required to operate together, while a level is one of a hierarchy of nested subsets, ranging from minimal (core) functionality to full (complete) functionally. A module is a collection of semantically related features that represents a unit of functionality, such as an image module. While modules can be implemented independently of one another, a module's definition and implementation may be explicitly dependent upon another or multiple modules. Discretionary items are those items in an implementation that have options in behavior, functionality, parameter values, error handling, and so forth. Deprecation and extensibility are self-explanatory. Deprecated features are marked as outdated and being phased out. Extensibility allows any developer to create extensions (additional features) beyond what is defined in the specification (Hazaël-Massieux & Rosenthal, 2005, ¶7).

Since the seven dimensions of variability are at the core of the definition of a specification's conformance model, there is "significant potential for negative interoperability impacts if they are handled carelessly or without careful deliberation" (Hazaël-Massieux & Rosenthal, 2005, ¶ 9). Generally, implementation variability complicates interoperability; identical implementations are better. However, there are cases when the net effect of conformance variability is not necessarily negative (Hazaël-Massieux & Rosenthal, 2005, ¶ 10). Consider the use of profiles. As mentioned earlier, profiles are developed for specific applications communities. Two different community profiles may not communicate well between themselves. However, if the two profiles are subsets of a large monolithic specification, each targeted at a specific application sector, then subdivision by profiles may actually enhance interoperability (Haaël-Massieux & Rosenthal, 2005, ¶ 10). Two areas where variability is dangerous are excessive variability in a specification or the multiplicative effect on variability when several dimensions are combined (Hazaël-Massieux & Rosenthal, 2005, ¶ 12).

Augmenting test documentation, metadata plays an important role in helping users to understand and execute the tests: "…Well-defined metadata can help in: (1) tracking tests during the development and review process; (2) filtering tests according to a variety of criteria — for example, whether or not they are applicable for a particular profile or optional feature; (3) identifying the area of the specification that is tested by the tests; (4) constructing a test harness to automatically execute the tests; (5) formatting test results so that they are easily understood" (Curran & Dubost, 2005, ¶2). By defining and using a minimal set of metadata elements (names, syntax, and usage), standardized tools are more likely to be developed since ambiguity is lessened.

Taxonomies

The emergence in recent years of digital libraries and of Internet-based communication applications have led some researchers to propose that the emerging data infrastructure of the Internet and digital libraries can be used to ease mining digital geospatial data across the Internet. Descriptive cataloging and metadata standards, as mentioned in Chapters IV and V, describe attributes of items to enhance the user's ability to restrict their research to appropriate content. As the amount of networked digital information continues to grow, the demand for "seamless" access also increases. However, for a single theory-neutral taxonomy to support granularity, language groupings, data classification, categorization, and linguistic annotation is asking a lot. For written and spoken language materials in minority languages, depending upon the person providing the description, for example, librarians, archivists, linguists, and so forth, existing ISO standards may not be granular enough to distinguish regional, social, or dialectical variations, or the standards may be too complex for the intended audience or describer. For example, ANSI/NISO Z39.53-200X provides "a standardized 3-character code to indicate language in the exchange of information is defined. Codes are given for languages, contemporary and historical. The purpose of this standard is to provide libraries, information services, and publishers a standardized code to indicate language in the exchange of information. This standard for language codes is not a prescriptive device for the definition of language and dialects but rather a list reflecting the need to distinguish recorded information by language" (National Information Standards Organization, 2001, p. 1). Although there are no standards for prescriptive or descriptive information for data elements or metatags (such as notes fields or subject headings), it is this type of work, common in librarianship, that may provide a necessary bridge to establish a common language/framework across standards, among users, and increase interoperability.

Geospatial data may incorporate numeric datasets of socioeconomic information, epidemiologic datasets, textual datasets, or use vector, raster, or tabular data. These data become digital objects when they are geocoded, or linked to a Cartesian co-ordinate system. These digital objects, whether tabular or thematic data, require description of their attributes, the geographic features of the extent of the area, and linkage to a unique identifier. Taxonomies for thematic data may use technical, specialized vocabulary, unfamiliar to the librarian, the person describing the data, or the end-user who is trying to acquire the data.

Brodeur et al. (2003) suggest that field of the interpersonal communication may provide a better framework with which to understand the issues in geospatial data interoperability. This communication process examines the interaction "between two agents, including the underlying internal representation of concepts along with encoding and decoding operations" (Brodeur et al., 2003, pp. 260-261) in both human-to-computer communication and computer-to-computer communication.

Harvey et al. (1999) also examine how to navigate differences in meaning. Their research takes a communications perspective since a central question in their research is "how people and social groups with different perspectives identify and possibly resolve their semantic differences" (p. 214). They suggest using a multidisciplinary approach using cognitive, computer science, and linguistic bases to examine semantic interoperability. Their cognitive approach, based on Fauconnier's general mappings among conceptual domains and Lakoff's metaphorical mappings, follows how partial mappings from multiple sources structure a target concept. This model also uses mathematically rigorous formalization and implementations. Their computer science approach, based on the work of Sheth's interoperable computer system semantics (Sheth, Avant, & Bertram, 2001), assesses context in semantic proximity (Harvey et al., 1999). Sheth's approach looks at vocabulary, content, and structure, not just database ontologies and declarative descriptions (Sheth, 1998; Sheth et al., 2002). Briefly, domains between two objects are mapped, and the contextual descriptions of the two objects are compared and described in a descriptive logic language that links the semantic and schematic level. The third framework is a linguistic framework that considers the processes social groups engage in to assure collaborative action or participatory design. It emphasizes the importance of involving relevant groups in articulating their differences in order to find common and viable solutions. Harvey et al. (1999) suggest, "If semantics are cultural agreements between independent agents observing the real world, then we expect that illuminating insights will come from the examination of the group processes that lead to 'accepted' understandings, and the role of language as the most fundamental way of finding and assuring agreement" (p. 228). They further suggest that information communities should conceptualize and articulate technical, organizational, and political semantic differences to resolve differences in consensual terminology and procedures.

Building Interoperable, Semantic Systems

In order to achieve semantic interoperability in heterogeneous information systems, systems must understand the meaning of the exchanged information, that is, the precise meaning of the data must be readily accessible and the system able to translate the data into an understandable form. Metadata is not just a description of the schema definition in a data set, but also a description of the conceptualization of the geospatial "reality." If semantics refers to user's interpretation of the computer representation of the world (Meersman, 1995), then metadata should contain a representation of the semantics of the data. Clearly, the dynamic data exchanges possible in GIS communities provide substantial advantages for sharing geographic information. However, to fully realise the advantages in heterogeneous, operational, and organizational environments requires developers and users to understand and resolve semantic differences (Harvey et al., 1999). While there has been substantial progress on technical interoperability, semantic interoperability remains a significant hurdle. Identifying and resolving semantic interoperability issues is especially pertinent for data sharing and considering future developments of standards.

Solutions to semantic interoperability involve three major frameworks: cognitive, computer science, and sociotechnical (Harvey et al., 1999; Miller & Han, 2000). Brodeur et al. (2003) suggest that a broader view of geospatial data interoperability is also in order. Using an ontology of geospatial data interoperability to refine the description of the conceptual framework, they identify the notions of concept, proximity, and ontology as fundamental to creating a new approach to geosemantic proximity (Brodeur et al., 2003, pp. 243-244).

Providing integrated access to data from a diverse, heterogeneous network requires a breadth of knowledge, not only about the structure of the data represented at each server, but also about the commonly occurring differences in the intended semantics of this data (Tawil, Fiddian, & Gray, 2001). The semantics of data are often couched in local schemas that may meet the needs of that user group, but lack the ability to be interoperable when searched by users in another setting. These semantically weak local schemas are a consequence of the limited expressiveness of traditional data models (Tawil et al., 2001). Stoimenov and Djordjevic-Kajan (2005) suggest the creation of architectures that can address semantic interoperability of distributed and heterogeneous GIS. A local community environment informs the perspective; mediation and ontologies define the architecture. First, they formally specify the meaning of the terminology of each community using local ontology. Then they define a translation between each community's terminologies, with an intermediate terminology represented by top-level ontology and common data model (Stoimenov & Djordjevic-Kajan, 2005).

If a *concept* consists of "the set of knowledge with the accompanying processes that an agent maintains about a phenomenon, which generate and recognize different representations of the concept" (Brodeur et al., 2003, p. 257), then domain-specific metadata is one possibility to upgrade the semantic level of local information systems. By considering an integrated framework, users receive better access and enhanced, semantically rich schema models. There are a number of frameworks, in development or currently available, that enrich the data definition language of resident servers. The schema's semantic knowledge is organised by levels of schematic granularity: database, schema, attribute, and instance.

Another possibility is geosemantic proximity, a framework that concurrently assesses the components of a geosemantic space (i.e., semantic, spatial, and temporal similarities) between a geospatial concept and a geospatial conceptual representation (Brodeur et al., 2003). Visually, these concepts and conceptual representations are segments on an axis made of an interior and a boundary. Geosemantic proximity is the intersection of their respective contexts. The interior of a concept "consists of its intrinsic properties that are components providing literal meaning (e.g., identification, attributes, attribute values, geometries, temporalities, and domain)" while the boundary of a concept "consists of its extrinsic properties that are components providing meaning through relationships with other concepts (e.g., semantic, spatial, and temporal relationships as well as behaviours)" (Brodeur et al., 2003, p. 260). Consequently, intersection between intrinsic and extrinsic properties illustrate attributes that can be assessed qualitatively taking into account the contexts of the respective representations (Kashyap & Sheth, 1996; Kashyap & Sheth, 1998).

Digital Libraries: Solutions and Possibilities

Dealing flexibly with differences among systems, ontologies, and data formats while respecting information sources' autonomy requires considerable thought. Adapting object-oriented digital libraries involves a number of solutions, such as mediation middleware, use and integration of Internet harvesting techniques, and new architectures based on object-oriented ontologies that affect the development of search modules and metadata description. Much of the research focuses on text-deciphering or linguistic equivalency algorithms or consensual activities on meanings of categories for resource description (Di Pasquale, Forlizzi, Jensen, Manolopoulos, Nardelli, Pfoser et al., 2003), interoperability of simple schemas with complex schemas, or designing frameworks for managing equivalencies between metadata models in different fields and languages (Baker & Klavans, 1998). Questions, such as best problem-solving practices or balance between human and machine for resource discovery, emerge. Since the transparent and integrated access to distributed and heterogeneous data sources is key to leveraging research, how would it be best to integrate data suitable for knowledge or information domains? Controlled annotation of semantic rich metadata with diverse types of data allows the use of sophisticated query schemes (Gertz & Sattler, 2003).

Information systems can address these questions by applying and extending metadata harvesting, and by building upon existing componentized frameworks (Ravindranathan, Shen, Gonçalves, Fan, Fox, & Flanagan, 2004). Two very important issues are (1) how to reconcile the diversity found within the harvested data to create a single, integrated collection view for the end-user and (2) how to create an integrated framework that addresses data quality, flexible and efficient search, and scalability (Gonçalves, France, & Fox, 2001). To provide technologies that improve the access to heterogeneous and distributed resources, several layers of metadata, related to users, communities, devices, and data sources, must be handled and efficiently used (Godard, Andrès, Grosky, & Ono, 2004). Further, in order to understand class and property hierarchies, support for inference should be available (Palmér, Naeve, & Paulsson, 2004). The quality of the information available to the information specialist to adequately address (matchmake) the semantics of requirements and resources is critical. The explicitness, structuring, and formality of this information can differ considerably leading to different types of matchmaking (Lutz et al., 2003).

Lutz et al. (2003) describes one sample framework for the analysis of practical problems. First, the information required for the matchmaking process is identified. Once identified, the required information is classified according to the qualities of explicitness, structuring, and formality. The information is assessed to determine the quality level of the required information that is appropriate for the task-at-hand. Finally, the different levels of explicitness, structuring, and formality can easily be associated to predefined scenarios that indicate possible implementation methods (Lutz et al., 2003). The research also has found that multiword terms provide the most effective snapshot of user searching behavior for query categorization. Using both approaches, researchers to classify their approach and judge whether the applied methods are appropriate for the task-at-hand (Yi, Beheshti, Cole, Leide, & Large, 2006). This is particularly important for the naïve user, since research shows that an information seeker's information need is identified through transformation of his/her knowledge structure (i.e., cognitive map, or perspective) (Cole, Leide, Beheshti, Large, & Brooks, 2005a; Cole, Leide, Large, Beheshti, & Brooks, 2005b; Yi et al., 2006).

Individual and Organizational End Users: Issues in Accessibility

Data sharing is defined as the "transfer of spatial data/information between two or more organizational units where there is independence between the holder of the data and the prospective user" (Calkins & Weatherbe, 1995, p. 66). Nine factors or conditions create a conducive environment for sharing: (1) sharing classes; (2) project environment; (3) the need for shared data; (4) opportunities to share data; (5) willingness to share data; (6) incentive(s) to share data; (7) barriers to sharing; (8) the technical capability to share; and (9) resources for sharing (Kevany, 1995). Users and the organizational constructs can create obstacles in the use of GIS data. These include variation in priorities among participants, differences in capacity to exploit GIS resources and services, dissimilarity in the level of awareness and spatial data handling skills; and inability to achieve agreements over access to information, leadership, data standards, equipment and training (Argentati, 1997; Masser & Campbell, 1995, p. 236). Others have seen obstacles to sharing as primarily behavioral factors as well as organizational dynamics (Longley, Goodchild, Maguire, & Rhind, 2001; Onsrud & Rushton, 1995). For individuals to be willing to engage in spatial data sharing, constructs, such as "attitude towards the behavior," "social norm," and "perceived control over the behavior," become important

(Wehn de Montalvo, 2003). The literature suggests that these constructs influence the structures, processes, and policies used in interorganizational relationships that facilitate building and sharing spatial databases. By examining the contextual factors that affect geographic information relationships, one can learn what mechanisms are effective in accomplishing database development and sharing.

Nedović-Budić (1999) suggest that "appropriate organizational motivation, attitudes, and structure are required for geographic data sharing to happen" (p. 190). For example, attitude toward data can significantly affect cooperative or sharing relationships. Limits on open access are contraindicated in facilitating sharing, but user expectations may not match with data restrictions/access. Economic, political, professional, or regulatory factors may also hinder incentives to share data. Structural and functional characteristics of organizations also affect user access, such as "new coordination mechanisms, communication channels, overarching bodies, responsibilities, and authority" (Nedović-Budić & Pinto, 1999, p. 191). Power equations between organizations regarding access will continue to change as the online environment becomes more pervasive. Initiatives, such as the NSDI and open standards development, encourage data sharing, infrastructure development, institutional arrangements, clearinghouse tools, enhanced metadata, and data transfer standards. However, to accommodate end-user needs and expectations, there must be clearly defined policies on access, cost recovery, data documentation, liability, interorganizational agreements, legal authority, and participant roles that accommodate the end user as well as the organizational structure (Calkins, 1992; Harvey & Tulloch, 2006; Tulloch & Shapiro, 2003). Policies should address the data itself (scope, extent, manipulation, handling); responsibility/ownership (redistribution and incorporation into new sets); cost; incentives; and formalization of the interorganizational relationship (Harvey & Tulloch, 2006; Nedović-Budić, 2001; Nedović-Budić & Pinto, 1999).

Although there has been increased acceptance and use of GIS and other digital data sets in private and public sector organizations, successful interorganizational GIS use is still problematic (Harvey & Tulloch, 2006; Nedović-Budić & Pinto, 2000). Characteristics that determine successful interorganizational GIS range from the intensity, quality, and interdependence of interorganizational relationships to resources and structure (Nedović-Budić & Pinto, 2000). Stability, culture, politics, and leadership affect success of GIS partnerships and collaborations. Nedović-Budić and Pinto (2000) address the importance of coordination mechanisms that manifest through established structures, processes, and policies. Structure is couched in a communication perspective, viewing the flow of information via channels, direction, and methods as well as level of shared components, which range from hardware and software to personnel and space. They firmly believe that coordination process is best undertaken through standardization, joint planning, or mutual adjustment, as formal or informal policies are established to address data-related issues, responsibilities, ownership, contributions, and incentives (Nedović-Budić, 2001; Nedović-Budić &

Pinto, 2000). As with all initiatives, outcomes must be evaluated on a number of criteria, including efficiency, effectiveness, decision-making impact, and equity. A combination of functionality, usability, and accessibility evaluation strategies applied iteratively to assess libraries from the perspective of patron needs also seems appropriate (Bertot, Snead, Jaeger, & McClure, 2006).

Harvey (2006) emphasizes the function of metadata in supporting trust in the collaborative process. He suggests that collective intentionality (the shared intent of a group) and status functions (i.e., rules, regulations, procedures, and standards) comprise an individual's trust in the institution's operating processes, or institutional reality. He posits that there are two types of trust, the type based on personal relations and the type based on the exchange of impersonal objective data. There is an interdependence and causality between the two, "[t]he reliance on objective data can undermine personal trust relationships in some cases" (Harvey, 2006, p. 146). Since technical solutions to data sharing can hinder the understanding of shared data, "data sharing must move one rung higher and become information sharing, which requires the collective recognition of status functions and the creation of collective intentionality" (Harvey, 2006, p. 145). Therefore, an environment of trust, at both the personal and data levels, is integral to successful information sharing. One way to assist the trust relationship is through the of rich, descriptive metadata that can create a collective understanding of the uses of GIS data (Harvey, 2006). Other important qualities include equity in the use of the data or other common resources; a fair decision-making process; ensuring users have an adequate control of common activities, organizational persistence to make the interorganizational arrangements succeed despite differing agendas and management styles, and coalition building, bargaining, and willingness to compromise (Nedović-Budić & Pinto, 2000).

Users may also be impacted by the lack of strategic information management, to institutionalize information and decision support tools, and to transfer the technology to planning settings (Nedović-Budić, 2001). Problems in these areas also affect users in the workflow process and in the creation of effective infrastructure needed for teaching, research, and technology transfer. Nedović-Budić (2001) suggests further research and policy development in education and technology transfer, database creation, data maintenance and access, standards development, and legislation and policy.

In summary, the use of GIS needs to mitigate organizational and political factors that "apparently offset, in many instances, the theoretical benefits to be obtained from structures which seek to promote information sharing" (Masser & Campbell, 1995, p. 247), which may have serious implications for end users in their use of geospatial information.

The Library's Role

The role of librarians as custodians and disseminators of information is not new. It has been suggested their role in the use of spatial data is increasingly challenging, "navigat[ing] emerging geospatial data standards, disparate data, and shifting technologies" (Schuurman, 2000), not to mention organizational and collaborative issues. Libraries prefer to provide their users transparent and integrated access to digital spatial information, and distributed, autonomous, and heterogeneous information services. How best to handle issues of interoperability and provide access to their users? Dataset format, semantics, scale, resolution, and geographic area can compound the delivery of geospatial information. Hunt and Joselyn (1995) offer several possible solutions: data conversion, data organization, intelligent retrieval software, and physical upgrade. Converting data to a software-dependent form may be feasible but will be labor intensive for library staff. However, the trade-off for end-user ease of access may be worth the additional training and product development. Data organization may involve physically partitioning or consolidating data by appropriate geographic units. It may also involve logically identifying, coding, or cataloging data to make their retrieval more intuitive to the user. Intelligent retrieval requires the patron to identify, access, and retrieve a specific subset of a given dataset at an appropriate scale or resolution in a timely fashion, again which may require more back-end coding of data or more front-end training of the patron on available applications (Hunt & Joselyn, 1995). As addressed in the chapter on collection development, the application software necessary to bring functionality to a dataset may determine if it will be accessible to the user as preformatted data or for use in an analytic GIS environment.

Another solution is the use of topic navigation maps, an international standard (ISO 13250) project. Much like library pathfinders, topic maps can assist in improved retrieval of online information, allowing users to define their own navigation strategies in electronic resources (Sigel, 1998). Further, topic maps can also assist in the creation and navigation of living documents and dataset repositories. Whether the resource is a structured information environment (such as a thesaurus or controlled vocabulary resource) or an unstructured information environment (no enhanced features, such as thesauri or other controlled vocabulary), topic maps provide "outside views," or a user-structured model, for navigating the resource. Again, this could become an extended skills set for catalogers as well as a revised skills set for subject bibliographers. According to Sigel (1998), building a topic map from a structured information resource is almost 100% automatic, easing burden on staff time and project completion. If there is no previous information structure, building a useful topic map will take more staff and user time. However, both results enhance resource discovery and create new metadata relationships between controlled and user vocabulary.

Stoimenov and Djordjevic-Kajan (2005) recommend using a hybrid ontology approach to resolve semantic heterogeneity of data sources. Not only is the meaning of the terminology of each community specified in the local ontology, they have created a methodology and software support for resolving semantic mismatches (conflicts) between terminologies. Although their methodology is computationally intense, work defining and mapping terminology is standard practice for librarians in establishing authority control. Their formal ontology, which consists of "definitions of terms, … includes concepts with associated attributes, relationships and constraints defined between the concepts and entities that are instances of concepts" (Stoimenov & Djordjevic-Kajan, 2005), could take the intellectual aspects of authority control to a higher level in the analysis and description of resources. As with any data-based resource, formal ontologies are "best for sharing, merging, and querying data, but not for reading and efficient processing"(Stoimenov & Djordjevic-Kajan, 2005, p. 217). However, perhaps the new model for library catalogs should consider the incorporation of data dictionaries, thesauri, and semantic rules, ontologies that are stored together as a knowledge base, and metadata that specify a common model (Stoimenov & Djordjevic-Kajan, 2005, p. 216). Semantic Web technologies such as these can bridge formal ontologies and natural language (Katz, Lin, & Quan, 2002).

In academia, personalization of services and resources in the support of learning activities is an emergent area in library services. As geospatial learning continues to become more interactive and collaborative, personal project spaces allow individuals to work in their personalized environment with a mix of private and public data and simultaneously share data with team members (Lim, Goh, Liu, Ng, Khoo, & Higgins, 2002; Lim, Sun, Liu, Hedberg , Chang , Teh *et al.*, 2004). Portability of resources, interoperability assurances, and enhanced resource discovery through multiple perspectives is critical. Librarians are particularly suited to work with users on a one-to-one and small group basis to establish a common knowledge base and navigation/discovery skills. In another example, scholars play the role of both consumer and contributor of intellectual works in the digital environment. How they seek, use, and create implicit and explicit implicit assemblages of resources provide a useful framework for the collection and organization of access resources in research libraries (Palmer, 2005).

As the roles of libraries continue to shift and mutate into new organizational constructs and services, the same should apply to how librarians see their roles in managing data-intensive information. Green (1998) suggests that four questions need to be addressed in information or knowledge management:

- "Is there a universal set of relationship types applicable across all contexts?
- "How can we build integrated knowledge organization schemes that reflect a multiplicity of relational views?

- "Is the incorporation of a relational approach to retrieval feasible, given the volume and diversity of material online?
- "How could we evaluate the impact of incorporating a relational approach to online retrieval?" (Green, 1998)

All four questions have relevance to the everyday work of librarians. Librarians determine the basic concepts and relations used in the description of an item. The current emphasis on catalog design reflects the importance of providing a multiplicity of views, based on user perspective, from the naïve user to the more sophisticated researcher, as well as the conceptual view of the librarian. The field also reflects the concerns over the feasibility and evaluation of indexing resources with many semantic relations and the effect on retrieval and resource discovery. How librarians integrate their knowledge and skills with other information specialists and database designers in the management of GIS data and resources is yet to be determined. The integration is a necessary and critical juncture in the field of librarianship.

Conclusion/Summary

Clearly, accessibility has many parameters that must be determined to achieve syntactical and semantic operability. There appears to be a merging of concepts, at least with respect to the geometries involved, which shows an evolution from "mere data exchange at the interface level to systems integration with common semantics" (Albrecht, 1999, p. 166). There are areas in GIS knowledge management, such as ontology design principles, quality issues surrounding the creation of subject metadata, limitations of subject analysis, or the concerns of decentralized versus centralized provision of resource description, discovery, retrieval, that require librarians to learn how better to handle these issues from the research in library and information science.

One of the future areas of research is the role that service discovery plays within the larger task of service composition, and what other subtasks play a role in ensuring semantic interoperability in service composition. For network and standards developers, as well as libraries, the task of service composition is an area worth exploring.

References

Albrecht, J. (1999). Towards interoperable geo-information standards: A comparison of reference models for geo-spatial information. *The Annals of Regional Science, 33*(2), 151-169.

Argentati, C. D. (1997). Expanding horizons for GIS services in academic libraries. *The Journal of Academic Librarianship, 23*(6), 463-468.

Baker, T., & Klavans, J. (1998). Metadata and content-based approaches to resource discovery. In C. Nikolaou, & C. Stephanidis (Eds.), *Research and advanced technology for digital libraries: Second European Conference, ECDL'98, Heraklion, Crete, Greece, September 1998, Proceedings* (pp. 737-738). Berlin, Germany: Springer-Verlag GmbH.

Batten, D. F., Karlsson, C., & Andersson, A. E. (1989). Knowledge, nodes and networks: An analytical perspective. In A. E. Andersson, D. F. Batten, & C. Karlsson (Eds.), *Knowledge and industrial organization* (pp. 31-46). Berlin; Heidelberg, Germany: Springer-Verlag GmbH.

Bertot, J. C., Snead, J. T., Jaeger, P. T., & McClure, C. R. (2006). Functionality, usability, and accessibility: Iterative user-centered evaluation strategies for digital libraries. *Performance Measurement and Metrics, 7*(1), 17-28.

Brodeur, J., Bédard, Y., Edwards, G., & Moulin, B. (2003). Revisiting the concept of geospatial data interoperability within the scope of human communication processes. *Transactions in GIS, 7*(2), 243-265.

Buehler, K., & McKee, L. (1996). *The openGIS guide: Introduction to interoperable geoprocessing: Part I of the open geodata interoperability specification (OGIS)* (2nd ed.). Wayland, MA: Open GIS Consortium, Inc., OGC Technical Committee.

Calkins, H. W. (1992). Institutions sharing spatial information. In P. W. Newton, P. R. Zwart, & M. E. Cavill (Eds.), *Networking spatial information systems* (pp. 283-292). London, England: Belhaven Press.

Calkins, H. W., & Weatherbe, R. (1995). Taxonomy of spatial data sharing. In H. J. Onsrud, & G. Rushton (Eds.), *Sharing geographic information* (pp. 65-75). New Brunswick, NJ: Center for Urban Policy Research.

Cole, C., Leide, J., Beheshti, J., Large, A., & Brooks, M. (2005a). Investigating the anomalous states of knowledge hypothesis in a real-life problem situation: A study of history and psychology undergraduates seeking information for a course essay. *Journal of the American Society for Information Science and Technology, 56*(14), 1544-1554.

Cole, C., Leide, J. E., Large, A., Beheshti, J., & Brooks, M. (2005b). Putting it together online: Information need identification for the domain novice user. *Journal of the American Society for Information Science and Technology, 56*(7), 684-694.

Crampton, J. W. (2001). Maps as social constructions: Power, communication and visualization. *Progress in Human Geography, 25*(2), 253-260.

Crampton , J. W. (2003a). How can critical GIS be defined? *GEO World, April.* Retrieved November 2006, from http://64.233.161.104/search?q=cache: Exjq8CSF2NEJ:geoplace.com/gw/2003/0304/0304cgis.asp+geoworld+cra mpton&hl=en&gl=us&ct=clnk&cd=3

Crampton, J. W. (2003b). *The political mapping of cyberspace.* Chicago, IL: University of Chicago Press.

Curran, P., & Dubost, K. (2005). *Test metadata* (W3C Working Group Note No. 14 September 2005). Cambridge, MA: W3C Quality Assurance Working Group. Retrieved from http://www.w3.org/TR/test-metadata/.

Di Pasquale, A., Forlizzi, L., Jensen, C. S., Manolopoulos, Y., Nardelli, E., Pfoser, D. *et al.* (2003). Access methods and query processing techniques. In G. Goos, J. Hartmanis, & J. van Leeuwen (Eds.), *Spatio-temporal databases: The CHOROCHRONOS approach* (*Lecture Notes in Computer Science, 2520*, pp. 203-261). Berlin, Germany: Springer-Verlag GmbH.

Elwood, S. (2006). Critical issues in participatory GIS: Deconstructions, reconstructions, and new research directions. *Transactions in GIS, 10*(5), 693-708.

Fonseca, F., Davis, C., & Cmara, G. (2003). Bridging ontologies and conceptual schemas in geographic information integration. *Geoinformatica, 7*(4), 355-378.

Fonseca, F., Egenhofer, M. J., Davis, C., & Cmara, G. (2002). Semantic granularity in ontology-driven geographic information systems. *AMAI Annals of Mathematics and Artificial Intelligence, 36*(1-2), 121-151. [Special Issue on Spatial and Temporal Granularity].

Frank, A. (2001). Tiers of ontology and consistency constraints in geographical information systems. *International Journal of Geographical Information Science, 15*, 667-678.

Frank, A. U. (2003). Ontology for spatio-temporal databases. In G. Goos, J. Hartmanis, & J. van Leeuwen (Eds.), *Spatio-temporal databases: The CHOROCHRONOS approach* (*Lecture Notes in Computer Science, 2520*, pp. 9-77). Berlin, Germany: Springer-Verlag GmbH.

Gertz, M., & Sattler, K.-U. (2003). Integrating scientific data through external, concept-based annotations. In S. Bressan, A. B. Chaudhri, M. L. Lee, J. X. Yu, & Z. Lacroix (Eds.), *Efficiency and effectiveness of XML tools and techniques and data integration over the Web: VLDB 2002 Workshop EEXTT and CAiSE 2002Workshop DIWeb, revised papers* (*Lecture Notes in Computer Science, 2590*, pp. 220-240). Berlin, Germany: Springer-Verlag GmbH.

Godard, J., Andrčs, F., Grosky, W., & Ono, K. (2004). Knowledge management framework for the collaborative distribution of information. In W. Lindner, M. Mesiti, C. Türker, Y. Tzitzikas, & A. Vakali (Eds.), *Current trends in database technology - EDBT 2004 Workshops: EDBT 2004 Workshops PhD, DataX,*

PIM, P2P&DB, and ClustWeb, Heraklion, Crete, Greece, March 14-18, 2004, Revised selected papers (*Lecture Notes in Computer Science, 3268.*, pp. 289-298). Berlin, Germany: Springer-Verlag GmbH.

Gonçalves, M. A., France, R. K., & Fox, E. A. (2001). MARIAN: Flexible interoperability for federated digital libraries. In P. Constantopoulos, & I. T. Sølvberg (Eds.), *Research and advanced technology for digital libraries: 5th European conference, ECDL 2001, Darmstadt, Germany, September 4-9, 2001, proceedings* (*Lecture Notes in Computer Science, 2163*, pp. 173-187). Berlin, Germany: Springer-Verlag GmbH.

Goodchild, M. F. (2004). GIScience, geography, form and process. *Annals of the Association of American Geographers, 94*(4), 709-714.

Goodchild, M. F., & Haining, R. P. (2004). GIS and spatial data analysis: Converging perspectives. *Papers in Regional Science, 83*(1), 363-385.

Green, R. (1998). *Challenges in knowledge organization "Structures and Relations in Knowledge Organization."* Panel conducted at the annual meeting of the *ISKO International Conference* Lille, France: International Society for Knowledge Organization. Retrieved November 2006, from http://index.bonn. iz-soz.de/~sigel/ISKO/Challenges.html

Gruber, T. (1992). *A translation approach to portable ontology specifications.* Stanford, CA: Knowledge Systems Laboratory, Stanford University.

Guarino, N. (1998). Formal ontology and information systems. In N. Guarino (Ed.), *Formal ontology in information systems* (pp. 3-15). Amsterdam, Netherlands: IOS Press.

Hannah, M. G., & Strohmayer, U. (2001). Anatomy of debate in human geography. *Political Geography, 20*, 381-404.

Harley, J. B. (1992). Deconstructing the map. In T. J. Barnes, & J. S. Duncan (Eds.), *Writing worlds: Discourse, text, and metaphor in the representation of landscape.* New York, NY: Routledge.

Harvey, F. (2006). Status functions, collective intentionality: Matters of trust for geospatial information sharing. In M. Raubal, H. J. Miller, A. U. Frank, & M. F. Goodchild (Eds.), *Geographical information science, Proceedings, 4th international conference, GIScience 2006, Münster, Germany, September 20-23, 2006* (*Lecture Notes in Computer Science, 4197*, pp. 145-152). Berlin, Germany: Springer-Verlag GmbH.

Harvey, F., Kuhn, W., Pundt, H., Bishr, Y., & Riedemann, C. (1999). Semantic interoperability: A central issue for sharing geographic information. *The Annals of Regional Science, 33*, 213-232.

Harvey, F., & Tulloch, D. (2006). Local-government data sharing: Evaluating the foundations of spatial data infrastructures. *International Journal of Geographical Information Science, 20*(7), 743-768.

Hazaël-Massieux, D., & Rosenthal, L. (2005). *Variability in specifications* (W3C Working Group Note No. 31 August 2005). Cambridge, MA: W3C Quality

Assurance Working Group. Retrieved from http://www.w3.org/TR/spec-variability/

Hunt, L., & Joselyn, M. (1995). Maximizing accessibility to spatially referenced digital data. *The Journal of Academic Librarianship, 21*(4), 257-265.

Kashyap, V., & Sheth, A. (1996). Semantic and schematic similarities between database objects: A context-based approach. *The VLDB Journal, 5*, 276-304.

Kashyap, V., & Sheth, A. (1998). Semantic heterogeneity in global information systems: The role of metadata, context and ontologies. In M. Papazoglou, & G. Schlageter (Eds.), *Cooperative information systems: Trends and directions* (p. 139–178). San Diego, CA: Academic Press.

Katz, B., Lin, J., & Quan, D. (2002). Natural language annotations for the semantic Web. In R. Meersman, Z. Tari *et al.* (Eds.), *On the move to meaningful Internet systems 2002: CoopIS, DOA, and ODBASE: Confederated International Conferences CoopIS, DOA, and ODBASE 2002, proceedings* (*Lecture Notes in Computer Science , 2519*, pp. 1317-1331). Berlin, Germany: Springer-Verlag GmbH.

Kevany, M. J. (1995). A proposed structure for observing data sharing. In H. J. Onsrud , & G. Rushton (Eds.), *Sharing geographic information* (pp. 76-102). New Brunswick, NJ: Center for Urban Policy Research.

Kwan, M.-P. (2004). Beyond difference: From canonical geography to hybrid geographies. *Annals of the Association of American Geographers, 94*(4), 756-763.

Lim, E.-P., Goh, D. H.-L., Liu, Z., Ng, W.-K., Khoo, C. S.-G., & Higgins, S. E. (2002). G-Portal: A map-based digital library for distributed geospatial and georeferenced resources. In G. Marchionini, & W. R. Hersh *Proceedings of the second ACM+IEEE joint conference on digital libraries (JCDL 2002), Portland, Oregon, USA, July 14-18, 2002* (pp. 351-358). New York, NY: ACM Press.

Lim, E.-P., Sun, A., Liu, Z., Hedberg , J., Chang , C.-H., Teh , T.-S. *et al.* (2004). Personalized issues in digital library: Supporting field study with personalized project spaces in a geographical digital library. In Z. Chen, H. Chen, Q. Miao, Y. Fu, E. Fox, & E. Lim (Eds.), *Digital libraries: International collaboration and cross-fertilization: 7th international conference on Asian digital libraries, ICADL 2004, Shanghai, China, December 13-17, 2004, proceedings* (*Lecture Notes in Computer Science, 3334*, pp. 553-562). Berlin, Germany: Springer-Verlag GmbH.

Longley, P. A., Goodchild, M. F., Maguire, D. J., & Rhind, D. W. (2001). *Geographic information systems and science*. Chichester, England: John Wiley & Sons Ltd.

Lutz, M., Riedemann, C. R., & Probst, F. (2003). A classification framework for approaches to achieving semantic interoperability between GI Web services. In W. Kuhn, M. F. Worboys, & S. Timpf (Eds.), *Spatial information theory*

(Lecture Notes In Computer Science, 2825, pp. 186-203). Berlin, Germany: Springer-Verlag GmbH.

Mark, D. M., Skupin, A., & Smith, B. (2001). Features, objects, and other things: Ontological distinctions in the geographic domain. In D. R. Montello (Ed.), *Spatial information theory: Foundations of geographic information science: International Conference, COSIT 2001 Morro Bay, CA, USA, September 19-23, 2001, Proceedings* (*Lecture Notes in Computer Science, 2205*, pp. 489-502). Berlin, Germany: Springer-Verlag GmbH.

Masser, I., & Campbell, H. J. (1995). Information sharing: The effects of GIS on British local government. In H. J. Onsrud, & G. Rushton (Eds.), *Sharing geographic information* (pp. 230-249). New Brunswick, NJ: Center for Urban Policy Research.

Meersman, R. (1995). An essay on the role and evolution of data (base) semantics. In R. Meersman, & L. Mark (Eds.), *Database applications semantics: Proceedings of the IFIP WG 2.6 Working Conference on Database Applications Semantics (DS-6), Stone Mountain, Atlanta, Georgia, U.S.A., May 30-June 2, 1995* 1st ed., (pp. 1-7). London, England: Chapman & Hall.

Miller, H. J., & Han, J. (2000). Reports from related meetings: Discovering geographic knowledge in data rich environments: a report on a specialist meeting . *ACM SIGKDD Explorations Newsletter, 1*(2), 105-107.

National Information Standards Organization. (2001). *Codes for the representation of languages for information interchange (Revision of ANSI/NISO Z39.53-1994)*. Bethesda, MA: National Information Standards Organization. Retrieved November 2006, from http://xml.coverpages.org/ANSI-NISO-Z39-53.pdf

National Institute of Standards and Technology. (1996). *Application portability profile (APP): The U.S. government's open system environment profile version 3.0* (NIST special publication No. 500-230). Gaithersburg, MD: NIST, Computer Systems Laboratory; Systems and Software Technology Division.

Nedović-Budić, Z. (2001). Geographic information science implications for urban and regional planning. *URISA Journal, 12*(2), 81-93.

Nedović-Budić, Z., & Pinto, J. K. (1999). Interorganizational GIS: Issues and prospects. *Annals of Regional Science, 33*(2), 183-195.

Nedović-Budić, Z., & Pinto, J. K. (2000). Information sharing in an interorganizational GIS environment. *Environment and Planning B: Planning & Design, 27*(3), 455-474.

Onsrud , H. J., & Rushton, G. (1995). *Sharing geographic information*. New Brunswick, NJ: Center for Urban Policy Research.

Palmer, C. L. (2005). Scholarly work and the shaping of digital access. *Journal of the American Society for Information Science and Technology, 56*(11), 1140-1153.

Palmér, M., Naeve, A., & Paulsson, F. (2004). The SCAM framework: Helping semantic Web applications to store and access metadata. In C. Bussler, J. Davies,

D. Fensel, & R. Studer (Eds.), *The semantic Web: Research and applications: first European semantic Web symposium, ESWS 2004 Heraklion, Crete, Greece, May 10-12, 2004, proceedings* (*Lecture Notes in Computer Science, 3053*, pp. 167-181). Berlin, Germany: Springer-Verlag GmbH.

Pickles, J. (1995). Representations in an electronic age: Geography, GIS, and democracy. In J. Pickles (ed.), *Ground truth: The social implications of geographic information systems* (pp. 1-30). New York, NY: The Guildford Press.

Ravindranathan, U., Shen, R., Gonçalves, M. A., Fan, W., Fox, E. A., & Flanagan, J. W. (2004). Prototyping digital libraries handling heterogeneous data sources: The ETANA-DL case study. In R. Heery, & L. Lyon (Eds.), *Research and advanced technology for digital libraries: 8th European conference, ECDL 2004, Bath, UK, September 12-17, 2004, proceedings* (*Lecture Notes in Computer Science, 3232*, pp. 186-197). Berlin, Germany: Springer-Verlag GmbH.

Reid, J. S., Higgins, C., & Medyckyi-Scott, D. (2004). Spatial data infrastructures and digital libraries: Paths to convergence. *D-Lib Magazine, 10*(5). Retrieved July 2006, from http://dlib.anu.edu.au/dlib/may04/reid/05reid.html

Schuurman, N. (2000). Trouble in the heartland: GIS and its critics in the 1990s. *Progress in Human Geography, 24*(4), 569-590.

Schuurman, N. (2003). The ghost in the machine: Spatial data, information and knowledge in GIS. *The Canadian Geographer/Le Géographe Canadien, 47*(1), 1-4.

Schuurman, N. (2006). Formalization matters: Critical GIS and ontology research. *Annals of the Association of American Geographers, 96*(4), 726-739.

Sheth, A. (1998). Changing focus on interoperability in information systems: From system, syntax, structure to semantics. In M. F. Goodchild, M. J. Egenhofer, R. Fegeas, & C. A. Kottman (Eds.), *Interoperating geographic information systems* (pp. 5-30). Dordrecht, Netherlands: Kluwer Academic Publishers.

Sheth, A., Avant, D., & Bertram, C. (2001). *System and method for creating semantic Web and its applications in browsing, searching, profiling, personalization and advertisement, Patent # 6,311,194*. Washington, DC: United States Patent Office.

Sheth, A., Bertram, C., Avant, D., Hammond, B., Kochut, K., & Warke, Y. (2002). Managing semantic content for the Web. *IEEE Internet Computing, 6*(4), 80-87.

Sigel, A. (1998). *Towards knowledge organization with topic maps*. Paper presented at the annual meeting of the SGML/XML Europe 98: "From Theory to New Practices" Paris, France: Graphic Communications Association (GCA). Retrieved November 2006, from http://www.infoloom.com/gcaconfs/WEB/paris2000/S22-02.HTM

Smart, P. D., Abdelmoty, A. I., & Jones, C. B. (2004). An evaluation of geo-ontology representation languages for supporting Web retrieval of geographical information. In *Proceedings of the GIS research UK (GISRUK) 12th annual conference* (pp. 175-178). Norwich, UK: University of East Anglia. Retrieved February 2006, from http://www.geo-spirit.org/publications/ psmart-gisruk04-final.pdf

Sondheim, M., Gardels, K., & Buehler, K. (1999). GIS interoperability. In P. A. Longley, M. F. Goodchild, D. J. Maguire, & D. W. Rhind (Eds.), *Geographical information systems: Principles, techniques, applications, and management* (pp. 347-358). New York, NY: John Wiley & Sons.

Stoimenov, L., & Djordjevic-Kajan, S. (2005). An architecture for interoperable GIS use in a local community environment. *Computers & Geosciences, 31*(2), 211-220.

Tawil, A.-R. H., Fiddian, N. J., & Gray, W. A. (2001). Domain-specific metadata a key for building semantically-rich schema models. In H. S. Kunii, S. Jajodia, & A. Sølvberg (Ed.), *Conceptual modeling: ER 2001: 20th international conference on conceptual modeling, Yokohama, Japan, November 27-30, 2001, proceedings* (*Lecture Notes in Computer Science, 2224*, pp. 225-240). Berlin, Germany: Springer-Verlag GmbH.

Taylor, P. J. (1990). Editorial comment: GKS. *Political Geography Quarterly, 9*(3), 211-212.

Tulloch, D. L., & Shapiro, T. (2003). The intersection of data access and public participation: Impacting GIS users' success? *URISA Journal, 15*(APA II), 55-60.

Wehn de Montalvo, U. (2003). In search of rigorous models for policy-oriented research: A behavioral approach to spatial data sharing. *URISA Journal, 15*(1), 19-28.

Yi, K., Beheshti, J., Cole, C., Leide, J. E., & Large, A. (2006). User search behavior of domain-specific information retrieval systems: An analysis of the query logs from PsycInfo and ABC-Clio's Historical Abstracts/America: History and Life. *Journal of the American Society for Information Science and Technology, 57*(9), 1208-1220.

Chapter VII

Reference Services

Ardis Hanson, University of South Florida Libraries, USA

Introduction

Geographers often define the spatial parameters of different environments by integrating diverse data sets with locational coordinates to create an attribute-rich digital geospatial dataset. From these datasets, researchers can observe and record phenomena and create new geographic metaphors in describing the geographic spaces of places, physical or virtual. One challenge for librarians is to be cognizant of not only the spatial extent of geospatial data, but to have the ability to interpret the socioeconomic milieu, which characterizes the attribute data and the environment in which it describes and to make it accessible for the user community. A second challenge is how to reinterpret traditional patron interactions in an increasingly online service environment and the best use of applications that must be used to "push" information to the patron. A third is how to market the library's GIS services and resources to our patrons, both old and new.

A fourth challenge is how instruction and instructional support must be reconfigured to meet the needs of a variety of users, including librarians, with a range of skills and knowledge levels. The final challenge is instructing patrons to achieve appropriate levels of information competencies (Abbott & Argentati, 1995). This chapter will examine the "new world" that reference librarians find themselves in today, how they may approach the challenges of geospatial reference, and inculcate information competencies and lifelong learning skills in their patrons.

Spatial Thinking

Any discussion on reference services must begin with a definition of what it means to work with geospatial data. To work effectively with geospatial data requires the librarian and the user to understand spatial concepts, tools used to create representations of geospatial data, and the cognitive processes used to frame questions in a geospatial manner. The Geographical Committee of the National Research Council suggests, "It depends on understanding the meaning of space and using the properties of space as a vehicle for structuring problems, for finding answers, and for expressing solutions. By visualizing relationships within spatial structures, we can perceive, remember, and analyze the static and, via transformations, the dynamic properties of objects and the relationships between objects. We can use representations in a variety of modes and media (graphic [text, image, and video], tactile, auditory, and kinesthetic) to describe, explain, and communicate about the structure, operation, and function of those objects and their relationships" (National Research Council, Geographical Sciences Committee, 2006, p. 3). Many of the elements identified in this passage are currently activities done by reference librarians, such as "structuring problems" are a function of the reference interview. However, for many librarians, providing reference services using geospatial data is a new, sometimes overwhelming experience.

There are different levels of GIS services libraries can offer. These range from high-level, which require a full GIS set-up, to mid-level, which utilize Web-based GIS applications and require user input, to low-level, which uses online static maps (Kowal, 2002). In 1997, the Association of Research Libraries (ARL) surveyed their research library members who participated in the GIS literacy project. Of the 72 respondents, 64 reported that they provide GIS services, either at the library or at academic departments offering GIS (Association of Research Libraries, 1999). Among libraries that offer GIS services but do not administer them, the most common activity is offering guidance in finding appropriate data sets. For those libraries that do not have a GIS unit (generally located in government documents or map libraries), GIS services are offered at the general reference desk (Association of Research Libraries, 1999). Librarians use a variety of hardware (computers, printers, digitizers, external storage devices, large format plotters, or scanners); platforms (Windows, MacIntosh, UNIX, and DOS); and software (ArcView, predominantly). The training offered by the Literacy Project was aimed primarily at documents and map librarians (Association of Research Libraries, 1999). Most of the respondents also had training by GIS software providers or GIS coursework. Technical support for GIS hardware or software was provided by library staff at 51 institutions (80%) (Association of Research Libraries, 1999). GIS patrons were primarily students and faculty/staff, although there were a number of users from community, business, and government (Association of Research Libraries, 1999).

A 2005 survey examined the use of GIS implementation within 168 academic libraries in institutions classed as Master's Colleges and Universities I and II (2000 Carnegie Classification) (Kinikin & Hench, 2005a). Almost a third of the respondents either currently support (13%, twenty-two libraries) or plan to implement GIS services in the future (16%, twenty-seven libraries) (Kinikin & Hench, 2005a). Only half of the libraries surveyed have staff devoted to GIS part time. Librarians had to have knowledge of a range of hardware (computers, printers, scanners, servers, GPS units, and plotters), platforms (Windows, Macintosh, Unix), software (ArcView, LandView IV, ArcGIS, ArcInfo, MapInfo, GRASS, Community 2020, MapArt/Adobe Illustrator, Spatial & Image Analyst (ArcView Extensions), ArcExplorer, and Idrisi/Erdas). Further, training is variable, from no training, to training outside of the library, to on-site training that itself ranged from one-on-one instruction to the use of group or individual tutorial programs (Kinikin & Hench, 2005a). Further, the collection resources ranged from Internet resources, software package data, reference databases with spatial data, archives, and institutional research (Kinikin & Hench, 2005a). Staffing levels were widely divergent, ranging from full-time employees who assist with GIS in addition to other duties to student workers who were knowledgeable about GIS. Finally, academic patrons' interests, specialties, and areas of interest included geography, geology, natural resources, business, engineering, sociology, political science, environmental science, biology, landscape architecture, planning, history, and nursing (Kinikin & Hench, 2005a).

The findings of these, and other surveys, touch on four of the five challenges mentioned in the introduction to the chapter: (1) interpreting the socioeconomic milieu, which characterizes the attribute data and the environment in which it describes and making it accessible for the user community, (2) reinterpreting the reference interview, (3) reconfiguring instruction and instructional support, and (4) instructing patrons to achieve appropriate levels of information competencies.

Interpreting the Milieu of Spatial Data

Traditional reference is "a mediated, one-on-one service that intervenes, and stands ready to intervene, at the information seeker's point of need. That need ... is every information seeker's universal predicament of wanting to move forward (cognitively) but being unable to progress until some missing information is found" (Lipow & Schlachter, 1997, p. 126). Today's information seekers want and need that gap filled with as little interruption as possible, so they can continue where they left off. Further, users want support in *their use of information* instead of *their seeking of information*. Today's users of geospatial data also require simple yet stable means of data distribution, innovative data discovery and interpretation mechanisms, and enhanced spatial content publishing, including the means to create and propagate spatial data beyond the academy while retaining local control of the data content. Multilingual support is becoming a necessity with geospatial content crossing boundaries and users forging new, global partnerships.

For this new patron group, librarians need to be well versed in geospatial and geographic data discovery tools and support. A quick review of job opportunities illustrate minimum requirements for librarians working with geospatial data: develop and manage spatial and numeric data resources and services; support teaching, research, and engagement; provide consultation to faculty and students; develop instructional modules; handle and create data documentation and research resources; and establish configurations for GIS workstations. This has implications for the traditional roles of librarians in collection development, which now has an intensive data focus; cataloging and creating metadata; current awareness and selective dissemination of information; scholarly activity (publications and presentations), and university service via committees, task forces, and teams.

Reference services now range from finding and selecting data to providing technical support, from troubleshooting application- and network-based questions to facilitating licenses and software installations. Instruction is also increasing, from comprehensive group workshops and trainings to individual, intensive instructional events. The content of these workshops range from highlighting new resources to targeting the reference/technical support continuum. Outreach services are also in demand, as libraries market their resources and services to very diverse clientele across colleges and disciplines. Outreach also creates new partnership opportunities, as data-intensive labs merge with libraries to form joint-use facilities. New functionality in geospatial data use, through new protocols, such as simple object access protocol (SOAP) and application program interfaces (API), continue to create new opportunities for librarians in the provision of reference services. Expanding reference services to research services is an emerging area, as librarians become essential members of research collaborations and multiuniversity partnerships. What is new for reference librarians is the emphasis on product development, a focus on intensive and in-depth consultation, and the transition to being a member of a research project.

Traditional libraries provided reference service for paper-based information resources, such as maps, statistical data, books, journals, and archival collections. Although maps are frequently housed in government documents or a map library, other materials that have a connection to geographic or geospatial data are dispersed throughout the library's collections (physical locations, such as reference, documents, general, special, and virtual locations on servers, desktops, laptops, and phones). Since today's researchers often have a geographic or geospatial component to their reference/research questions, reference and research consultation assistance with queries involves GIS, spatial, and numeric data, and traditional cartographic materials as well as "fugitive" print materials in books, documents, and journals. A question on assessing a neighborhood, for example, may involve demographic and outcomes data, service data, geographic data, infrastructure data, and resource data[1], which may be in numeric datasets, reports, maps, or even in photographs.

Reinterpreting Traditional Patron Interactions

If most questions asked are based on "what" and "how" in reference to a particular location and time, a GIS has two characteristics that allow it to answer these types of questions: (1) the ability to apply spatial operators to the data and (2) the ability to link data sets together (Hanson, 2001). For example:

The first question, "What is at...?", asks what exists at a particular location, say a neighborhood, a county, or a state. That place can be described in a variety of ways as mentioned above, e.g. place name, postal code, geographical co-ordinates, a local Cartesian co-ordinate system, or census block. The second question, "How do I get from ... to ...?", links two locations, such as the particular characteristics between point A and point B. The third question, "Where is this condition true [or not true]?", asks where is that place where certain conditions are satisfied, for example, is there a daycare center by bus route #37. The fourth question, "What has changed since ...?", may be based on the previous three questions but looks for the results for the two moments in time. Question five, "What are the pattern(s)?" determines patterns, such as clusters of mental or physical illness and the existence of service centers. Question six, "What if...?", tries to determine what will happen if something new is added. (Hanson, 2001, p. 50)

As librarians provide more resources, services, and technologies to patrons, the increase on more specialised, targeted reference questions and consultations also increase (Cardina & Wicks, 2004). Further, the range and depth of each reference interaction may differ significantly based upon patron, services offered, training provided, staffing, and the question asked. Kowal (2002) suggests that the type of patron experience and knowledge will require the reference librarian to gauge and match service to user need. For example, a high-level service user has a holistic understanding of his or her information needs, is computer and GIS-application literate, and is able to analyse and identify information from primary and raw data. A mid-level service user uses dynamic, Web-based GIS resources to answer specific questions. A low-level service user uses static maps generated off Web-, CD-ROM-, or DVD-based resources.

Morris (2006) describes the importance of content issues in the performance of a reference interview. The data extent, or coverage of the study area, is often the first question, followed by thematic content (Morris, 2006). Currency of the information and its relationship to other data/information used in the project is an important factor in selecting content. The availability of the attributes of the data, such as percent of street addresses available for a specific project, or if the data can be easily geocoded, is also key (Morris, 2006). The format of the data determines if it can be used "as is." It also determines if conversion is necessary or if there will be unacceptable data loss when converted. Coordinate systems may need to be reprojected while data may need conversion (Morris, 2006). Is there information accompanying the dataset that describes if these activities need to be performed or how best to perform them? Ease of access is also a basic question, from one large download to numerous extractions of data that are then recompiled (Morris, 2006). Licensing and pricing concerns may also require examination of redistribution rights for faculty and/or researchers (Morris, 2006). Additional questions may concern types of service, such as image, feature, or geocoding, access protocols, and reliability and uptime (Morris, 2006). Also, for all beta or demonstration content/services, *caveat emptor*.

Kinikin and Hench's survey of smaller academic libraries (Carnegie Classification Master's Colleges and Universities I & II) reported significantly different results from that of larger academic libraries. Of the 11 smaller academic libraries surveyed, 8 of the libraries indicated fewer than two users per week with only one library reporting more than five users per week (Kinikin & Hench, 2005). Houser (2006) describes the joint Library and GIS/Data Lab services at the University of Kansas Libraries. After the integration of the Map Library and GIS lab services, the number of data consultations with library staff increased from 82 to almost 160 interactions during 2003-2004, with the 2005 monthly averages showing an increase of 16% over the previous year (Houser, 2006). As with the Kinikin and Hench study, more granular statistics were not available. However, a more detailed, longitudinal study from the Yale Map Collection GIS Service indicated that the average amount of time spent per consultation with a GIS librarian is about 4 hours (Parrish, 2006). The Yale statistics also indicate that faculty and doctoral students typically average about twice as much time per consultation than undergraduate or master's level students. The Yale Map Collection GIS Service staff average approximately 155 consultations a year, with faculty and students generally making four visits each, most face-to-face (Parrish, 2006). However, it is not uncommon during complex, long-term projects for the librarians to have multiple consults with a single patron. The Yale statistics show that 30 consults between one patron and a GIS librarian (Parrish, 2006).

Abbott and Argentati (1995) saw the provision "of access to spatial information and software tools and facilitating intellectual access to GIS concepts and applications" as a "natural evolution and extension of existing library services and expertise" (p. 251). However, researchers and librarians have also warned about the complexity and technical difficulty of GIS reference. For example, a staff person may need a minimum of at least 20 hours to complete a GIS tutorial that will allow him or her to begin working on complicated research problems (Deckelbaum, 1999). One also must be practical. Finding or processing spatial data usually takes twice as long as initially expected (Sweetkind-Singer & Williams, 2001).

Basic cartographic elements and their placements require a knowledge of map display and rudimentary statistics to insure that appropriate statistical measures are used to calculate and display information (Deckelbaum, 1999). Reference librarians also need to be aware that paper maps may themselves be essential in determining preliminary data patterns or trends, for collecting additional data, or may themselves be repurposed to digital formats (Deckelbaum, 1999).

Visualization and analysis of digital spatial data is in itself a complex process, involving the understanding and use of legacy materials in other formats, and requiring the reference librarian to delve into new, uncharted areas or to think in new ways about content and context of information. Questions may be asked at all stages of a research process or arise serendipitously.

In addition to the traditional interview, reference librarians will need to know other service parameters for their GIS users. These include basic access questions, such as whether they have access to the necessary data or what are the rules regarding access to a GIS laboratory (Houser, 2006). Other issues include knowing the appropriate software and hardware for that researcher, if there is enough data storage capacity, and how data may be distributed (Houser, 2006). Reference librarians may have to determine if there is a need to obtain GIS assistance from campus staff or other expert users and the types and levels of training and related resources required for the project (Houser, 2006). Librarians may be asked about purchasing and licensing of GIS software, if there are campus GIS listservs, and what commonly used datasets are housed and maintained by the library (Houser, 2006). The availability of grant sources to support GIS development and use in research projects, collection development, staff training, and new services will be a frequently asked question by doctoral students and research and teaching faculty.

Traditional maps and statistical publications are static resources, with no capability for the researcher or librarian to merge, create, or modify data spontaneously. GIS adds an opportunity for librarians to become producers of information (Pfander & Carlock, 2004) as well as enter into collegial, scholarly relationships with fellow academic researchers. However, entering into new publishing or scholarly relationships require reference librarians to understand not just the traditional reference question but also the acquisition and use of geospatial resources. A clear understanding of intellectual property, data redistribution rights, and derivative rights are critical in the reuse or manipulation of geospatial data, whether public domain, public sector, or leased proprietary data. Opportunities and challenges presented by new information technologies redefine intellectual property issues around the world, almost on a daily basis. Since researchers have responsibilities to promulgate their research, retaining rights over how their results are used is essential, especially for long-term or multiyear studies. Intellectual property issues can influence their ability to distribute their products. Reference librarians who work with researchers need to be current to be effective.

For example, national and international projects may be affected by confidentiality clauses that may exist in software licences, use and access agreements for geospatial data, project documents, or funding arrangements. For librarians who are acquiring, supporting, or providing access to researchers, contract terms and/or license language will need to accommodate different countries or different legal jurisdictions. Longhorn *et al.* (2001) suggest that "[b]oth staff and institutions should recognize their rights and responsibilities in such cases, and stated policies should be in place, including appropriate non-disclosure and confidentiality agreements and forms, both in contracts of employment and perhaps even on a project basis" (p. 29). Since university policies vary, the reader is asked to consult the General Counsel's Office of their university as well as to review the policies of the corporate entity owning rights to the original datasets.

Ethics also come into play. Researchers conducting research with groups of people must go under an institutional review board (IRB). Any study involving observation of or interaction with human subjects that originates at an academic or research setting, including course projects, reports, or research papers, must be reviewed and approved by the IRB for the protection of human subjects in research and research-related activities. The IRB ensures that data about study participants are kept confidential, secure, and, if health related, that the data complies with the parameters of the federal Health Information Portability and Accountability Act. Further, in academic libraries that host human subjects datasets, researchers are responsible for obtaining approval of their research projects from the appropriate local or private IRB responsible for assuring protection of human research subjects. For example, if a library hosts research data from a local or national study, and students need to access it as part of a class project, IRB approval may be needed before access to the system is provided. Librarians should be familiar with what requirements must be formally filed and approved, as in the IRB, for access to certain datasets or even the GIS, if bought for that project.

Understanding how to operate within current intellectual property and copyright concerns is also crucial in developing resource management policies and procedures as part of a research team when working with multiple private and public sector vendors. If the reference librarian has part of his or her time assigned to a research project, he or she may be considered a research collaborator. Either as an official or unofficial member of a research project, or as a reference guide, the librarian may be held accountable should liability issues arise. As a result, librarians seek to adapt their behavior, resources, and services to better serve their clientele and define their own roles.

Evaluating and Assessing Services

Library measurement has been defined as "the collection and analysis of objective data describing library performance on which evaluation judgments can be based" (Van House, Weil, & McClure, 1990). It has been suggested that librarians use effectiveness, cost effectiveness, cost benefit, and performance measurement as measures for evaluation which can be related to inputs, outputs and outcomes (Lancaster, 1992; Vickery & Vickery, 1987). Inputs are system resources, such as library budgets and finances, staffing, and print and digital resources. Outputs are the activities the system provides, such as transactions, hours of access, the availability, use, and usability of resources. Although there have been numerous reinventions of the role of librarians, the quantitative and qualitative measures by which they determine how well they are performing have not changed. Input measures, such as volumes held, serial subscriptions, expenditures, and staffing are not correlated to output measures, typically transactions (reference circulation, interlibrary loan, and bibliographic instruction) (Kyrillidou, 2002). This is compounded by counting data downloads that can only measure what the library does have, not what it does not. More rigorous market analysis may be required to adequately assess use, identify shortcomings, and predict demand.

When creating a new service area or services, quantitative and qualitative measures allow library management to determine the implications of a major change in order to deal with all of the elements that must be addressed (Arsenault, Hanson, Pelland, Perez, & Shattuck, 2003). Using quantitative indicators, such as frequencies of interactions, and qualitative indicators, such as effectiveness, libraries can "increase their visibility, restructure to meet the needs of their users, achieving their objective of remaining the preeminent source of information within the academy" (Dewhurst, 1985, p. 1838). In addition, statistics and measures are powerful planning tools that show emerging trends that require additional resources, workflow processes, training needs, staffing changes, and entrepreneurial opportunities. They also allow comparison with peer institutions and aid areas planning a new service to project staff, instructional, and technology demands.

Problems with library measures are legion. Traditional counts of productivity may underestimate the actual volume of work performed, as the actual volume of business is becoming more complex to gauge (Hernon & Altman, 1998). Ways of counting are problematic. Few libraries use their reference counts to determine frequently asked questions or monitoring changes in the types and complexity of questions asked. In the interest of accumulating higher numbers, these qualitative aspects are often lost. Even if the question comes through an "Ask-A" service, there is no systematic coding or review of the questions, recording the length of time spent on answering it, or the number of and types of resources consulted to answer a particularly exotic question. Repurposing and measuring existing operations may provide more effective library measures. For example, Steinhart's (2006) description of the benefits of sharing data provides additional measures that could address institutional and academic support as well as outreach and partnerships. The number of interorganizational activities and reuse of geographic data by external organizations certainly count as outreach or academic-community partnership activity. Data error correction can be counted as response to user feedback and the use of geospatial data may be counted as part of public distribution requirements (Steinhart, 2006). Training and education also measure professional development as well as meeting institutional or organizational competencies in the use of geospatial data.

Definition of quality of services also varies and, in the library literature, the concepts of quality and satisfaction are used interchangeably, even though the two are not necessarily the same (Hernon & Nitecki, 2001). Further, instruments meant to measure service quality are often better measures of satisfaction. However, service quality is seen as the antecedent of satisfaction, that is, without service quality, satisfaction cannot be measured (Cronin & Taylor, 1992). Assessing customer satisfaction may be more critical if there are other service providers or competing resources for users than just the library's GIS services and resources. Sometimes, the difference is never clarified on the questionnaire or survey.

Take the example of the service quality measure, LibQual+™, created in 1999 as a standardized measure of library service quality by the Association of Research Libraries (ARL) in collaboration with Texas A&M University. LibQual+™, measures library users' evaluations of service quality on four dimensions: "Affect of Service," "Library as Place," "Personal Control," and "Access to Information." The structure of LibQual+™ mirrors that of SERVQUAL™, an industry- based service performance measure. Although LibQual+™ is described as a measure of service quality rather than satisfaction, users of the questionnaire frequently consider it a measure of the latter. In addition, many library measures have a problem with validity and reliability. For example, performance-only assessment is the most valid framework for gauging customer satisfaction when using LibQual+™ or SERVQUAL™ (Landrum & Prybutok, 2004; Roszkowski, Baky, & Jones, 2005). By eliminating the difference scores in these instruments, the problems with reliability and validity are eliminated and respondents will need less time to complete them (Landrum & Prybutok, 2004). Another suggestion is to use a shorter version of the LibQual+™ with just performance-only ratings, as in the case of the SERVPERF, the performance-only variant of SERVQUAL (Roszkowski, Baky, & Jones, 2005).

Libraries may also choose to use assessment guidelines in higher education as a model for library or service assessment. Dow (1998) describes how the University of Rochester, a Carnegie 1 research university, used the American Association of Higher Education's (AAHE) *Principles of Good Practice for Assessing Student Learning*, as part of its strategic assessment. The principles identified included effective ongoing assessment, asking questions that people cared about, attentiveness to outcomes, and gathering evidence as to how and to what degree its programs influenced the quality of effort that students and faculty invested in learning experiences (Dow, 1998). This was a departure from traditional resource-based input and output measures to define quality and impact.

The ARL New Measures agenda examines learning outcomes, research outcomes, institutional outcomes, and personal control or electronic service quality issues. Smith (2000) examined how accreditation agencies were identifying student outcomes while tying them back into institutional objectives. He suggests that a library should understand its students, not just what their knowledge and skills are when they arrive, but also what challenges they face in the future as to learning. He also suggests that a library should understand what learning outcomes are required for student success and on which dimensions of learning it should focus its efforts on. He further suggests that a library effectively measure the extent to which outcomes are achieved as well as those that are not achieved (Smith, 2000). After all, without understanding where problems lie, there is no opportunity to correct or change them. Finally, after correcting the problems or making identified changes, is the library meeting its defined outcome? (Smith, 2000).

Franklin (2001) examines the relationship between a library and sponsored research activities within university settings through a review of 20 years of KPMG (an internationally known consultancy) reports. Using three indicators, total research and development funding at a university (total R&D funding); total library expenditures (total Library dollars), and library expenditures in support of sponsored research as a percentage of total library expenditures (percent of Library dollars), there was a high correlation between total R&D funding and total library expenditures (Franklin, 2001). Little or no correlation was found between total library expenditures and percent of library expenditures in support of sponsored research. Little or no correlation was also found between an institution's R&D funding and percent of library expenditures in support of sponsored research (Franklin, 2001). This may be due to the fact that few libraries track expenditures other than by departmental/college codes and do not track by contract or grant codes.

Reconfiguring Instruction, Training, and Instructional Support

In 1966, four skills were considered critical for understanding GIS: literacy (text), articulacy (spoken words), numeracy (mathematical notation), and graphicacy (visual communication) (Balchin & Coleman, 1966). All four skills enhance one's ability to communicate effectively and understand those relationships that cannot be expressed solely with text, spoken words, mathematical notation, or visually. Text, mathematics, graphics, and discourse complement one another, and all are necessary in various contexts for the successful comprehension and exchange of ideas. Balchin and Coleman argue "neither words nor numbers nor diagrams are simpler or more complex, superior or inferior. They are only more suitable or less suitable for particular purposes" (1965, p. 25). For example, users may have difficulties in conveying their information needs to learn content-based image retrieval systems (Yang, 2004) or to deconstruct a map. Compounding communication is recent research that suggests only a very small percentage of the general population prefer to learn by reading (Weiler, 2005). Clearly, understanding when a particular communication mode is more appropriate than another is and how to use each mode effectively become essential knowledge-sets and skill-sets for librarians.

Creating Patterns: Information-Seeking Models

Digital geospatial data archives and repositories are directly related to specific question-answering research, which "reflect[s] the information needs of their creators" and "tend[s] to be tailored to the specific cultures of research communities, rather than adhering to existing blueprints for access resources" (Palmer, 2005, pp. 1142-1143). Since disciplines play an important role, especially in the influence of context and search task on information seeking (Solomon, 2002), the impact of the different types of conceptual models for information seeking and retrieval will impact how librarians themselves learn and instruct others in the use of geospatial information (Järvelin & Wilson, 2003). Such models serve different research purposes. What kind of models are there and in what ways may they help librarians to better instruct, train, and support GIS research and reference activities? Are different types of models needed for various purposes? Differences in discipline affect faculty and student information seeking skills. Individuals in the hard sciences, such as engineering and math, use a search strategy based on a specific problem-solving process while individuals in the humanities are more likely to be based on a topical interest that is extensively researched (Palmer, 2005). Understanding diverse analytical models of task-based information seeking can contribute to the development of a research area, that is, "Conceptual models may and should map reality, guide research and systematise knowledge, for example, by integration and by proposing systems of hypotheses" (Järvelin & Wilson, 2003, ¶9).

One such critical model on information-seeking behaviours was developed by Ellis based on Glaser and Strauss's grounded theory for data analysis (Ellis, 1989). He determined six generic information behaviors: *starting, chaining, browsing, differentiating, monitoring,* and *extracting.* Each of these activities can involve a librarian. For example, *starting* is comprised of the initial steps to finding information, such as a review of indexing and abstracting sources or consultation with a librarian in starting a literature review. *Chaining* is following "chains of citations" (referential connections) between sources identified, such as following references from an initial source or following other sources that refer to an original source. *Browsing* includes scanning of published journals and tables of contents, as well as references and abstracts of other literature reviews or bibliographies. *Differentiating* is qualifying sources of information to ensure relevance in one's information search. *Monitoring* is a current awareness activity in which one regularly examines core journals, conferences, new books, and catalogs of resources via hand searching or by using current awareness services. *Extracting* selectively identifies and acquires relevant material from resources (e.g., sets of journals, series of monographs, collections of indexes, abstracts or bibliographies, and computer databases). Ellis (1989) notes "… *the detailed interrelation or interaction of the features in any individual information seeking pattern will depend on the unique circumstances of the information seeking activities of the person concerned at that particular point in time" (p. 178).*

Similar to Kuhlthau's model in terms of the various types of activities or tasks carried out within the overall information-seeking process (Kuhlthau, 1988; Kuhlthau, 1991; Kuhlthau, 1993), Ellis's empirically based model is considered authoritative in its classification of types of information seeking behaviors. Building on Ellis' work, four new areas have been added to his model: *accessing, verifying, networking,* and *information managing* (Meho & Tibbo, 2003). With the increasing dependence on online resources, Meho and Tibbo suggest that, "Although not all of these new features are information searching or gathering activities, they are tasks that have

Figure 1. A process chain of Ellis's information seeking activities

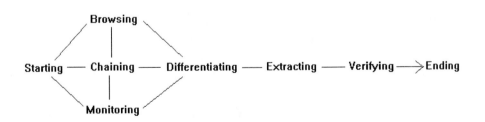

Figure 2. Meho and Tibbo's expanded model based on Ellis information activity model

SEARCHING
Starting, Browsing, Chaining, Differentiating, Monitoring, Extracting, Verifying, Networking

ACCESSING
Decision-making

♫ return to search
↳ forward to Processing

Information-Managing

PROCESSING
Chaining, Extracting, Differentiating, Verifying, Information-Managing, Syntheses And Analyses,
♫ return to Search, Processing
↳ forward to Ending

ENDING
Research Product Completed

Constant Motion Between Stages As Information Needs Are Identified or Change

significant roles in enhancing information retrieval and facilitating research" (p. 583). *Accessing* is a critical component of information seeking because without access, one may miss essential pieces of data to answer one's question. In addition, given the importance of archival materials and other forms of primary sources in geospatial data, librarians may need to investigate why access to information is denied and to intervene to obtain the items. *Verifying* are those activities that establish the accuracy and authority of the data or information (Meho & Tibbo, 2003). Librarians and researchers should consider bias and lack of reliability and accuracy in materials produced by governmental agencies as well as private sector organizations. Detailed metadata on lineage of data can alert users to discrepancies in data and in use of such information (see Chapter IV on metadata and VIII on collection development). Data comparison with other sources can also point out errors in data. *Networking* are those activities associated with communicating or maintaining relationships with fellow researchers or governmental and non-governmental organisations, or individuals who could provide access to or verify information or data (Meho & Tibbo, 2003). The final information-seeking activity is *information managing*, which emphasizes the importance of filing, archiving, and organizing the information used to facilitate one's research (Meho & Tibbo, 2003).

Ellis, Meho, and Tibbo conclude that the activities identified in their studies may not be entirely or always sequential. However, Meho and Tibbo (2003) suggest that the information-seeking behaviors of social scientists are divided into four interrelated stages, *searching, accessing, processing,* and *ending* (pp. 584-585), that then contain a number of activities. For example, during the *searching* stage, all of Ellis' activities and Meho and Tibbo's networking activity begin. During the *accessing* stage, decision-making to continue on to *processing* or to return to *searching* is based on acquisition of requested materials. It is the author's belief that *information managing* also begins in this stage to document acquired items. During the *processing stage*, chaining, extracting, differentiating, verifying, and information-managing activities begin or continue as do syntheses and analyses. There will be constant motion between the searching, accessing, and/or the processing stages until the *ending* stage, when the project is completed (Meho & Tibbo, 2003, pp. 584-585).

Ellis' existing and extended models, which examined primarily social scientists, are expanded upon by Palmer (2005), who suggests that scientists and humanists use primarily four research activities: confirmation searching, discovery searching, collecting, and consultation. Further, there are important distinctions in source selection, decision making, and how materials are used (Palmer, 2005), which can impact reference services and support to research faculty. To act as a consultant, reference staff need to increase significantly their own understanding of the different research processes and information-seeking behaviors.

In addition to activity models, reference librarians also need to be familiar with various cognitive models for information seeking. Ingwersen's cognitive model examines how the search process itself affects information retrieval. He suggests that the all facets of the information-seeking system, from the user's questions, documents, or objects retrieved, interaction with intermediaries, such as search engines, and, by extension, librarians, are the result of explicit or implicit cognitive models of the domain of interest at that particular point (Ingwersen, 1992; Ingwersen, 1996). Ingwersen also suggests that the user experiences various cognitive transformations from where he or she begins the information-seeking process to the point where a search is successful in identifying objects to answer the user's questions (Ingwersen, 1992; Ingwersen, 1996). Therefore, users usually have implicit models of their information need or work task that they can tell individuals or query systems. These cognitive structures and their transformations need to be effectively communicated throughout the information system as identified previously, again which includes the reference librarian. What is even more crucial is the cognitive framework of the reference librarian, since his or her knowledge structures "are determined by the individual and its social/collective experiences, education, etc." (Ingwersen, 1982, p.168). The librarian and, ultimately, the user, are influenced by his or her environment "which may possess social conventions, preferences, and collective cognitive structures adhering to domains" (Ingwersen, 1992, p. 17). This extends to the transformative aspect of knowledge and the importance of effective communication, since "elements of a communicated message must be perceived or recognized, in order to allow the message to transform into a new state of knowledge. This transformation does not necessarily produce a simple accumulation of categories or concepts, but can be seen as a reconfiguration, a restructuring or a compression in part of the recipient's knowledge structures" (p. 17-18).

Järvelin and Wilson (2003) expand on a cognitive model and address how user-perceived tasks must be considered. They suggest that individuals interpret the same objective task differently, particularly for what is perceived to be a complex task. Task complexity entails a number of variables: "repetition, analysability, *a priori* determinability, the number of alternative paths of task performance, outcome novelty, number of goals and conflicting dependencies among them, uncertainties between performance and goals, number of inputs, cognitive and skill requirements, as well as the time-varying conditions of task performance" (Järvelin & Wilson, 2003). How the user perceives the task is the basis for the actual performance of the task, the interpretation of his or her information needs, and the decision making to ensure satisfaction, or successful completion of the task (Järvelin & Wilson, 2003). Järvelin and Wilson consider five classes of tasks. These are automatic information processing tasks; normal information processing tasks; normal decision tasks; known, genuine decision tasks; and genuine decision tasks (¶31). These range from simple computational tasks that are *a priori* completely determinable and structured to decision-making tasks, where information, result, or process is unknown, and the first task must be to create the information-seeking structure.

Byström and Järvelin (1995) examined what types of information are sought through which types of channels, from what kinds of sources, in which kinds of tasks. They determined that as task complexity increases the needs for domain information and problem-solving information increases. Complexity also increases the proportion of general-purpose sources (e.g., experts and personal collections) and decreases that of problem- and fact-oriented sources. Finally, complexity decreases the success of information seeking, decreases the internality of channels, and increases the number of sources (Byström & Järvelin, 1995).

There are a number of studies that show clear differences in student and faculty information-seeking behavior across disciplines (Whitmire, 2002) and in how faculty and students teach and learn (Neumann, Parry, & Becher, 2002). Cole *et al.* (Cole, Leide, Large, Beheshti, & Brooks, 2005) found that domain (subject area) novice users in the beginning stages of researching a topic find themselves searching for information before they have clearly identified their information need. Data also supported their hypothesis that the user's information need is identified through transformation of his or her knowledge structure (i.e., the seeker's cognitive map or perspective on the task for which information is being sought) (Cole et al., 2005). Heinström (2006) investigated whether information-seeking patterns relate to discipline differences, study approaches, or personality traits. Her quantitative study of master's thesis students (n=305) found that both the level and scope of information students wished to retrieve and the way they searched for it determined if searches were exploratory or precise (Heinström, 2006).

The most common patterns seen in users' information seeking can be described along a continuum of broadness (holistic) vs. specificity (serialist). Users with a broad divergent learning style typically start their learning process through an overall understanding of a topic, addressing several aspects of the information need at the same time, relating new information, much like a "big picture" approach. This style is often used to become acquainted with a new area, to build basic knowledge, or to understand causal relations. There can be redundancy in these types of searches, which may not be strictly relevant to the user's information needs (Ford, 1995). This type of information-seeking behavior generally uses a broad search pattern throughout project stages (Edwards, 2003). Users with a convergent rational, problem-focused style use a different learning approach, concentrating on one item at a time. They often generate precise searches, mastering one component of a topic before moving on to the next. For these users, the broad, or big, conceptual picture emerges late in the search (Ford, 1995). Users with this type of "serialist" learning style also generate controlled and systematic seeking with precise goals. Both types of users may perform "effortful information search and processing," believing that the investment of additional cognitive effort in an information processing task is the best way to achieve accuracy (Heinström, 2006). For most geospatial-based queries, a problem-based learning model would also be appropriate. Similar to students using a problem-based learning model (Rankin, 1999), GIS users also tend to be frequent library users, use a wider variety of resources, and ask more complex reference questions. Both types of students also show more independence in information gathering and generally acquire information skills sooner than their peers (Rankin, 1999).

Learning styles have implications in the development of instructional materials for users as well as for librarians. Using instructional materials structured to suit individualistic learning styles is significantly superior for both types of learners, and serialist instructional materials resulted in overall better learning performance and efficiency than did holist materials (Ford, 1995). Adaptive training methodologies have the potential to accommodate individual differences within required time limits to avoid instructional failures, ensure minimal proficiency outcome, and maximize individual proficiencies. Creating instructional models that capitalize on student strengths and match content and structure of training events to the student's conceptual structure provides librarians more insight into the information-seeking and resource-discovery process. Further, the development of adaptive instructional systems, an emerging trend in education, addresses learner styles with personalized training schemes, with a focus on lifelong learning, just-in-time training delivery, and integrated training across campus settings (Rosenberg, 2001). With personalization, the learners' objectives, current status of skills/knowledge, and learning style preferences are targeted to better deliver content as well as to monitor progress in skills sets (Rosenberg, 2001). Another advantage of personalization is the repurposing and scaffolding of materials to create reusable learning objects. This is a critical component in a complex knowledge area such as GIS. Tancheva (2003) suggests that tutorials be problem-based, teach concepts, and be anchored to class-based instruction support or discipline. This is essential since students often lack sound mental models of databases or datasets. Conceptual models ensure the transferability of competencies across systems, and improve synthesis and evaluative thinking (Tancheva, 2003). Interactivity is an attribute highly valued by generation Xs and Ys. It provides active and situated learning through simulations as well as opportunity for online collaboration with other learners and instructors (Rosenberg, 2001). Interactive mapping and direct manipulation of data to create graphic display mirrors many simulation and real-time activities with which users have grown up. Geospatial data is also very media rich, with multiple formats and output styles and indexing or descriptive metadata. Adaptive learning systems permit a variety of presentation styles, and also allow users to index and organize data for their own use (Rosenberg, 2001). Users will wait to the last possible moment to learn a new system for a class project or paper. Just-in-time, or point-of-need, delivery provides training at the exact time and place that it is needed for the user to complete a specific task (Rosenberg, 2001; Tancheva, 2003). Adaptive instructional methods emphasize user-centric environments. In this type of environment, the learner takes responsibility for his or her own learning (Rosenberg, 2001) and the librarian is seen more as an expert guide and facilitator. However, tutorials are best seen as complementary or as supplementary instruction rather than as an effective stand-alone teaching tool (Tancheva, 2003). Gutierrez and Wang (2001) found that "students who regularly used the library benefited the most" from literacy instruction. Their study reinforces the argument that research skills require practice, that is, a single, traditionally conducted library research class will not significantly improve the ability of the user to perform research.

In addition to user education, librarians have important roles in curriculum development (Kiran, 2004). Serving on a curriculum committee provides reference librarians a stronger voice in the planning and implementation of a curriculum's information management component and information resources by students (Kiran, 2004). Librarians also become more involved as a resource person for faculty, instructional support staff, and students, ranging from basic assistance or instruction at the reference desk to resource management and utilization to consultation with faculty and students.

How librarians conceptualize government information should be changed. Cheney (2006) suggests that librarians "teach faculty and student researchers why, not just how, to use government information rather than agency organisation and its function" (p. 307). This concept reduces the emphasis on government information as "a separate specialization, based on its provenance, rather than its subject content" (Cheney, 2006, p. 307). This approach clearly favors the range and breadth of GIS as both a subject within a discipline as well as a discipline in its own right. Understanding the user's information-seeking need, based on identification of the user's discipline and question, can create better use of government information as well as better users of that information. Users generally do not care how the information is organized as to agency, division, or department, just "does it answer my question?" This applies not just to search skills, but also to our "how to use the library on X" lectures, instructions, workshops, and tutorials.

Both broad exploration and specific searching have their particular benefits dependent on the stage of the research process, or a preferred way of learning, which also affect instruction (Ford, 1995). Järvelin and Wilson (2003) emphasize the use of conceptual models to "map reality, guide research and systematise knowledge, for example, by integration and by proposing systems of hypotheses" (¶10). Teaching and research models developed by reference staff should focus on developing problem solving skills, critical thinking skills, and effective information-seeking behaviors. By focusing on how "… information contributes to their critical thinking within the context of a discipline" (Cheney, 2006, p. 307), librarians will be able to create effective instructional and training materials for their users.

Achieving Information Competencies

Whether one calls it information skills (Rankin, 1999) or information literacy competencies (Cunningham & Lanning, 2002), these skills are essential to the learning process. Using geospatial data requires the user to have a range of skills, from determining the nature and extent of the information needed to using information effectively to accomplish specific purpose. Solving problem scenarios and answering geospatially based questions requires proficiency in a number of competency standards that must be completed successfully. Compounding requisite skills sets is the evaluative component of technology and its use. Users must develop skills to judge the reliability and validity of information as well as emerging technologies and methodologies (Tedesco, 1999). Adding to the discussion of literacy is the American Library Association's requirements for information literacy, which has four components. These components focus on the user's ability to recognize when information is needed; the ability to locate the needed information; the ability to evaluate the suitability of retrieved information; and the ability to use effectively and appropriately the needed information (Association of College and Research Libraries, 2000).

However, information professionals must also consider new models of reasoning, new systems or applications, or work processes from the perspectives of both naïve and experienced users (Wells & Hanson, 2003). Shapiro and Hughes (1996) suggest seven dimensions when designing a curriculum that promotes comprehensive information literacy. These dimensions begin with tool and resource literacy, that is, the ability to use print and electronic resources and software and to understand their form, format, and access issues (Shapiro & Hughes, 1996). These are followed by literacies in the context of information in academia, such as the import of discipline and scholarly publications Social-structural literacy, research literacy, and publishing literacy address how information is socially situated and produced, how tools are used to conduct research, and publication and production of research

Table 1. The "information literacy competency" taxonomy

	Knowledge	Comprehension	Application	Analysis	Synthesis	Evaluation
Hardware	What are the hardware components of a system?	What do the components of a hardware system do?	When would the hardware suit my needs?	How does this piece of hardware work?	How would I build this hardware?	What improves hardware design?
Software	What are the software components of a system?	What is the role of software is in a system?	When would the software fit the situation?	How does this software work?	How would I build this software?	What conditions produce quality software?
Data	Where can I get data?	What does this data mean?	When would I use this data?	How is this data interpreted?	How do I appropriately gather the data?	What factors increase the value and reliability of data?
Procedure	What actions can be taken?	What is the purpose of an action?	When would an action occur?	What are the steps of the action?	How would I define the steps of the action?	Which aspects of an action are necessary and which are sufficient?
People	Who are the stakeholders?	What are the roles and relationships of individuals in a situation?	When should an individual become involved?	How is the person responding?	How can the individuals have their responses changed?	What significance does an individual have to the progress of a system?

Reprinted with permission of publisher. Vitolo, T. M., & Coulston, C. (2002). Taxonomy of information literacy competencies. Journal of Information Technology Education, 1(1), 43-51. Retrieved from http://jite.org/documents/Vol1/v1n1p043-052.pdf.

results (Shapiro & Hughes, 1996). The literacies end with emerging technology literacy and critical literacy, which address the adaptation, adoption, and evaluation of information technology in terms of their intellectual and social capital, as well as their benefits and costs (Shapiro & Hughes, 1996).

Considering the intensive computational and technology skills sets necessary to work with geospatial data, additional approaches to literacy may be necessary. Vitolo and Coulston (2002) suggest that analytical and model-based reasoning consider the understanding of relationships among objects, the application of ordering principles to the objects, and the use of basic computational tasks/relevant operations relevant to the relationships and ordering, critical skills for working with geospatial data. Therefore, information literacy, analytical reasoning, and model-based reasoning all begin with the ability to gather data about an environment, to understand cause and effect relationships, and to do deductive reasoning within an environment (Vitolo & Coulston, 2002). They suggest that a new information literacy competency taxonomy can be created by mapping the six levels of the educational objectives of Bloom's Taxonomy (1984, 1956) to the five fundamental units of information systems (Shelly, Cashman, & Rosenblatt, 1998), pp. 1.4-1.6). This provides an expanded way of thinking about not only the skills, but also how one may assist a user to acquire and hone those skills (Vitolo & Coulston, 2002), p. 47). Devised to express educational objectives, "intended behaviors which the student shall display at the end of some period of education" (Bloom, 1984, p. 16), Bloom's Taxonomy ties these educational objectives to larger information literacy competencies, "e.g. intended behaviors *in the context of information literacy* which the student shall display at the end of some period of education." (Vitolo and Coulston, p. 46). This taxonomy allows the librarian to identify those areas in which he or she needs to augment his or her skills, while honing current skills to perform more effectively in the delivery of reference services using GIS.

This new way of looking at skills and competencies makes sense in a statistically intensive, computing-intensive, and quickly evolving field such as GIS. The evolution of the National Spatial Digital Infrastructure has required librarians to deal with multi-institutional standards and the dynamic nature of the geospatial knowledge. It has also required that GIS librarians not only learn, but also understand the theoretical and methodological choices used in geospatial and geoscience research. The end products of GIS research efforts are "large, interdependent databases of specific statistics, high-resolution images, textual data sets, and complicated algorithms that must be sustained in a dynamic computer environment that supports their easy migration from one research application to another" (Shuler, 2003, p. 329). Librarians are natural partners in research collaborations. They are the individuals who seek out datasets, help define information queries, assist and instruct users in resource discovery, and create collaborative arrangements with other academic and research libraries. Shuler (2003) suggests the new geospatial literacy "demands that librarians work in partnership with their community among a new set of knowledge choices, and then preserving a credible record of how well (or badly) those choices were made" (p. 329).

Conclusion/Summary

There are a number of issues that are still of concern in offering reference GIS services. In addition to establishing standards for GIS centers and services to reduce variance in user needs and expectations across academic libraries (Badurak, 2000), reference librarians have several other key opportunities in collection development, instruction on accessing spatial information and software tools, and facilitating user knowledge of GIS concepts and applications (Abbott & Argentati, 1995).

GIS is an example of an interdisciplinary tool that is used across colleges, departments, and academic administration. Regardless of the size of the college or university, libraries may be the point-of-access for students, faculty, and community users. Further, since GIS is data centered and makes use of many secondary and tertiary products (monographs, reports, maps, etc.), the library is a natural choice to establish a GIS service point. More importantly, GIS offers libraries a new venue to create, market, and distribute specialized content.

Finally, as the level of reference services using GIS data increases, so does the level of complexity and base knowledge the librarian must have to interact effectively with advanced users or participate on research teams. This has implications for continuing professional development and library education. However, the rewards of working in a more dimensional manner with students and faculty, as well as highlighting the library's support of research at the university level, are well worth the time and effort involved.

References

Abbott, L. T., & Argentati, C. D. (1995). GIS: A new component of public services. *The Journal of Academic Librarianship, 21*(4), 251-256.

Arsenault, K., Hanson, A., Pelland, J., Perez, D., & Shattuck, B. (2003). Issues for library management when implementing large-scale programmatic change. In A. Hanson, & B. L. Levin (Eds.), *Building a virtual library* (pp. 165-179). Hershey, PA: IDEA Group Publishing.

Association of College and Research Libraries. (2000). *Information literacy competency standards for higher education*. Chicago, IL: ACRL. Retrieved from http://www.ala.org/ala/acrl/acrlstandards/standards.pdf

Association of Research Libraries. (1999). *The ARL Geographic Information Systems Literacy Project: A SPEC kit* (ARL SPEC Kit No. 238). Washington, DC: Association of Research Libraries, Office of Leadership and Management Services.

Badurak, C. (2000). Managing GIS in academic libraries. *WAML Information Bulletin, 31*(2), 110-114.

Balchin, W., & Coleman, A. (1966). Graphicacy should be the fourth ace in the pack. *The Cartographer, 3*(1), 23-28.

Bloom, B. S. (1984). *Taxonomy of educational objectives. Handbook 1: Cognitive domain*. New York, NY: Longman.

Byström, K., & Järvelin, K. (1995). Task complexity affects information seeking and use. *Information Processing & Management, 31*(2), 191-213.

Cardina, C., & Wicks, D. (2004). The changing roles of academic reference librarians over a ten-year period. *Reference & User Services Quarterly, 44*(2), 133-142.

Cheney, D. (2006). Government information collections and services in the social sciences: The subject specialist integration model. *The Journal of Academic Librarianship, 32*(3), 303-312.

Cole, C., Leide, J. E., Large, A., Beheshti, J., & Brooks, M. (2005). Putting it together online: Information need identification for the domain novice user. *Journal of the American Society for Information Science and Technology, 56*(7), 684-694.

Cronin, J. J., & Taylor, S. A. (1992). Measuring service quality: A reexamination and extension. *Journal of Marketing, 56*(7), 55-68.

Cunningham, T. H., & Lanning, S. (2002). New frontier trail guides: Faculty-librarian collaboration on information literacy. *Reference Service Review, 30*(4), 343-348.

Deckelbaum, D. (1999). GIS in libraries: An overview of concepts and concerns. *Issues in Science and Technology Librarianship, 21*(Winter). Retrieved November 2006, from http://www.istl.org/99-winter/article3.html

Dewhurst, W. T. (1985). *Input formats and specifications of the National Geodetic Survey data base*. Rockville, MD: Federal Geodetic Control Committee: U.S. Dept. of Commerce, National Oceanic and Atmospheric Administration.

Dow, R. F. (1998). Using assessment criteria to determine library quality. *Journal of Academic Librarianship, 24*(4), 277-281.

Edwards, J. A. (2003). The interactive effects of processing preference and motivation on information processing: Causal uncertainty and the MBTI in a persuasion context. *Journal of Research in Personality, 37*(2), 89-99.

Ellis, D. (1989). A behavioural approach to information retrieval design. *Journal of Documentation, 46*(3), 318-338.

Ford, N. (1995). Levels and types of mediation in instructional systems: An individual differences approach. *International Journal of Human-Computer Studies, 43*(2), 241-259.

Franklin, B. (2001). Academic research library support of sponsored research in the United States. In *Proceedings of the 4th Northumbria International Conference on Performance Measurement in Libraries and Information Services* (pp. 105-111). Washington, DC: Association of Research Libraries. Retrieved November 2006, from http://www.libqual.org/documents/admin/franklin.pdf

Gutierrez, C., & Wang, J. (2001). A comparison of an electronic vs. print workbook for information literacy instruction. *The Journal of Academic Librarianship, 27*(33), 208-212

Hanson, A. (2001). Community assessments using map and geographic data. *Behavioral and Social Sciences Librarian, 19*(2), 49-62.

Heinström, J. (2006). Broad exploration or precise specificity: Two basic information seeking patterns among students. *Journal of the American Society for Information Science and Technology, 57*(11), 1440-1450.

Hernon, P., & Altman, E. (1998). *Assessing service quality: Satisfying the expectations of library customers*. Chicago, IL: American Library Association.

Hernon, P., & Nitecki, D. A. (2001). Service quality: A concept not fully explored. *Library Trends, 49*, 687-708.

Houser, R. (2006). Building a library GIS service from the ground up. *Library Trends, 55*(2), 315-326.

Ingwersen, P. (1982). Search procedures in the library analysed from the cognitive point of view. *Journal of Documentation, 38*, 165-191.

Ingwersen, P. (1992). *Information retrieval interaction*. London, England: Taylor Graham. Retrieved November 2006, from http://www.db.dk/pi/iri/#the_whole_book

Ingwersen, P. (1996). Cognitive perspectives of information retrieval interaction: Elements of a cognitive IR theory. *Journal of Documentation, 52*(1), 3-50.

Järvelin, K., & Wilson, T. D. (2003). On conceptual models for information seeking and retrieval research. *Information Research: An International Electronic Journal, 9*(1), Article No.163. Retrieved November 2006, from http://informationr.net/ir/9-1/paper163.html

Kinikin, J., & Hench, K. (2005a). Survey of GIS implementation and use within smaller academic libraries. *Issues in Science and Technology Librarianship,* (42), [n].

Kiran, K. (2004). Emerging roles for the librarian in problem based learning. *Journal of Problem-Based Learning, 2*(2). Retrieved December 2006, from http://eprints.rclis.org/archive/00007450/

Kowal, K. C. (2002). Tapping the Web for GIS and mapping technologies: for all levels of libraries and users. *Information Technology & Libraries, 21*(3), 109-114.

Kuhlthau, C. C. (1988). Developing a model of the library search process: Cognitive and affective aspects. *RQ, 28*(2), 232-242.

Kuhlthau, C. C. (1991). Inside the search process: Information-seeking from the user's perspective. *Journal of the American Society for Information Science, 42*(5), 361-371.

Kuhlthau, C. C. (1993). *Seeking meaning: A process approach to library and information services.* Norwood, NJ: Ablex.

Kyrillidou, M. (2002) From input and output measures to quality and outcome measures, or, from the user in the life of the library to the library in the life of the user. *Journal of Academic Librarianship, 28*(1-2), 42-46.

Lancaster, F. W. (1992). *If you want to evaluate your library* (2nd ed.). Champaign, IL: Graduate School of Library and Information Science, University of Illinois.

Landrum, H., & Prybutok, V. R. (2004). A service quality and success model for the information service industry. *European Journal of Operational Research, 156*(3), 628-642.

Lipow, A. G. & Schlachter, G. A. (1997). Thinking out loud: Who will give reference service in the digital environment? *Reference & User Services Quarterly, 37*(2), 125-129.

Longhorn, R.A., Henson-Apollonio, V., & White, J.W. (2002). *Legal issues in the use of geospatial data and tools for agriculture and natural resource management.* A Primer. Mexico, DF International Maize and Wheat Improvement Center (CIMMYT). Retrieved November 2006, from http://www.cimmyt.org/Research/NRG/map/research_results/tech_publications/IPRPrimer.pdf

Meho, L. I., & Tibbo, H. R. (2003). Modeling the information-seeking behavior of social scientists: Ellis's study revisited. *Journal of the American Society for Information Science and Technology, 54*(6), 570-587.

Morris, S. P. (2006). Geospatial Web services and geoarchiving: New opportunities and challenges in geographic information services. *Library Trends, 55*(2), 285-303.

National Research Council, Geographical Sciences Committee. (2006). *Learning to think spatially GIS as a support system in the K-12 curriculum.* Washington, DC: National Academies Press. Retrieved from http://darwin.nap.edu/books/0309092086/html

Neumann, R., Parry, S., & Becher, T. (2002). Teaching and learning in their disciplinary contexts: A conceptual analysis. *Studies in Higher Education, 27*(4), 405-417.

Palmer, C. L. (2005). Scholarly work and the shaping of digital access. *Journal of the American Society for Information Science and Technology, 56*(11), 1140-1153.

Parrish, A. (2006). Improving GIS consultations: A case study at Yale University Library. *Library Trends, 55*(2), 327-339.

Pfander, J., & Carlock, D. (2004). The Arizona electronic atlas: A new reference and instructional tool. *Issues in Science and Technology Librarianship, 41*(Fall). [Special topic issue: Nontraditional Reference Services].

Rankin, J. A. (1999). *Handbook on problem-based learning.* New York, NY: Medical Library Association.

Rosenberg, M. (2001). *e-Learning: Strategies for delivering knowledge in the digital age.* New York, NY: McGraw-Hill.

Roszkowski, M. J., Baky, J. S., & Jones, D. B. (2005). So which score on the LibQual+™ tells me if library users are satisfied? *Library & Information Science Research, 27*(4), 424-439.

Shapiro, J. J., & Hughes, S. K. (1996). Information literacy as a liberal art: Enlightenment proposals for a new curriculum. *Educom Review, 31*(2). Retrieved November 2006, from http://www.educause.edu/pub/er/review/reviewArticles/31231.html

Shelly, G. B., Cashman, T. J., & Rosenblatt, H. J. (1998). *Systems analysis and design* (3rd ed.). Cambridge, MA: Course Technology.

Shuler, J. (2003). On and off the grid: Geographic information science & technology and academic libraries. *The Journal of Academic Librarianship, 29*(5), 327-329.

Smith, K. R. (2000). *New roles and responsibilities for the university library: Advancing student learning through outcomes assessment.* Washington, DC: Association of Research Libraries. Retrieved November 2006, from http://www.arl.org/stats/newmeas/outcomes/HEOSmith.html

Solomon, P. (2002). Discovering information in context. *Annual Review of Information Science and Technology, 36*, 229-264.

Steinhart, G. (2006). Libraries as distributors of geospatial data: Data management policies as tools for managing partnerships. *Library Trends, 55*(2), 264-284.

Sweetkind-Singer, J., & Williams, M. (2001). Supporting the information needs of geographic information systems (GIS) users in an academic library. *Information and the Professional Scientist and Engineer, 16*, 175-191.

Tancheva, K. (2003). *Online tutorials for library instruction: An ongoing project under constant revision.* Paper presented at the annual meeting of the Association of College and Research Libraries Charlotte, NC: ACRL. Retrieved December 2006, from http://www.ala.org/ala/acrl/acrlevents/tancheva.PDF

Tedesco, L. A. (1999). Responding to educational challenges with problem-based learning and information technology. In J. A. Rankin (Ed.), *Handbook on problem-based learning* (pp. 113-120). New York, NY: Medical Library Association.

Van House, N. A., Weil, B. T., & McClure, C. R. (1990). *Measuring academic library performance: A practical approach.* Chicago, IL: American Library Association.

Vickery, B., & Vickery, A. (1987). *Information science in theory and practice.* London, England: Butterworth.

Vitolo, T. M., & Coulston, C. (2002). Taxonomy of information literacy competencies. *Journal of Information Technology Education, 1*(1), 43-51. Retrieved 01/12/02, from http://jite.org/documents/Vol1/v1n1p043-052.pdf

Weiler, A. (2005). Information-seeking behavior in generation Y students: Motivation, critical thinking, and learning theory. *Journal of Academic Librarianship, 31*(1), 46-53.

Wells, A. T., & Hanson, A. (2003). E-reference. In A. Hanson, & B. L. Levin (Eds.), *Building a virtual library* (pp. 95-120). Hershey, PA: IDEA Group Publishing.

Whitmire, E. (2002). Disciplinary differences and undergraduates' information-seeking behavior. *Journal of the American Society for Information Science and Technology, 53*(June), 631-668.

Yang, C. C. (2004). Content-based image retrieval: A comparison between query by example and image browsing map approaches. *Journal of Information Science, 30*(3), 254-267.

Endnote

[1] *Demographic and outcomes date* describes the prevalence of conditions, such as poverty rate, household composition, employment, crime, etc. *Service data* captures what services currently available to residents. *Geographic data* maps the spatial location of the neighborhood by zip code, census tract, local neighborhood boundaries, or other specific designations. *Infrastructure data* describes transportation systems, storm water and sewage facilities, streets and sidewalks, the age of physical facilities, and other aspects of the physical environment of the neighborhood. *Resource data* captures the governance and financing systems that control community resources (Hanson, 2001).

Chapter VIII

Collection Management Issues with Geospatial Information

John Abresch, University of South Florida Libraries, USA

Ardis Hanson, University of South Florida Libraries, USA

Peter Reehling, University of South Florida Libraries, USA

Introduction

Among the most challenging aspects of GIS are identifying needs, acquiring re-sources, and managing the collection, a process that involves decision making in a dynamic and changing environment. Libraries that have traditionally collected maps have a good grounding in many of the issues, yet even they must learn new approaches, new technology, and think beyond the needs of traditional map users. Librarians will find challenges throughout the life of geospatial information, from its acquisition to its disposition, especially as a library collection migrates from a primarily print format to a focus on digital formats.

Building a digital geographic data collection from scratch and acquiring computer software and hardware systems in which to manage and display the geographic in-formation is becoming standard practice. In a study conducted on the integration of GIS in academic libraries that are a part of Carnegie Classification Master's Colleges and Universities I and II, there was significant interest in the use of implementing

GIS services to support academic endeavors (Kinikin & Hench, 2005). However, building such a collection does not mean that librarians must build it in a void. An extensive tradition of collecting geographic materials exists in both public and academic libraries for outlining strategies to build new collections (Larsgaard, 1998; Ristow, 1980). How GIS is used throughout higher education has prompted libraries to examine the issue of providing digital geospatial data resources and services. This chapter will address collection development issues for geospatial data, including establishing a collection development policy, determining user needs and their relationship to resource development, building digital and print geospatial collections, issues in collecting data at the local, state, and federal levels, archival concerns, and legal and licensing considerations.

Creating a Collection Development Policy

Central to the planning process of either adding geospatial information to an existing library collection or assembling a primarily digital geospatial information collection is the creation of a collection development policy. A collection development policy is the instrument a librarian creates and then utilizes that not only defines the collection, but also is a guide to the ongoing management of the collection. Evans and Saponaro (2005) define collection development defined as a "process of making certain the library meets the information needs of its service population in a timely and economical manner, using information resources produced both inside and outside of the organization" (p. 70). Further, "effective collection development requires creating a plan to correct collection weaknesses while maintaining its strengths" (Evans & Saponaro, 2005, p. 70). Therefore, an effective collection development policy is an action plan that is used to assist staff in the acquisition and decision-making process (Evans & Saponaro, 2005).

In their discussion of collection management issues and electronic collections, Pettijohn and Neville (2003) add that "collection development represents not just the acquisition of information, but a strategic investment in knowledge" (p. 21). They feel that the guiding principles, goals, and strategies of this process are formally stated in collection development policies (Pettijohn & Neville, 2003). Further, these policies are based upon an understanding of the strengths and weaknesses of the collection, the availability of shared resources, and the information needs of the community. To define subject coverage, depth, level, and scope, librarians emphasize or exclude specific subject areas, languages, formats, and genres (Evans & Saponaro, 2005). Existing collection development policies may be adapted for use in selecting electronic resources or revised to consider additional formats, features, and evaluative criteria. Policies must consider the virtual library from a dual perspective; it is both a dynamic collection in its own right and a hybrid collection created by merging the virtual and physical libraries (Manoff, 2000). These hybrid collections contain multiple formats, iterations, and archival concerns, with paper maps, atlases, globes, satellite imagery, numeric datasets on CD-ROMs, DVDs, and server-based digital information, archived on a variety of software and hardware.

In creating a collection development policy for geospatial information, librarians need to recognize and synthesize two critical factors that shape the parameters of the policy. These factors are an understanding of the environment in which the collection will be housed and understanding of the needs of the user community. For example, in an academic setting, the user community may vary depending on the mission of the college or university. Librarians also need to be cognizant of what publications, data, media, computer hardware, and software components can populate a potential collection and the extent of its related subject area in academic environments. The concept of library collections extends beyond static locations, as library catalog may involve information resources and collections from a wider virtual arena. Or, as Pettijohn and Neville (2003) suggest, "Ultimately, the goal of collection development in academic libraries is unchanged: to meet the immediate and anticipated information needs of users and to serve the research and teaching missions of the university. This is accomplished through strategically selecting, sharing, retaining, duplicating, divesting, archiving, and facilitating access to intellectual content" (p. 21).

Steps in Creating a Collection Development Policy: The Library Environment

An important first step in planning for a geospatial information collection or any library collection is to gain an understanding of the library environment in which the collection will be integrated. An appropriate place to begin is to examine the diverse factors that shape a collection development policy. Various sources in the library literature offer techniques and procedures for taking an objective and evaluative overview of an academic library's collection and services. Using published guides about collection management from the American Library Association as source material, Evans and Saponaro (2005) posit that there are three major elements: the collection overview, details of subject areas and formats collected, and miscellaneous issues.

A library's mission, either academic or public, will influence how the collections would develop. An essential part of a library's collection development policy is an explanation of the institutional objectives of a library. The document should include a general description of the community the library serves (e.g., degree-granting programs and the level of degrees). It should also elaborate about the patrons of the library, such as the range of students (undergraduate to postdoctoral students), and faculty (teaching to research). A statement about the parameters of the collection identifies what subject fields the library will collect and the limitations, if any, to the formats and types of materials, including emphasis on research materials for academic library collections or provisions for inclusion of technology. The latter is important in the provision of technical access, resource finding, and virtual library services as the "role as information intermediaries demands a new sub-set of quasi-technical skills and awareness. Librarians must not only identify and facilitate access to electronic information resources; we also must educate library patrons about how to access them and when to use them" (Kovacs & Elkordy, 2000, p. 335). Once the statements about the extent of a digital geospatial collection and the environment of which it will be part of can be made, the collection can be further defined by subject area.

In determining the composition of a collection's scope and subject area, librarians have a number of options to set levels of collection intensity, for example, the American Library Association has a five-level collection intensity guideline, ranging from comprehensive, research, study, basic and minimal. Other libraries, such as the Research Library Group and the Association of Research Libraries, have similar guidelines within a multiuse conspectus that identifies collecting levels (Evans & Saponaro, 2005). A useful tool in creating a collection development policy, the conspectus model requires librarians to perform subject analysis, which is particularly useful in an emerging field such as geographic information science and/or content areas of geospatial data. Larsgaard (1998) uses the conspectus model developed by Mosher and Pankake (1983) as the basis to a collection development policy for geospatial data. This model has 6 levels of collection: 0 - out of scope, 1- minimal information level, 2 - basic information level, 3 - study or instructional support level, 4 - research level, and 5 - comprehensive level, with a numerical value assigned to each subject area for the current collecting level and existing collection strengths (Mosher & Pankake, 1983).

Since a collection development policy establishes authority on what individuals or groups have selection privileges, and to what type of materials are to be selected for the collection, a framework or mechanism for resource evaluation of materials for possible inclusion for a particular library collection is essential. For new librarians assigned to geography and/or geospatial-dependent collections, it can be difficult to understand how to apply the conspectus model to the range of print and digital geospatial data. Larsgaard (1998) advocates a two-pronged approach: work with other librarians experienced with collecting geospatial data and become familiar with notable collections of geospatial data hosted by various institutions and libraries. Using the collection development policy as a guide, librarians can create a process of assessing the needs of their user community, as well as other factors, such as university mission, library strategic planning, and budget.

As discussed, the determination of collection level scope and goals of building a collection involves the integration and analysis of a number of factors. The librarian may have to examine relevant publications and existing collection development policies from other map libraries to understand the practical functions of geospatial collections, how they relate with the user community, and how they integrate with the library organization. An initial step can involve examining the rich history of map librarianship in the United States and Europe, which has essential facts relevant to the librarian involved in managing the disposition of geospatial data in both print and digital formats in academic library collections. Map librarianship also offers some interesting guidance on assessing user needs *vis à vis* academic collections as a whole.

User Needs

The library literature offers a rich, historical background in the development of map librarianship and the delivery of geospatial information services in academic libraries (Larsgaard, 1998; Nichols, 1982; Parry & Perkins, 2001; Ristow, 1980). In addition, it discusses parameters that librarians can use to gauge the knowledge of geospatial data users, as well as defining skill sets and knowledge needed by librarians to manage GIS and geospatial collections

Marley (2001) determined from interviews with users of GIS software applications that basic interpretive skills related to maps are still relevant to librarians and users. Library staff assigned to work with geospatial data need to be very familiar with "concepts such as scale, projection, symbolization, grids, geodesy and direction, and to be cognizant of the many different types of maps, their subject matter, and methods of reproduction" (Marley, 2001, p. 17). Basic map skills include knowing how to examine a map carefully, understanding the relationship between maps and gazetteers, or how to understand and interpret from a single map to an individual series of maps. In assessing potential and actual users of geospatial data, Marley suggests a methodology proposed by Winearls (1974, cited in Marley, 2001) in assessing user needs for a university map collection.

- Many people, the majority perhaps, know little about maps, which makes it difficult to provide information at an appropriate level and in a useful format. The layperson may have simple requirements such as locating parks in a city.
- Then there are students who use maps so infrequently that they are not familiar with the library layout or indexes. They hope that the information desired will be provided on one map, preferably measuring 8.5 X 11 inches, to be easily incorporated into a paper. This type of user is ill prepared to consult several different maps in order to synthesize information.
- On the next level are academics, specialists who know their subject matter, but are unfamiliar with maps as information sources. The academic is often

prepared to do a certain amount of work to achieve goals and is usually easy to work with.

- Finally there is the subject/map expert who really only needs assistance to know how to use the library, to find uncataloged materials and to be referred to sources that may not be in a particular map collection" (Winearls, as cited in Marley, 2001, p. 17).

Marley (2001) emphasizes that contemporary users of geospatial information collections still have the same questions and inquiries even though they usually deal with a digital environment. User questions are classed into locational questions, such as inquiries of the whereabouts of a particular place, and requests for a map of particular place (Marley, 2001). Winearls' approach of grouping user queries about map collections is still a useful strategy in assessing user needs of a digital geospatial information collection. Essential to his methods was the use of survey tools to gather data from users. Surveys about map libraries and their use by large academic and research libraries use a variety of different types of methods to gather data, for example, noting circulation records or recording patron-generated statistics. Variation in survey responses can appear because of the relative location of a map collection, the number of proficient Geography or Geosciences users with easy access to a map collection, or a broad map collection accessed by large numbers of general users (Marley, 2001, p. 18). Surveys are delivered in a variety of ways, from low-tech pencil and paper surveys to Web-delivered questionnaires to determine use of digital geospatial data and of GIS software. Whatever the method, the survey will most probably indicate a wide range of GIS and digital geospatial data users, especially in an academic campus setting. For example, asking the nature of the data use may uncover new classroom or research initiatives using digital geospatial information. In academic library settings, a survey of the instruction and research activities of faculty and staff may indicate a significant growth in recent years of the use of GIS software and digital geospatial information among faculty. The situation is reflective of the increased availability of powerful desktop computers and digital mapping software to academic researchers.

Building Geospatial Information Collections

Many academic libraries use an integrated approach in building collections of geospatial information. Librarians often rely on a number of both public and private sources for geospatial information, such as the Federal Depository Library Program, as well as digital geospatial data from government agencies and corporate sources. The use of geographic information systems in the federal government has extended beyond traditional users of spatial data, such as the Bureau of Land Management

(U.S.), and the Census Bureau (U.S.). Federal GIS users now include the military, Federal Emergency Management Agency (FEMA), the Justice Department, and the Library of Congress. Executive Order 12906, signed by President Clinton, requires agencies to base the geographic data documentation on federal metadata standards, post the data electronically, and join in industry standard activities. The current metadata standard issued by the Federal Geographic Data Committee (FGDC) determines the accuracy or quality of geospatial data and defines standards for data collection. A number of federal agencies participate in the National Geospatial Data Clearinghouse, accessible through the FGDC's home page on the World Wide Web http://www.fgdc.gov/.

Following the example of federal offices, state and local government agencies have integrated GIS technology into their organizational structure and business processes. The use of GIS illustrates a trend by researchers to apply computer technologies to departmental communication processes and problem solving tasks (Isaacs, Walendowski, Whittaker, Schiano, & Kamm, 2002; Safai-Amini, 2000). The use of geospatial data for daily transactions has created a voluminous record of electronic information within government agencies, which in turn has generated significant issues involving the management of such data, such as information collection, data storage, and public accessibility to information. Many governmental agencies at the county and local level use a variety of strategies that combine paper maps and GIS to maintain archiving standards for digital record preservation and to facilitate public finding aids. Collection development librarians who are creating a geospatial/geographic information collection will also use a number of strategies to obtain identified resources in both print and electronic formats.

Collecting Geospatial Information: Print Media

In acquiring print media for geospatial information collections, libraries in the United States can turn to the federal government for a variety of cartographic products. Virtually every sector of the government uses some form of the geospatial data for reference, analysis, and computation, such as the U.S. Geological Survey, the Department of Commerce, and the Department of Defense. The U.S. Geological Survey (USGS) is the civilian topographic mapping agency of the United States. It compiles and publishes the topographic map series of the United States, its protectorates, and territories. The main map publication of the agency is the topographic map, specifically the 1:24,000 scale series. The USGS also produces many other geological and mineralogical maps, such as the Geological Quadrangle Series, Miscellaneous Investigations Series, GeoPhysical Investigations Series, and Oil and Gas Investigations, to name but a few. As a contributor to the Federal Depository Library Program, many thousands of USGS maps are distributed to libraries in the United States each year.

The Department of Commerce, which houses the U.S. Census Bureau and the National Oceanic and Atmospheric Association (NOAA), is also a major map producer. The Census Bureau compiles demographic data that is conflated with geographic coordinates of a particular reference grid, such as a census tract, city block, or even street area. The Census Bureau produces a series of maps that portray the different enumeration regions of the decennial census. The series includes county subdivisions, urbanized area map series, census tract outline maps, county map series, place map series, metropolitan map series, and block statistics maps. The use of computing mapping software in conjunction with the data compiled by the Census Bureau engenders the possibility of creating a large number of specialty maps beyond the scope of its decennial series maps (Larsgaard, 1998). The National Oceanic and Atmospheric Administration has, under its jurisdiction, a number of agencies that produce myriads of spatial data products. The National Weather Service produces a variety of maps and charts, the National Federal Aviation Administration utilizes aeronautical charts, and the U.S. Coast and Geodetic Survey produces aircraft positioning charts, jet navigation charts, coast charts, harbor charts, small-craft charts, and offshore mineral leasing area maps. To understand better the nature of charts that are created, librarians are advised to consult the *Catalog of Aeronautical Charts and Related Publications*.

The cores of many map collections in depository libraries often contain maps produced by the Department of Defense. The Department of Defense contains the National Imagery & Mapping Agency (NIMA), which produces maps and charts for the various military agencies in the federal government. Although NIMA produces a variety of aeronautical, topographic, nautical, and miscellaneous items, its main endeavor is producing imagery. Charts and maps relating to navigation and flood plains are produced by the Army Corps of Engineers and the Department of Homeland Security, respectively. The vast cartographic output of NIMA and its predecessor agencies have long contributed maps to the federal depository program. The practice originated in the late 1940s when the Army Mapping Service established a formal map depository program for distributing its surplus stocks to participating libraries. Larsgaard (1998) notes that during the peak of the program in the 1960s there were over 200 participating libraries, with each library receiving over 300 maps a year. In 1984, the USGS and NIMA joined the federal depository library program enhancing the number of cartographic materials that participating libraries had access to.

A good source of information for librarians in attempting to navigate the vast number of federal map agency publications is to use the various indices, or guides, that the agencies produce. The USGS, for example, produces a number of indices and catalogs that identity its various map series. The publication entitled *Index to topographic and other map coverage* compiles listings of maps for a particular state, while the *Geological Map Index* lists many items for individual states. Since the catalog lists include numerous materials (state publications, periodical articles, and theses and dissertations, etc.), it can be difficult to comprehend for collection development purposes. However, there are academic and professional organizations with information resources that can offer assistance in building a map collection. One such source is the Map and Geography Round Table (MAGERT) of the American Library Association.

MAGERT is an organization that facilitates a community of individuals with an interest in map and geography librarianship. The MAGERT Web site is a useful resource for librarians needing guidance in building their cartographic collections or in building a print geospatial information collection. MAGERT provides a number of support functions for librarians, such as advocacy and educational programs on map librarianship and cartographic literacy programs. It also acts as a forum for the exchange of ideas by individuals working with or interested in map and geography collections. MAGERT also publishes several guides that are helpful in building print geospatial information collections. These include *Guide to Cartographic Products*

Table 1. Map vendors and publishers

East View Publications, Inc.	http://www.eastview.com
GeoCenter ILH (Germany)	http://www.geokatalog.de
Globe Corner Bookstore	http://www.globcorner.com
Gone Tomorrow	http://www.gonetomorrow.com
Hema Maps (Australia)	http://www.hemamaps.com.au
Hereford Map Centre (United Kingdom)	http://www. themapcentre.com
International Travel Maps (Canada)	http://www.itmb.com
Intercarto (France)	http://www.intercarto.com
Latitudes Map and Travel Sore	http://www.latitudesmapstore.com
National Geographic Map Store	http://www.nationalgeographic.com
MapLink	http://www.maplink.com
A Map Solution, Inc.	http://www.maps911.com
The Map Store	http://www.themapstore.com
Map World (New Zealand)	http://www.mapworld.co.nz
Map World (United States)	http://www.mapworld.com
Masons Maps (Australia)	http://www.masonsmaps.com
MetroData (Puerto Rico)	http://www.metropr.com
Mexico Maps (Mexico)	http://www.mexicomaps.com
Michelin (France)	http://www.michelin.fr
Milwaukee Map Service, Inc.	http://www.milwaukeemap.com
Omni Resources	http://www.omnimap.com
Outstanding Maps (United Kingdom)	http://www.mapsonline.co.uk
NetMaps, SA. (Spain)	http://www.netmaps.es
Platts Maps	http://www.platts.com
Rand McNally & Co.	http://www.randmcnally.com
Turinta Maps (Portugal)	http://www.turnita.pt
Universal Map	http://www.universalmap.com
United State Geological Survey Store	http://www.usgs.gov
William & Heintz Map Co.	http://www.whmap.com
World of Maps (Canada)	http://www.worldofmaps.com

of the Federal Depository Library Program, base line: a newsletter of the Map and Geography Round Table, an *Occasional Paper* series, and the *Guide to U.S. Map Resources*. Both *baseline* and the *Guide* are available from the MAGERT Web site (http://magert.whoi.edu/pubs/FDLPguide.html).

Map library Web sites are another resource for librarians building a geographic/ geospatial collection. The University of Oregon maintains a resource page on map publishers and distributors and remote sensing image suppliers (http://libweb.uoregon.edu/map/mappublink.html). An additional resource for librarians interested in ordering federal, state, foreign, or commercial maps can be found on the University of Nevada-Reno Library Web site (http://www.delamare.unr.edu/Maps/order.html). The Western Association of Map Libraries maintains a virtual Map Librarian's Toolbox that contains a number of links to print map publishers, cartographic suppliers, and vendors, as well as guides for map cataloging and map storage (http://www.waml.org/maptools.html).

Without a doubt, many historical geospatial materials will be in print. Retrospective conversion of all existing print maps and gazetteers is simply not feasible at this time. The uneven quality of map production at the federal government level offers challenges to librarians in developing storage techniques for the geospatial information. Common factors affecting preservation among the different map types include heat, light, moisture, dust, and biological agents.

Preservation and Archival Issues

Participants in the Federal Depository Library Program, depository libraries, such as larger public libraries and academic libraries, receive regularly scheduled shipments of government publications and generally duplicate the archival modes and methods for geospatial information stored by different federal agencies. Flat filing cases are generally used to store map collections. Typically, storage techniques for paper maps involve the use of horizontal or vertical filing equipment (Laarsgard, 1998). Most libraries use horizontal filing cases. These cases have shallow drawers, usually five to a unit. The units stack on top of one another, so the height of cabinet is limited by the ceiling height and load bearing limits of the floor on which they are situated. The drawers of such cases are usually about 2-inches deep and may hold as many as 300 packed sheets per stack, but only of same-sized sheets, such as sheets of topographic maps. Horizontal filing cases are constructed of steel or of wood. Unlike horizontal, (flat) files, vertical files hold maps in a vertical position using different methods of suspension. Maps in a suspension file configuration can hang from hooks, be held by some form of rod-like binder or clasp, filed in racks, troughs, or suspended folders, which can slide on rails along the sides of the cabinet (Laarsgard, 1998). Remote sensing and advanced digital imaging technologies, such as aerial photographs, and satellite photos, are often stored in protective plastic sleeve, put in archival boxes, and placed upon standard shelving. All the storage cabinets and boxes are placed in climate-controlled environments with temperature and humidity levels that stay at a constant rate.

Moving Beyond Print Map Collections: Digital Geospatial Information

As noted in Chapters II and III, the rapid adoption of telecommunications hardware and communication technologies by industries of the information economy has led to the increased use of digital geospatial data across the public and private sectors. A common economic trend of the information sector is rapid development. Research and development investment in the various communication hardware and computer components that comprise the sector's infrastructure often lead to new methods of information delivery. Internet-based interactive mapping products similar to the older CD-ROM products contain more functionality and data. Companies offer options and products that present the user with different levels of access to digital geospatial information. Many of the digital mapping products range in detail from simple screen captures of their print counterparts to more complex GIS packages that have relational data and have advanced analytical tools. Complex digital mapping products allow users to initiate spatial queries while referencing a base map with a variety of locational data and relational data attributes. Making sense of these resources to develop a state-of-the-art geospatial collection can be daunting. However, a solid understanding of print geospatial information, the existing collection, library environment, and user needs can assist in building a focused and useful collection for research and teaching.

An essential step in creating and integrating GIS services and collections in an academic library is in creating a sound collection-development policy. As discussed earlier, a number of factors, such as user needs, available budget, technological infrastructure, and staff development programs, are important factors in constructing a policy. In this collaborative effort, the responsible geographic information librarian would probably need to seek input from and work with their libraries IT department or Systems Librarian, Government Documents Librarian, Reference Services Librarians in order to plan for GIS services. Outreach to the community of GIS users who would potentially be library GIS service patrons is essential. In determining the scope of the GIS services and collections, the librarian may pursue a detailed examination of the use of digital geospatial information and geographic information system software among users in their library user community. As noted by Pettijohn and Neville (2003), "Usage statistics theoretically offer a quantitative method for evaluating the use of electronic resources" (p. 28). They also add that usage statistics are used in basic cost-benefit analyses to determine cost-per-use of a resource and/or to justify its expense. This is critical since geospatial data is dependent upon software and hardware for access and analysis.

In creating a collection development policy for GIS services, librarians can incorporate elements of a needs assessment into their workflow to help organize the various types of information elements they collect. Needs assessments are commonly performed across government organizations and private firms in planning for new information services. The financial aspects of a needs assessment is especially important in setting up a GIS service or collection. Most academic users of GIS will most probably be using desktop versions of a GIS program that usually have a specific number of licensed users. Listed in Table 2 are a number of commonly used GIS software programs used by campus researchers. In an environment with diverse

Table 2. Commonly used GIS software in academic environments

ARCGIS http://www.esri.com	"**ArcGIS** is an integrated collection of GIS software products for building a complete GIS for your organization. ArcGIS enables users to deploy GIS functionality wherever it is needed—in desktops, servers, or custom applications; over the Web; or in the field."
CLARK LABS http://www.clarklabs.org	"IDRISI Andes is an integrated GIS and Image Processing software solution providing over 250 modules for the analysis and display of digital spatial information."
ER MAPPER http://www.ermapper.com	"ER Mapper Professional is a powerful, yet simple to use, geospatial imagery processing application. ER Mapper Professional enhances your geographic data to make it more meaningful. It allows you to extract quantitative information and solve problems."
GRASS http://grass.itc.it/index.php	"Geographic Resources Analysis Support System GRASS, this is a Geographic Information System (GIS) used for geospatial data management and analysis, image processing, graphics/maps production, spatial modeling, and visualization. GRASS is currently used in academic and commercial settings around the world, as well as by many governmental agencies and environmental consulting companies."
INTERGRAPH http://www.intergraph.com	"GIPS enables the efficient collection of geospatial data and provides tools to ensure that the resulting product meets its intended specification. The GIPS encompasses: GeoMedia Topographer: Data Capture and Image Processing GeoMedia Curator: Data Integration and Management GeoMedia Cartographer: Product Generation GeoMedia Export Services: Data Dissemination"
MAPINFO http://www.mapinfo.com	"*MapInfo Professional* is a powerful Microsoft Windows-based mapping application that enables business analysts and GIS professionals to easily visualize the relationships between data and geography."

GIS software users, it may benefit the library to enter in a collaborative agreement with the users to help defray costs, which can be expensive for the Library if there are a large number of users for the software. Besides licensing issues, other factors in selecting a GIS software package for a library include functionality, expandability, and usability. With the advent of library Web sites and online catalogs, the GIS software should be able to interface with the library management software. The software should also be adaptable to accommodate changes in industry and data formats. As noted earlier in Chapter III, use of standards in data description are slowly becoming more widespread among GIS data producers, which is easing problems of access to digital geospatial information.

In determining the usability of a digital geospatial product, such as MapInfo Professional or ArcExplorer, there are a number of concerns. Hands-on trials by a range of users are critical. When considering usability with geographic information software packages, usability is more than a product's analytical capabilities. Ease of use, ADA compliance, graphic design features, navigability, and intuitiveness of the interface play a role in user choice and satisfaction. Basic and advanced GIS options should be available and easily located. Available geospatial information and services should be clearly indicated on the screen. Help menus should be well marked and have clear, easy to understand information. Usability also plays a major role in determining the number of clicks, views, and errors in database usage logs. In addition, the graphics intensive, high bandwidth nature of server-based GIS sites require certain computer hardware capabilities but also server capacity. Since users' access to resources varies upon hardware, software, and network connections, GIS resources must be evaluated using a mix of operating systems and browsers. The hardware, software, resources, and browsers used should be noted on the evaluation instrument as well as the tasks performed by the user.

Webware, Hardware, and Software Evaluation

A new area for collection management is the evaluation of software, hardware, and Webware. Any GIS should be evaluated strictly in terms of the potential user's needs and requirements in consideration of their work procedures, production requirements, and organizational context. In addition, CD librarians examining data requirements and costs also must evaluate compatibility of the database design with existing system(s) and initial data loading requirements and costs. If acquiring a new GIS system, costs include system acquisition and installation, as well as system life cycle and replacement costs. Getting the system and data is not the end of the evaluation process. What will be the day-to-day operating procedures and costs of the item? There will also be costs in staffing requirements and user training (including library staff). Finally, there may be application development and costs to integrate this new software, hardware, or Webware into existing systems. Finally, the timetable of these new systems must also be examined. Turnkey options are very rare the further one moves away from a " MapQuest moment" to complex, multidisciplinary research. This puts more of a burden on the evaluation process and on current staff. It may be that CD staff turn to expert help outside the library or to a professional GIS consultant to determine platform, communications options, software, and viability of product for the heterogeneous GIS community found in academic settings.

Librarians involved in establishing a geospatial collection may often focus on supporting the focused interests of some academic researchers. Many academic researchers often request advanced features on mapping software packages to enable their analytic research. While it is important to fulfill the needs of a community of researchers, librarians should also consider the spatial information needs of the beginning user (Longstreth, 1995). For example, Harvard, as a rule, purchases datasets that will be used more frequently for teaching and research as opposed to very specific, time-dependent datasets (Florance, 2006). Taking the time to determine one's users is a key component of building a successful geospatial collection. Larsgaard (1998) suggests observing user needs and requests from several months to up to a year before determining any significant changes in a collection development policy. Additional information and advice about GIS software selection can be found in trade Web sites and publications such as: *Directions Magazine* (http://www. directionsmag.com/), *Geoinformatics* (http://www.geoinformatics.com/), *Geospatial Solutions* (http://www.geospatial-solutions.com/geospatialsolutions/), GeoPlace http://www.geoplace.com/), and GIS Lounge (http://gislounge.com/). Some scholarly publications that publish geographic-information-system-related articles include: Annals of the Association of American Geographers (ISSN 0004-5608, Blackwell), Cartographica (ISSN: 0317–7173; University of Toronto Press), Computers & Geosciences (ISSN 0098-3004; Elsevier), Geographical Analysis (ISSN 0016-7363; Blackwell); Geoinformatica (ISSN 1384-6175; Springer), International Journal of Geographical Information Science (ISSN 1365-8816; Taylor and Francis), Journal of Geographical Systems (ISSN 1435-5930; Springer), Transactions in GIS (ISSN 1361-1682Blackwell), and The URISA Journal (ISSN 1045-8077; URISA).

The increased use of mapping software technology has created a veritable explosion of GIS data, much of which is available to consumers in a variety of digital formats, such as CD-ROM, or accessed via online protocols, such as FTP. In addition to spatially referenced data, there are a number of other digital applications of geographic information as well. Products may contain a digital version of their print antecedents, for example, dictionaries, encyclopedias, travel guides, and gazetteers. Other products may be historical in nature with scanned images of historical maps. Yet other products may be part of a multimedia package, such as an educational toolkit prepared for instruction in GIS principles. In terms of selecting digital geospatial information products for a library collection, Parry (2001) advocates considering "map related CD-ROM's mainly in terms of their thematic and regional content and their primary function" (p.77). He also suggests an ad-hoc scheme that includes "multi-media electronic atlases (regional, national, international), route planners, street level atlases, topographic map collections, historic map collections, demographic data sets, earth science and environmental data sets, and gazetteers" (Parry, 2001, p. 77). The more basic spatial data products, such as an electronic atlas, will most likely have a wider audience, while the more advanced GIS products will serve needs of a more focused community of users.

Microsoft's electronic encyclopedia, *Encarta,* offers an electronic atlas, with geo-political, climatic, and topographical maps integrated in the content of some 42,000 articles. Newer versions of *Encarta* have Internet capabilities as well. The *2007 Encyclopædia Britannica Ultimate DVD-ROM or CD-ROM* contains over 100,000 articles supplemented by 2,523 maps. The map images integrated into electronic atlases tend to be scanned raster maps depicted with related census or other numeric data. Interactive spatial data are also found in travel and route-planning software products.

Delorme's *Street Atlas USA 2007* DVD contains spatial data representing over 3.4 million roads in both the United States and Canada. In relation to the road level data that is contained on the DVD product, the *Street Atlas* contains a number of unique socioeconomic data, phone listings, and other data types, such as specific geographic features and imagery. National Geographic's *Back Roads Explorer* CD-ROM set functions much like a digital road atlas of the United States. According to the product's Web site, the product is created from nationwide seamless, 1:100,000-scale (1:250,000 in Hawaii and Alaska) USGS topographic maps, with an overlay of streets and roads. Related spatial data on the set includes over a million geographic points of interest. Other functionality includes the ability to customize and print photo-quality map and a 3-D imaging capability. *The Back Roads Explorer* is also compatible with GPS systems. Another digital mapping product that also integrates advanced imagery is *Mountain High Maps* from Digital Wisdom publishers. A four-disc CD-ROM set for either Windows or Macintosh computers, *Mountain High Maps* contains 78 maps that contain a number of spatial data layers and geographic features. The data layers on the CD-ROM set include physical elevation, vegetation, and political information that are available in three colorized formats. Other features are depicted in grayscale. Each map in the CD-ROM set includes features such as outlines, borders, country names, rivers, cities and towns, physical features, linear scales, and lines of latitude and longitude. These products indicate a growing level of sophistication and comfort with daily use of geospatial data in very specific applications designed to be non-technical and very user-friendly, which may have implications when reviewing more complex GIS software applications across formats/networks.

While acquiring digital geospatial data in CD-ROM or DVD formats can be cost-effective in building a geographic information library rather quickly, librarians need to be cognizant of a number of issues, such as format, when acquiring datasets for their library collection. Given the rapid development in computer hardware and storage devices, the CD-ROM format is no longer used exclusively by many digital geospatial data publishers in distributing their information. As noted by the earlier descriptions of electronic atlases and other mapping products, publishers often offer mixed format options for their products. Further, many commercial mapping products have secured Internet access for a wider selection of data. It is with the Internet that librarians can find myriad mapping applications and commercial products that can extend the capabilities of their CD-ROM based digital geographic information collections. Yet the wide variability of the different types of maps available on the Internet provides different challenges in deciding how to choose maps.

The ubiquitous, online environment of the Internet, with data transfer capabilities, and Web browser technology, is a particularly fertile environment for the display of geographic information. It is much more cost effective to distribute maps in an online format than in a traditional print or CD-ROM format. The ease of updating information in a Web browser environment also allows for regular maintenance of geographic information posted online, a distinct advantage over print maps and CD-ROM products (Peterson, 2003).

Most library users are familiar with interactive trip-planning functions. In recent years, competition between Web search engine companies and subsequent upgrades in information options have led to the incorporation of mapping functions on popular Web browsers and indexes, especially, Google and Yahoo. The interactive Web sites allow users to create customizable street maps between specific street addresses. The mapping functions are usually coupled with related information, such as business address, points of interest, landmarks, natural features, and imagery. Some companies have expanded beyond simple display of raster graphics of gif images, and offer detailed imagery for their maps and cover most of the United States.

From a collection development standpoint, search engines are a good initial step in finding maps across the Internet. The search engine results for maps, though, are only as good as the metadata description is for the returned hits. Most results pages from search engines return only lists of textual information that contain less information of a graphical nature. While searching the Internet, there a number of Web sites that maintain lists of available sources for online mapping sources. The ESRI corporation supports the Geography Network (GN) (http://www.geographynetwork.com/aboutus/index.html), which is a global network of geographic information users and providers. GN provides the infrastructure needed to support the sharing of geographic information among data providers, service providers, and users around the world. Many types of geographic content, including dynamic maps, downloadable data, and advanced Web services, are available. Another Internet resource with a comprehensive listing of hyperlinks, with all manner of resources for maps and cartography, is Odden's Bookmarks (http://oddens.geog.uu.nl/index.php). Created by Roelof Oddens, the curator of the map library of the faculty of GeoSciences, at the University of Utrecht, the site has over 22,000 links.

Table 3. Digital cartography-software vendors, GIS-GPS developers

American Digital Cartography, Inc.	http://www.adci.com
Articque (France)	http://www.articque.com
Avenza Systems Inc.	http://www.avenza.com
Claritas	http://www.claritas.com
Clary-Meuser Research Network	http://www.mapcruzin.com
Core Software Technology	http://www.coresw.com
DigiAtlas (Spain)	http://www.digiatlas.com
Digital Earth (Australia)	http://www.digitalearth.com.au
East View Cartographic	http://www.cartographic.com
EuroCartrographie (Netherlands)	http://www.eurocartographie.nl
GeoConcept SA (France)	http://www.geoconcept.com
GeoFrameworks	http://www.geoframeworks.com
GIS Dynamics	http://www.gisdynamics.com
Harvard Design and Mapping	http://www.hdm.com
High Country Software (United Kingdom)	http://www.mobilemaps.com
IMGS (Ireland)	http://www.imgs.ie
Intergraph Corporation	http://www.intergraph.com
Klynas Engineering	http://www.klynas.com
Magellan Navigation, Inc.	http://www.promagellangps.com
MainStreetGIS	http://www.mainstreetgis.com
Mapcom Systems	http://www.mapcom.com
MapFrame	http://www.mapframe.com
Map Solute (Germany)	http://www.mapsolute.com
StreetMap (United Kingdom)	http://www.streetmap.biz
TeleAtlas	http://www.teleatlas.com
Telemorphic, Inc.	http://www.telemorphic.com
ThinkGeo	http://www.thinkgeo.com

With establishment of digital collections by many academic libraries, digital map libraries are another good source of online cartographic and GIS materials. One notable example of digital libraries of this type is the Alexandria Digital Library (ADL) (http://www.alexandria.ucsb.edu/). As documented on its Web site, the ADL is a "distributed digital library with collections of georeferenced materials. The library includes the operational library, with various nodes and collections, and the research program through which digital library architectures, gazetteer applications, educational applications, and software components are modeled, prototyped, and evaluated. The ADL also provides HTML clients to access its collections and gazetteer, and provides specific information management tools, such as the Feature Type Thesaurus for classing types of geographic features, as well as downloadable software code."

Other online map collections include the Perry-Castaneda Library Map Collection of the University of Texas Library System (http://www.lib.utexas.edu/maps/) and the online Yale Map Collection (http://www.library.yale.edu/MapColl/online.html), which contains many digital images of historical maps and globes. Many user interfaces build customizable maps online, such the Geospatial & Statistical Data Center from the University of Virginia (http://fisher.lib.virginia.edu/collections/gis/). They create maps based on public domain data produced by different federal agencies. As with paper maps, the U.S. federal government is a good resource in acquiring digital geospatial data as discussed below.

Collecting Governmental GIS Data

In many instances, librarians assigned to collect digital geospatial data can rely on an extensive network of data sources that can range from clearinghouses such as the FGDC to the tremendous amount of publicly available GIS data output. Most of the publicly available data is from government agencies at the federal, state, and local levels. One aspect of publicly created GIS data sources is the ubiquitous nature of certain data types, such as shapefiles, that are available for download from many government data sites. The ease of download and interoperability of many such government data files, such as U.S. Census topography, has led to the widespread integration of similar data files across different GIS coverages. Different government agencies often tend to share data, such as a file that contains a boundary or topology information, to help avoid duplication when working with similar geographic extents in projects. In some instances, GIS technicians located in the same geographical area, though in different city and county agencies, may use the same base coverage in building their specific GIS applications. An example would be technicians in a city planning agency and county environmental agency using the same vector map to display their information. The situation presents a number of challenges in collecting descriptive information in determining the scope and lineage of the GIS data for processing. Librarians would need to collect descriptive information about the GIS data for cataloging purposes and in preparing other library services for the GIS data, like access and reference services.

In most instances, the government agencies and departments will have produced metadata of varying degree to accompany their GIS data products. Some municipalities produce metadata layers according to FGDC standards, but for others librarians may have to search related documentation for description information of the data. The geographic data that is integrated into a spatial database often exists in other formats, such as specialized reports, tables, graphs, charts, spreadsheets, or even maps. In researching the lineage of a particular set of geospatial data, the librarian would have to delve into the local government codes concerning the archiving of spatial data records in both print and electronic formats. This approach in evaluating geospatial data sources becomes important in creating a historical archive of digital geospatial data. The ease of importing GIS data files and in creating new data views using existing ones can greatly limit the permanency of a particular dataset. A particular GIS dataset could be replaced rather quickly by other data. By taking a holistic approach to digital geospatial data collecting, the librarian will be in a favorable position to answer quickly the common inquiry of having good data from seasoned practitioners of GIS applications. The ability to create a richer data description will make the GIS data more understandable to novice users as well as offer additional descriptive variables to researchers.

Federal Data

The distribution of federal government produced GIS data across the United States was a factor in many state, county, and city governments using digital geospatial data in support of management functions, such as with tax assessment, planning, utilities management, environmental and social services. The use of digital geographic data and GIS software in different agencies has resulted in a tremendous output of GIS datasets. Recently, the federal government has established a Web site to facilitate access to digital geospatial data. Geodata.gov, also known as the Geospatial One-Stop, is a geographic information system public gateway portal to improve access to geospatial information and data under the Geospatial One-Stop E-Government initiative. Geospatial One-Stop is one of 24 E-Government initiatives sponsored by the Federal Office of Management and Budget (OMB) to enhance government efficiency and to improve citizen services. The portal is a catalog of geospatial information containing thousands of metadata records (information about the data) and links to live maps, features, and catalog services, downloadable data sets, images, clearinghouses, map files, and more. The metadata records are submitted to the portal by government agencies, individuals, and companies, or by harvesting the data from other geospatial clearinghouses. Other readily available digital geospatial data sources include TIGER/Line files produced by the U.S. Census Bureau. TIGER/Line files are extracts from the Census Bureau's TIGER (Topologically Integrated Geographic Encoding and Referencing) to support the mapping and related geographic activities required by the decennial census and sample survey programs. (http://www.census.gov/geo/www/tiger/tiger2005se/TGR05SE.pdf). The geographically referenced TIGER/Line files are used for a variety of socioeconomic and spatial analyses. Using commercially available geographic information system software products, researchers can import the TIGER/Line files and enhance its attribute characteristics by linking its geographic spaces or points with other datasets. With publicly available datasets, private software developers also import data from the public domain and create a new, proprietary, geospatial depiction or coverage of an area.

As in collecting print media, librarians will find that the federal government is an essential and complex source for libraries in the acquisition of digital geospatial data. Similar to the distribution practices of private firms like ESRI, many government agencies produce GIS datasets in different formats to enable compatibility with different GIS software platforms, although the ESRI shapefile format is usually the file mode of preference of the government agencies. The federal GIS datasets are generally accompanied by detailed metadata that aids in the selection, cataloging, and description processes for a library collection. In some cases, however, librarians may have to process federally produced digital geospatial data that is not packaged for easy processing in a library setting. In such cases, the record management documentation of the agency will provide guidelines in deriving descriptive information about the GIS data.

The schedules for the management of digital geospatial data contain essential information about the lineage, quality, and extent of digital geospatial data, three critical components in assessing data for acquisition for research and instruction purposes. Since federal geospatial data is created and used in complex policy or decision-making processes or daily activities specific to an agency, the data may create very diverse collection depths or breadths for a library. Librarians may need to examine a wide array of documentation and related data that provide additional perspectives in describing the context of the data being considered for a collection. This can be illustrated by examining the directives and policies used by the Federal Bureau of Land Management in managing GIS data.

The Federal Bureau of Land Management contains a number of schedules for the retention, disposition, and processing of records. Schedule 20 covers the handling and processing of disposable electronic records created or received by Federal agencies. This includes records created by computer operators, programmers, analysts, system operators, and all personnel with access to a computer. Disposition authority is provided for specific master files, including some tables that are components of database management systems, and some files created from master files for specific purposes as well as some types of disposable electronic records produced by end-users in office automation applications. This is important since digital geospatial data is often comprised of files that have components of a master file nature and may contain some disposable aspects as well. Since the schedule identifies the production of spatial information and illustrates the various information components that can be related to the data as it is processed within an organization, understanding the schedule allows librarians to describe the data for cataloging purposes, and to link the data to other relevant datasets and documents.

Schedule 20 specifically defines a geographic information system (GIS) as an auto-mated system designed to capture, store, process, analyze, and display graphically referenced data. The GIS data are used for a wide variety of human and environ-mental related analyses, which is defined by the analysis of the geographic distribu-tion of data. Products include graphic images (plots), hard copy maps, displays of statistics on data, cartographic products, and any forms or combinations of these products in reports. These GIS products may be simple screen displays, outputs of analyses, and copies of tabular files, maps, or files for use in other computer sys-tems. The basic GIS data are retained electronically and are continually updated. Local GIS data is locally stored and managed by the Bureau of Land Management field offices. The GIS systems contain information uniquely collected by federal offices/agencies must conform to preservation and descriptive metadata standards as defined by the Federal Geographic Data Committee. Similar schedules exist at the state and county level.

State Level Data: The State of Florida

The Florida Department of State contains a number of general record schedules that define the nature of records and prescribes their disposition. General Record Schedule G12 for Property Appraisers recognizes and addresses the handling of geospatial sources of a physical and print format, such as maps. As with federal maps and geospatial data, maps are integral components in various land manage-ment procedures at the county and city government level, and are considered to have administrative value. Maps then must be retained as long as they are considered to have value. Maps scheduled for retention include maps for government land offices, highway maps, original maps, sectional maps, subdivision plats, and sales maps. These maps need to be retained for 1 year or until they have lost their value and are considered obsolete.

 The handling and disposition of certain aspects of digital geospatial data that is compiled by state, county, and city government agencies is discussed in chapter 1B-26 of the *Florida Administrative Code (FAC)*. This chapter outlines standards and requirements for electronic record keeping, which are applicable to all state agencies. The rules establish minimum requirements for the creation, utilization, maintenance, retention, preservation, and disposition of master copies of electronic records. Electronic records include numeric, graphical, aural (sound), visual (video), and textual information, which are recorded or transmitted in digital form. The rules apply to all electronic record keeping systems, including microcomputers, mainframe computers, and image recording systems.

As in the schedules used by federal agencies to manage and archive geospatial data, the State of Florida requires state agencies to create a record management system concerning geospatial data. The system that is implemented for the geospatial data must be integrated into the management of other records and information resources of the agency. The standards specified by the *FAC* follow federal schedules in that computer hardware and software must be compatible with computer systems used by other government agencies and data providers. Each agency must also specify the location and medium in which electronic records are maintained to meet retention requirements, establish and document security controls for the protection of the records, and maintain inventories of electronic record keeping systems to facilitate information disposition.

Digital geospatial data should also meet state provisions in conforming to public access to public records, which addresses description and attributes. The *FAC* requires that any electronic record keeping system that manages the digital geospatial have technical documentation that specifies technical characteristics necessary for reading or processing of records. For digital geospatial data, the documentation must include a data dictionary, a quality and accuracy report, and a description of the graphic data structure, such as recommended by the Federal Geographic Data Committee. Thus, the *Florida Administrative Code,* while specifying that a record management system be established to be able to process digital geospatial data, references federal practices in terms of maintaining data quality and control of the information. Local county and city government agencies, while conforming to the standards set forth in the Florida Code concerning the disposition of digital geospatial data, use different techniques and procedures to archive digital geospatial data and to make the data accessible to users.

Collecting Data at the County Level: Hillsborough County, Florida

Collecting digital geospatial data at the county and local level of governments is similar in scope to collecting data for state and federal agencies. Often the management of geospatial data is integrated with existing record management practices of different offices. For example, in Hillsborough County, Florida, geospatial data that are part of official transactions are managed and archived according to different provisions of both state and federal schedules concerning the archiving of geospatial data. The Tax Collector's Office and the Office of the Property Appraiser of Hillsborough County both make use of geospatial data in both print and digital formats. The graphic representations of various land parcels are kept in the form of plat maps that outline different land areas across Hillsborough County. The plat maps of official record are printed on varying grades of paper and are kept according to established federal and state practices in flat files within the county archives facilities. Copies of the plat maps are kept in other county offices, such as the Metropolitan Planning Organization for long-range transportation planning for Tampa,

Temple Terrace, Plant City & Hillsborough County, or the Hillsborough County City-County Planning Commission.

Related records that contain geospatial data are kept in the Records Library of the Clerk of the Circuit Court. The records library provides access to filmed and imaged recorded documents, with the earliest records dating back to 1836. The documents that are contained within the library include deeds, mortgages, liens, judgments, satisfactions, military separations, plat books, and tax rolls. Since many of the items are available in microfilm format and are searchable in an online index, county government documents have both online and physical representations that are reflective of the different government schedules regarding archiving of data. Since federal archiving rules do not consider geospatial data stored in an online accessible format as being an official record, other data formats are retained, for example, plat maps, as official land records. The county government does use GIS applications in a publicly accessible search engine that allows users to perform queries on land records. The online GIS search engine ensures public access to digital geospatial in accordance with the *Florida Administrative Code*. Thus, the librarian collecting local digital geospatial data at the county level would probably find a variety of data recorded on a number of formats in a number of places.

Managing Geospatial Data Records

Managing and archiving digital geospatial data records can present a number of challenges to academic librarians. In determining guidelines to process such data, the Federal Geographic Data Committee (FGDC) Content Standards for Digital Geospatial Metadata outlines parameters that can be adapted by libraries. The Content Standard specifies required elements for capturing information about lineage, processing history, sources, intended use, status of the data, and other types of information available through a clearinghouse. The clearinghouse concept is a central component of a National Spatial Data Infrastructure (NSDI) initiative (see Chapters II, III, & V in this volume). The NSDI included initiatives to improve public access to and use of geospatial data and to implement content standards for metadata.

The FGDC also advocates preservation of relevant geospatial data. Based on National Archive and Records Administration (NARA) directives, the committee emphasizes three areas of consideration when evaluating geospatial database systems:

1. **Records Retention:** Every data set, record, or file in the system should have a designated retention period. Temporary records should be deleted or transferred to alternative storage media or facilities for temporary records only at specific times according to an approved records retention schedule.

2. **Records Preservation:** Geospatial data creators are required under 44 U.S.C., chapter 29 to preserve permanent records, both the data and appropriate documentation. When the designated permanent records are transferred to NARA at the predetermined date, the transfer will be in the format and on a media acceptable to NARA at the time of transfer.

3. **Records Integrity:** The hardware and software systems design must ensure data integrity. This can be accomplished by using passwords and audit trails, by restricting when records can be edited and by maintaining a "history" file in a meaningful format of all changes, when appropriate." (National Spatial Digital Infrastructure, 2003, p. 2).

In elaborating on the NARA guidelines for geospatial data preservation, the FGDC recommends diverse types of geospatial data for preservation purposes. Their preservation policy recommendations include records in geospatial data base systems that provide evidence of the organization, policies, programs, decisions, procedures, operations, or other activities of an agency of the Federal Government. A broader body of geospatial data may be preserved because of the value of the information it contains. Storage media for the recommended data may include magnetic tape, floppy and hard disks, and optical cards and disks. The electronic records may include geospatial data files and databases with a national scope or those at the project or operations level. An important aspect of data integrity of the digital geospatial data is with the originating software that produced the geospatial data. Hardware and software system capabilities need to ensure appropriate retention and disposition as required by law. Information resource managers need to ensure that the information resources meet all legal requirements as outlined by NARA directives.

Most government agencies produce information management policies that follow NARA directives, yet meet the needs of their respective departments. Recognizing that significant geospatial data and GIS software data will be used in transactions that are part of preparing land-use plans, the Federal Bureau of Land Management has outlined extensive recommendations in archiving digital geospatial data used in land record management. The policies illustrate applications of the NARA and FGDC directives at the federal agency level.

Consider that land-use decisions are often the outcome of an integrated evaluation process that involves various information sources and input from personnel with diverse roles, such as data stewards, GIS specialists, and records administrators. Land-use planning data can include, but is not limited to, GIS layers and products, word processing files, studies, resource inventories, memoranda, e-mails, photographs, images, maps, and charts. Once the planning data is codified into a complete land-use plan, with a signed record of decision, the plan becomes a dynamic knowledge base. The planning data continues to be updated as activities are defined, permits are reviewed, and questions are received and answered. In addition to need to archive the active planning data, there is also a continued need to track and periodically archive completed planning data.

Archiving planning data saves a permanent copy of data for future retrieval. Data may be retrieved for responding to public questions as well as for reference to historic resource decisions. Unlike other archiving activities, which delete data from the system, land-use planning data remains present on existing systems for ongoing use, analysis, and reference. After identification by a planning team, the local system administrator would carry out the archiving of digital data. For example, during the life of the land-use plan, planning data would ideally be archived at the end of a fiscal year. Within each year, regular 30-, 60-, or 90-day backups would accommodate most needs to retrieve digital planning data that has been used, processed, and changed. Paper documents would be kept with the case file throughout the life of the plan and archived yearly. Land-use plan data archives are considered true copies of the original data used in making planning decisions. The local office manager usually certifies the archived data as being a complete copy of the land-use plan. One important caveat is that Internet-housed land-use plan data are not considered as permanent archives.

The physical process of archiving geospatial data in paper form that do not have a digital equivalent involves a number of steps. The archiving of such materials begins with the retention of the original paper copy of each land-use planning document, map, chart, and photograph, used in planning decisions on site at the local office with the case file. A duplicate copy is made of each printed document, map, chart, and photograph used in decisions. The duplicate copy is stored at a secure off-site location, and an annotation is made to indicate that electronic versions of the information are not available. Offices using geospatial data in paper form are not required to convert physical data products to electronic formats, but can carry out such conversions when feasible and practical.

Digital storage formats for electronic planning data involves using widely available formats, such as Adobe Acrobat and Microsoft Word, for processing documents. Archiving digital geospatial data involves the retention of its original data format and data directory structure, as it was compiled in the GIS software. Digital planning files that are the result of conversion from paper documents would be a duplicate of the corresponding paper documents. Once the land-use planning data has been identified and archived, there are specific storage options for the data. Planning text files, word processing files, and geospatial data used in the planning process are copied to CD-ROM media. For data sets up to two gigabytes in size, three CD-ROMs would suffice. For data sets that require more than two gigabytes of storage space, a tape backup system is considered to produce a permanent copy of the digital planning data used in the decisions. Documentation of the digital data include a text file and an archive metadata file that describe the name of the land-use plan, date of archiving, phase of the plan at the time of archiving, directory of data structure, software compression processes, index to archived information, and any other unique identification number for the individual dataset.

The written documentation and applied methods used by federal agencies concerning the disposition and preservation of geospatial data in both print and digital format are a record that can be referenced by librarians involved in processing digital geospatial data. Selected guidelines and directives for handling geospatial data both in print and digital formats that have been codified and applied to federal government agencies have been followed by both state and local government agencies, such as taxing authorities, and agencies that are involved in land-use regulation, and planning. For example, the state of Florida has referenced different federal schedules concerning the disposition and archiving of electronic records and geospatial data in the *Florida Administrative Code* concerning records management. The *Florida Administrative Code* in turn serves as a guide and reference for local governmental departments at the county and city level in processing and archiving geospatial data.

Legal Considerations in Collecting Geospatial Information

In addition to cost and availability of geospatial information, licensing, and distribution of data as well as applications come into play as more academic users want access to primary data and its tools. However, data size, format complexity, and potential restrictions are important issues to review in the acquisition of geospatial data. These restrictions may be due to copyright, access, or license agreements created by either public or private data producers. Intellectual property rights, liability issues, distribution methods, and data management practices must be understood by librarians from the dual perspectives of librarian and vendor. Changes in access to government information after the passage of the Homeland Security Act, for example, may affect content and access.

Public Domain vs. Public Sector Data

The concepts of "public domain" and public sector data are not interchangeable. Public domain is a legal status, that is, items in the public domain are copyright free. Public sector information is not necessarily in the public domain. Since it is not in the public domain, it may or may not be publicly accessible. As mentioned earlier, access to public sector information may be governed by constitutional, federal, or state law. Further, although an individual may consider an item in the public domain as "free," the legal definition implies that no property rights or restrictions are associated with the product and there is often an explicit disclaimer of copyright. Public domain material, however, can be modified, giving the person who did the modification intellectual property rights as well as copyrights for the modification, not the original product. "Public sector data" are data produced by a public sector body, which may be in the public domain or be protected data, depending on governmental and institutional policies, which vary with country.

Copyright

Copyrights protect the form of expression of an idea, concept, method or formula, and not the idea itself. National and international laws must account for changes in the nature of information and technology. One such area is the concept of "related rights," including rights for the electronic version of a manuscript or a database (Longhorn, Henson-Apollonio, & White, 2002). In the United States and Europe, researchers can generally use copyrighted material because of the "fair use" exception, based on exceptions permitted in the Berne Convention (World Intellectual Property Organization, 2001). Although fair use does not permit large portions of copyrighted material to be copied or transferred to third parties, the scope of "their application is sufficiently uncertain, however, that, where possible, parties should contract for anticipated uses rather than rely on fair use doctrine or other uncertain legal doctrines to sanction the licensee's activities" (Committee on Licensing Geographic Data and Services, 2004) p. 110).

Since neither the Berne Convention for the Protection of Literary and Artistic Works (WIPO, 2001), the U.S. Supreme Court automatically allows copyright for a database; all countries in the European Union now have separate database protection laws (Hugenholtz, 2001).

When determining lineage of the data, European content may require additional investigation as to educational and research uses.

Infringement, Accountability, and Liability Issues

There are certain legal responsibilities for anyone who creates, uses, or disseminates spatial information and tools, or services based on the data and tools. Legal risks relating to geodata and GIS include failure to secure or infringement of intellectual property rights, which include access to geodata or tools that result in illegal use (Cho, 2005). Legal risks also address accountability, such as failure to secure accountability for defective data or GIS tools, such as models, methodologies, and services based on the data and tools (Cho, 2005; Onsrud, 1999). One example is when defective geospatial data are used in decision making that have consequences at a planning or population-based level of policy or practice.

Librarians also need to be aware of the legal risk if confidentiality or privacy obligations are breached (Cho, 2005). Vendors of spatial information and GIS tools often invoke confidentiality restrictions that allow the purchaser to use the data or tools (software or models), but prohibit the purchaser from disclosing any details to a third party (Longhorn et al., 2002). Privacy issues are also a concern (Cho, 2005). Coupling descriptive data to precise location data is the key to many types of spatial analyses that have an epidemiologic or socioeconomic framework. However, when locations are linked to identities of individuals, there is a potential for violating personal privacy. An example is the use of health information that must abide by use regulations set forth by the Health Information Portability and Accountability Act of 1996. Typical requirements for using personal data are that the data be obtained with informed consent, only be held as long as required for the authorized use, and de-identified data only are used in the final product.

Outsourcing also poses a legal risk for librarians. This includes the tasks of geospatial data collection, processing, and dissemination, regardless if the outsourcing is performed by a government agency or for such an agency or private enterprise (Cho, 1998; Cho, 2005). Much of this involves the difference in cultural and legal laws in a country, including laws on intellectual property, laws governing international contracts, and foreign courts and their use of international contracts. Since laws are constantly undergoing legal amendments and evolving regulatory updates, constant supervision of the legal issues involved in the outsourcing process is vital. For example, contracts signed with foreign agents or involving foreign law might not be honored, creating a breach of contract. The ramifications of such a breach may be significant and require time spent in foreign courts to litigate the case. Another example is the difference between jurisdiction on intellectual property law, which may fall under international law while private international law governs the use of contracts. A foreign court may determine which nationality's law will govern the contract, and which court will have jurisdiction. Other issues involved with outsourcing includes labor, compliance with regulations, taxation, document management, and disaster recovery. For many academic librarians, this is indeed strange territory, and perhaps is best dealt with in concert with the Office of General Counsel or outside legal experts.

Librarians will need to understand licensing and liability issues in the acquiring and use of geospatial data, not only for the library's protection, but also for the researchers who plan to use the data. Since many academic researchers now work on national and international projects, all partners should be aware of any confidentiality clauses that may exist in software licences, use and access agreements for geospatial data, project documents, or funding arrangements. For librarians who are acquiring or providing access to researchers, contract terms and/or license language will need to accommodate different countries or different legal jurisdictions. Longhorn *et a.l* (2001) suggest that "[b]oth staff and institutions should recognize their rights and responsibilities in such cases, and stated policies should be in place, including appropriate non-disclosure and confidentiality agreements and forms, both in contracts of employment and perhaps even on a project basis" (p. 29).

Managing Legal Risks

Factors that can mitigate liability include how much care was exercised in developing a product or service, how much was charged, and whether appropriate disclaimers were provided. Licenses and contracts that provide such information are a key means of limiting liability, and allow libraries to manage the strange new worlds in which geospatial data collection and use exist. For example, for some projects, more than one copyright owner will provide data. A health services researcher may use epidemiologic data collected by the U.S. Public Health Service, Medicaid/Medicare data collected by the state of Florida's Agency for Healthcare Administration, a database copyrighted by a private managed care company, and transportation data from another academic research project located within another college. Collection development librarians, and administration, need to be aware of the administrative burden of negotiating terms with each user and provider of data, particularly for databases compiled from several sources.

ESRI offers several examples of how to handle licensing/contract agreements. Its "Digital Chart of the World for use with ARC/INFO® software" states clearly that it "is a license and not an agreement for sale" and addresses duration of the agreement, warranties, and export regulations. The ESRI Master License Agreement includes a "scope of use" table showing what types of licenses apply to ESRI software and data products while a separate ESRI Data License Agreement addresses permitted and not permitted uses, redistribution rights for derived data sets, no warranty given for quality, limitation on liability, and export restrictions (Longhorn et al., 2002).

One important area for librarians to review in licenses for use in research projects is "redistribution rights for derived data sets," which prohibit transfer of data to unlicensed third parties. Since external datasets are often integrated with data collected during a research project, limitations on distribution rights for the "derived data sets" can be crucial, especially if the dataset is a base set for a longitudinal, or expanded, study. Librarians should ensure that the data vendor does allow redistribution rights for derived data sets for the researcher, even if at an additional cost. The cost of the dataset and this right may be borne by the granting or funding agency, if written into a research grant or amended with notification to the funding agency. Librarians should also be aware of the restrictions placed on data or tools made available to educational institutions, that is, "noncommercial use only." A topographic base map for classroom instruction clearly meets the non-commercial use, however, a researcher in a private-public venture would face potential infringement. Since librarians may be actively involved in identifying, locating, and acquiring datasets for research, it is their responsibility to be informed of the terms and conditions of any external data sources or applications used and to document this information in writing to the researcher. Finally, librarians should be aware of any rights restrictions based on access to previously purchased data or software should there be updates or upgrades, respectively. Librarians should always examine the software upgrades and associated licenses to ensure that all the original purchase/ lease terms still apply.

Longhorn., Henson-Apollonio, and White (2002) suggest, at a minimum, the following policies regarding the management of geospatial and numeric datasets. For each project, a laboratory or project notebook should document "data sources, data created, enhancements to data, all software used or created, and any transfers of data or software among research groups or institutions" (p. 9). The notebook should clearly "indicate who did what when" and be updated and backed up regularly, preferably in real time as transactions occur in the lab or on the data. Staff and users should all read the license agreements when acquiring software packages or access to data sources. File maintenance is at two levels: physical and virtual. A physical file containing all data and software transfer agreements should be created and maintained. A virtual file, or metadata, should also be updated in real time and contain references to all data and software transfer agreements. A formal data or software transfer agreement should be created, to document terms of any interchange, checking that the terms in the distribution agreement do not conflict with other licenses. For example, data from third parties may have very different parameters surrounding their use. Longhorn, Henson-Apollonio, and White (2002) also suggest secure storage for all data (primary and secondary) for a minimum period of 10 years following closure of a project. They seriously reiterate the need for digital watermarks on all major datasets to allow data to be identified even it has been extensively modified (p. 9).

A Case Study of an Integrated Geospatial Data Collection: The Florida Geographic Data Library

The Florida Geographic Data Library (FGDL) is a collection of digital geospatial data that is warehoused and maintained at the University of Florida's GeoPlan Center, a geographic information systems (GIS) research and teaching facility. The FGDL was created to help support a variety of research endeavors among public and private institutions in North Central Florida. The mission of the FGDL is to operate as a mechanism for the distribution of a variety of satellite imagery, aerial photographs, and other digital geospatial data throughout the state of Florida and beyond. The FGDL offers a variety of GIS services and technical support that are structured around its digital data collections. The organization of the center and of its data collections reflect the structure and functions of regional spatial data centers such as the Alexandria Digital Library at the University of California Santa Barbara

Table 4. Agency data collected by the FGDL

Bureau of Indian Affairs (BIA)	Bureau of Transportation Statistics (BTS)
Environmental Protection Agency (EPA)	Federal Aviation Administration (FAA)
Federal Communications Commission (FCC)	Federal Emergency Management Agency (FEMA)
Florida Cooperative Fish and Wildlife Research Unit	Florida Department of Environmental Protection (FDEP)
Florida Department of Revenue (FDOR)	Florida Department of Transportation (FDOT)
Florida Division of Emergency Management (FDEM)	Florida Division of Historical Resources
Florida Fish & Wildlife Conservation Commission	Florida Marine Research Institute (FMRI)
Florida Natural Areas Inventory (FNAI)	Florida Resources & Environmental Analysis Center (FREAC)
Florida's 5 Water Management Districts	Land Processes Distributed Active Archive Center
National Climatic Data Center (NCDC)	National Oceanic and Atmospheric Administration (NOAA)
Office of Economic and Demographic Research (Florida Legislature)	Space Imaging Earth Observation Satellite Company (EOSAT)
SPOT Image Corporation	Subsurface Evaluations, Inc.
Tallahassee-Leon County GIS (TLCGIS)	University of Florida GeoPlan Center
US Census Bureau	USDA Forest Service
USDA - Natural Resources Conservation Service (NRCS)	US Department of Agriculture
US Fish & Wildlife Service (FWS)	US Forest Service (USFS)
US Geological Survey (USGS)	

(Hill, Carver, & Larsgaard, 2000) and the Idaho Geospatial Data Center (Jankowska & Jankowski, 2000a). The FGDL incorporates geographic query techniques and specialized user interfaces used at these institutions. :The FGDL is an important information asset at the regional level and as it develops its data collections and services can be an important component of the National Spatial Data Infrastructure in the United States.

The FGDL has over 350 layers of digital geospatial data that describes a wide range of physical and cultural aspects of Florida. The core of the FGDL data collections are comprised of a series of digitized topographic maps, aerial photographs, and other remote-sensed imagery of the State of Florida. The images comprise a broad layer of base maps of Florida over which a variety of GIS data can be arrayed. The FGDL has collected many GIS data files of geological and demographic significance to Florida that have been produced by a number of government agencies at the federal, state, and local level.

Most socioeconomic information that resides in the library is conflated with U.S. Census Bureau derived topology. A key component of the collections are the U.S. Census Bureau Tiger Line Files that portray demographic data in Florida and are enumerated at the county level and block level, which are produced by the U.S.

Census Bureau. Significant topographic and geological layers of Florida information in the collection are digital raster graphics representing the 1:24,000 map series. Other GIS data layers produced by Federal Agencies that are in the collection include a variety of specialized and diverse information, such as Bathymetric contours for the State of Florida and Surrounding Areas, Cancer Mortality in the State of Florida, Waterway Networks, Critical Habitat of Endangered Species, and Coastal Management Emergency Flood Data. The contributing agencies include the National Oceanic and Atmospheric Association, U.S. Geological Survey, U.S. Army Corps of Engineers, U.S. Fish and Wildlife Service, and the Federal Emergency Management Agency (http://www.fgdl.org).

GIS data that was produced at the state level include information compiled by Florida State Government Agencies, like the Florida Department of Environmental Protection, Florida Department of Transportation, Florida Marine Research Institute, and Florida Department of Revenue. GIS data from these agencies is as diverse as federally produced data layers and include summarized county tax records, aquifer statistics by county, brownfield locations, bus transit routes, railroad crossings, and surface water classification boundaries.

The GIS data collections that have been selected for the FGDL have a Florida focus, with some regional focus in North Central Florida, especially with data produced by University of Florida researchers (http://www.fgdl.org). Locally produced GIS data in the collection includes information that had been derived for projects by the GeoPlan Center at the University of Florida. The GIS data that has been produced involves planning and analysis activities that have a geographic focus on Florida. Some of the projects facilitated by the GeoPlan Center include an environmental analysis and national pollutant discharge elimination system (NPDES) database project, EPA southeastern landscape ecological analysis project, Florida statewide greenways systems planning project, recreational trail map series for the Florida Trail Association, and Ichetucknee Springs protection study.

Access and Organization

Users can access digital geospatial data through the FGDL homepage at http://www.fgdl.org. Upon entering the site, users interact with a simply designed homepage that is composed of some windows and hyperlinks to further information. One of the windows offers a menu of options, which describe the mission and scope of the FGDL and its digital data collections. The menu options also describe the types of software needed to view and use the GIS data that is available for viewing. It includes comprehensive instructions to view specific GIS files and to download the files. The FGDL has a site license for a suite of GIS software from ESRI Corporation, which includes different extensions of ArcMap, ArcView, and ArcExplorer. Other software that is used on the site is MrSID GeoViewer. The FGDL has developed its own software applications based on ArcView to view specific types of GIS files that it has in its collection.

:The FGDL offers as its main search tool a Web-based Metadata Explorer application. The application queries the contents of an ArcIMS Metadata Service. The ArcIMS Metadata Service contains FGDC metadata corresponding to all data layers in the FGDL. When users query the data using the metadata explorer, they are searching the contents of the metadata. Another hyperlink from the main menu on the homepage offers links to extensive technical documentation of the GIS data, which includes information such as data source, projection, and datum of GIS coverages. The Metadata Explorer offers the option for users to download data, In addition, users can simply browse through the metadata and read about FGDL data layers. The technical documentation is linked to an extensive metadata file that can be read either online or downloaded in an ASCII text format. The metadata file describes all of the FGDL GIS coverages and is essential in finding specific GIS data that the researcher may need.

- The FGDL offers data distribution options to users in both CD-ROM format and DVD formats. The data that is offered in these sets is diverse, for example, some DVD's contain data sets that are enumerated at the state level of Florida, including coastal areas. The data sets contained on the DVDs can include: Regulatory or Governmental Boundaries, Regional Planning or Ecological Data, Institutional Locations, Coral Reefs, Mangroves, and Bathymetry. Other data products would be familiar to most government documents librarians since the products contain aggregations of U. S. Census Data in CD-ROM format.
- U. S. Census data on CD-ROM are compilations of data taken from the U. S. Census Bureau, and offered to users in ESRI's shapefile format. The CD-ROMs contain historic spatial and attribute data dating back to 1970. The spatial data for 1970 and 1980 is in the form of centroid points and is down to the block group level. The 1990 spatial data is represented with both centroids and block group boundaries. The spatial data for 2000 is much more comprehensive and contains data enumerated at the block level. Other related datasets are enumerated at the block groups, census tracts, places, and county levels. The basic metadata structure for each item in the FGDL data collection includes a general data description (e.g., data source, scale of original source map, date, and geographic extent of the information). Feature attributes of the GIS data, detailed map projection parameters, and extensive notes on the quality of the GIS data to ensure the proper interpretation of the GIS data are included.
- The FGDL integrates some important aspects of other digital libraries, such as the use of a specialized user interface with a flexible browsing tool to establish and perform queries (Jankowska & Jankowski, 2000). The digital collection in the FGDL does not return queries by geographic footprint, such as the Alex-

andria Digital Library search interface (Hill et al., 2000), but by information taken from metadata files that can be accessed by Web browsers such as at the Cornell's University Geospatial Information Repository (Herold, Gale, & Turner, 1999). As a regional data center the FGDL has a fairly comprehensive collection of digital geospatial data. As a digital library, the FGDL meets the definition of a managed collection of information, with associated services, where the information is stored in digital formats and is accessible over a network (Arms, 2001).

Conclusion/Summary

Geospatial data has become an important part of contemporary socioeconomic processes, political activities, and academic research. It is integral to the functioning of geographic information systems (GIS), which are widely used by a community comprised of both government and private sector users. Management of geospatial data at the federal level has been guided by directives and recommendations from NARA and the FGDC. Selected federal guidelines and directives for handling geospatial data, both in print and digital formats, have been followed by both state and local government agencies, such as taxing authorities, and agencies that are involved in land-use regulation, and planning in Hillsborough County, Florida. A survey of the implementation and management of geospatial data at the federal, state, and local levels indicate that it is becoming part of the official record. Even though directives for archiving geospatial data have been written concerning documentation standards, standard archiving procedures for digital geospatial data remain very broad and require more definition, especially as more and more government agencies begin to use digital geospatial data as part of official transactions. Two critical issues in the collecting and archiving of geospatial data must address diversity of data, in content and format and the complexity of data as shown in the use of geospatial data as part of larger policy decision-making processes or daily work activities. Other critical issues address the continued growth in the use of geospatial data and the development of new software applications, and the obsolescence of formats and data.

Collection development activities for geospatial data and the facilitation of associated digital geospatial data collections have an impact on the establishment of GIS services in libraries. Expectations for services and resources can put unique demands on the library staff. The rapid development of GIS technology requires that library staff participate in regularly scheduled instruction to understand complex resources, improve computer skills, and learn new procedures about digital resources and applications in libraries (For a more thorough discussion on the geographic information literacy and training, see Chapters VII and IX). A number of researchers advocate a collaborative approach with other disciplines that use GIS, which would assist libraries in building digital geospatial collections and in preparing Web services (Boxall, 2002; Hyland, 2002).

- A raging debate on access to, and use of, public sector information is underway in the United States and in Europe. Several key points of the debate focus on definition, appropriate use, cost, and responsibility. Defining the public sector and what public sector information is key (Longhorn *et al.*, 2002). The repercussions of repurposing of traditional public sector data, including scientific data, by the private sector is yet to be determined. We have not yet determined answers to what the cost of collecting geospatial information is and who should bear that cost. A better question may be to maintain, if not increase, access to public sector information. One possibility is with collaborative partnerships among libraries. Several examples in the library literature provide information on exemplar data management and distribution policy builders, such as the Cornell University Geospatial Information Repository (CUGIR) the Harvard Map Collection, and the State University System of Florida (Aufmuth, 2006; Florance, 2006; Steinhart, 2006).

The Committee on Committee on Licensing Geographic Data and Services (2004) suggests that individuals involved in geospatial acquisition and collection share "contract negotiation experiences and techniques" and refresh "their understanding of data acquisition and dissemination options and user needs" (p. 1). The Committee also supports "unambiguous, standardized, and automated licensing," as a way "to improve coordination of data acquisitions" (p. 1). Librarians appear to be their best to ensure that these goals are met.

References

Arms, W.F. (2001). *Digital libraries.* Cambridge, MA: MIT Press.

Aufmuth, J. (2006). Centralized vs. distributed systems: Academic library models for GIS and remote sensing activities on campus. *Library Trends, 55*(2), 340-348.

Boxall, J. (2002). Geolibraries, the global spatial data infrastructure and digital Earth: A time for map librarians to reflect upon the moonshot. *INSPEL, 36*(1), 1-21.

Cho, G. (1998). *Geographic information systems and the law: Mapping the legal frontiers.* Chichester, England: J. Wiley & Sons.

Cho, G. (2005). *Geographic information science: Mastering the legal issues.* Hoboken, NJ: J. Wiley & Sons.

Committee on Licensing Geographic Data and Services. (2004). *Licensing geographic data and services*: National Research Council. Retrieved November 2006, from http://www.nap.edu/catalog/11079.html

Evans, G., & Saponaro, M. (2005). *Developing library and information center collections* (5th ed.). Englewood, CO: Libraries Unlimited.

Florance, P. (2006). GIS collection development within an academic library. *Library Trends, 55*(2), 222-235.

Herold, P., Gale, T. D., & Turner, T. P. (1999). Optimizing Web access to geospatial data: the Cornell University Geospatial Information Repository (CUGIR). *Issues in Science & Technology Librarianship, 21*(Winter), Article 2.

Hill, L. L., Carver, L., & Larsgaard, M. L. (2000). Alexandria Digital Library: User evaluation studies and system design. *Journal of the American Society for Information Science, 51*(3), 246-259.

Hugenholtz, P.B. (2001, April 19-20). The new database right early case law from Europe. Paper presented at the annual conference on International IP Law & Policy, Fordham University School of Law, New York. Retrieved November 2006, from http://www.ivir.nl/publications/hugenholtz/fordham2001.html

Hyland, N. C. (2002). GIS and data sharing in libraries: considerations for digital libraries. *INSPEL, 36*(3), 207-215.

Isaacs, E., Walendowski, A., Whittaker, S., Schiano, D. J., & Kamm, C. (2002). The character, functions, and styles of instant messaging in the workplace. In E. F. Churchill, J. McCarthy, C. Neuwirth, & T. Rodden (Eds.), *Proceedings of the ACM Conference of Computer Supported Cooperative Work (CSCW'2002)* Vol. 2002 (pp. 11-20). New York, NY: ACM.

Jankowska, M. A., & Jankowski, P. (2000a). Is this a geolibrary? A case of the Idaho Geospatial Data Center. *Issues in Science & Technology Librarianship, 19*(1), 4-10.

Kinikin, J., & Hench, K. (2005). Survey of GIS implementation and use within smaller academic libraries. *Issues in Science and Technology Librarianship,* (42), [n].

Kovacs, D. K., & Elkordy, A. (2000). Collection development in cyberspace: building an electronic library collection. *Library Hi Tech, 18*(4), 335-361.

Larsgaard, M. L. (1998). *Map librarianship: An introduction* (3rd ed.). Englewood, CO: Libraries Unlimited.

Longhorn, R. A., Henson-Apollonio, V., & White, J. W. (2002). *Legal issues in the use of geospatial data and tools for agriculture and natural resource management: A primer.* Mexico, DF: International Maize and Wheat Improvement Center (CIMMYT). Retrieved November 2006, from http://www.cimmyt.org/Research/NRG/map/research_results/tech_publications/IPRPrimer.pdf

Longstreth, K. (1995). GIS collection development, staffing, and training. *The Journal of Academic Librarianship, 21*(4), 267-274.

Manoff, M. (2000). Hybridity, mutability, multiplicity: Theorizing electronic library collections. *Library Trends, 48*(4), 857-876.

Marley, C. (2001). The changing profile of the map use. In R. B. Parry, & C. R. Perkins (Eds.), *The map library in the new millennium* (pp. 12-27). Chicago: American Library Association.

Mosher, P. H., & Pankake, M. (1983). A guide to coordinated and cooperative collection development. *Library Resources and Technical Services, 27*(4), 417-431.

National Spatial Digital Infrastructure. (2003). *Managing historical geospatial data records: Guide for federal agencies.* [s.l.]: NSDI. Retrieved November 2006, from http://www.fgdc.gov/library/factsheets/documents/histdata.pdf

Nichols, H. (1982). *Map librarianship* (2nd ed.). London: Bingley.

Onsrud, H. J. (1999). Liability in the use of GIS and geographical datasets. In P. Longley, M. Goodchild, D. D. Maguire, & D. Rhind (Eds.), *Geographical information systems: Management issues and applications* (p. 643–652). New York, NY: John Wiley & Sons.

Parry, B. (2001). Offline digital maps. In R. B. Parry, & C. R. Perkins (Eds.), *The map library in the new millennium* (pp. 72-87). Chicago; London: American Library Association; Library Association Pub.

Parry, R. B., & Perkins, C. R. (2001). *The map library in the new millennium.* Chicago; London: American Library Association; Library Association Pub.

Peterson, M.P. (2003). *Maps and the Internet.* Cambridge, England: Elsevier Scientific Press.

Pettijohn, P., & Neville, T. (2003). Collection development and acquisitions. In A. Hanson, & B. L. Levin (Eds.), *Building a virtual library* (pp. 20-36). Hershey, PA: IDEA Group Publishing.

Ristow, W. W. (1980). *The emergence of maps in libraries.* London, England: Mansell Publishing.

Safai-Amini, M. (2000). Information technologies: Challenges and opportunities for local governments. *Journal of Government Information, 27*, 471-479.

Steinhart, G. (2006). Libraries as distributors of geospatial data: Data management policies as tools for managing partnerships. *Library Trends, 55*(2), 264-284.

World Intellectual Property Organization. (2001). *Berne Convention for the Protection of Artistic and Literary Works (1886-1979).* Berne, Switzerland: WIPO. Retrieved November 2006, from http://www.wipo.org/treaties/ip/berne/index.html

Chapter IX

Geographic Information and Library Education

John Abresch, University of South Florida Libraries, USA

Ardis Hanson, University of South Florida Libraries, USA

Peter Reehling, University of South Florida Libraries, USA

Introduction

"I invite all of you to become geographers, if not by vocation then by avocation. GIS is about thinking geographically. Beyond being an essential component of GIS, geography also opens new avenues of examining and analyzing the world around us. More importantly, it provides us with totally new appreciation of everyday life and the environment in which we live it" (DeMers, 1997, p. 199). This quote sets the tenor for this chapter, in which we examine the educational requirements for librarians in the provision of GIS services. Implementing GIS services in academic libraries and facilitating associated digital geospatial data collections can be a daunting task for the librarian assigned these duties. The technical knowledge and computer skill-sets alone involved in understanding how GIS software operates are accompanied with a high learning curve. The research literature emphasizes collaboration with academic departments with the expertise in using GIS software. This chapter will cover the types of services that GIS users need for a prototypical GIS literacy project and basic geographic literacy for librarians. It will examine competencies in academic librarianship and geographic information literacy and offer a sample curriculum that meets the needs of geolibraries, librarians, and their patrons. A brief discussion of preservice and in-service issues, such as mentoring and communities of practice, follows. The conclusion discusses implications for library science in the preparation of new librarians and the professional development of practicing librarians.

Preparing Librarians for a Paradigm Shift

The new spatial paradigm is clear: "maps are data - numbers first and pictures later. They tell us where it is (inventory), and they provide insight into how it could be (analysis). In this context map analysis has become as an emerging discipline, recognizing fundamental map analysis operations independent of specific applications" (Parihar, 2002, ¶3). This paradigm shift has implications for skill-sets for practicing librarians as well as future graduates of library and information science programs. Since the bulk of GIS materials and resources to access GIS materials are electronic, or digital, in nature, a review of digital libraries seems appropriate. Choi and Rasmussen (2006) suggest that aligning digital library applications with traditional library collections and services requires "staff with new expertise that adds another dimension to library practice" (¶3). In their study of digital librarians[a] at ARL Libraries, the most frequently mentioned responsibilities were Web site-related tasks (35% of participants), policies and procedures (28%), collaboration (28%), supervision (26%), overall responsibility for digital projects/initiatives (26%), monitoring of technical standards and practices (21.7%), and writing and administrating grants (21.7%) (Choi & Rasmussen, 2006). Other researchers have also dealt with the role of the digital librarian, with core competencies and skills, depending upon the range of duties required (Arms, 2001; Chowdhury, 2002; Chowdhury & Chowdhury, 2003). Now, in addition to their traditional library skills and knowledge, professional librarians are expected to possess additional knowledge and skills required for work within the digital information world.

Basic Skills for Librarians

Librarians must have a variety of skills, ranging from the ability to engage in critical reflection to knowledge of different learning styles and teaching methods, including coaching and facilitation. Conceptual shifts must also occur, as librarians in academic and research settings move from the use of specific information tools towards viewing information as a holistic educational outcome based on transferable concepts and skills. Understanding the context, terminology, and political background that impact library services and staff is an essential skill, as librarians are required to develop new services and initiatives, create resources, and engage in increasingly non-traditional roles. There is also a responsibility to develop, deliver, and effectively evaluate these programs. The increased involvement in student-centered teaching and services requires librarians to understand academic requirements through physical and online classroom environments. Concurrently, librarians must be able to collaborate with faculty in designing learning activities that promote studentcentered learning and foster lifelong learning. In addition, librarians, especially those working within a GIS environment, will need a solid foundation in automated systems, computer and network technology, database searching, microcomputer applications, Internet searching, metadata, and resources in non-textual formats. Project management and supervisory skills and managing contract law, negotiation, and licensing may be paramount for the solo librarian as well as for librarians assigned to research and/or administrative/managerial roles.

Then there are the personal traits that foster confidence and success in the academic environment. These include the ability to learn continuously and quickly well, flexibility, creativity, risk-taking, facility in fostering and managing change, good interpersonal and communication skills, negotiation and conflict resolution skills, and the ability to work both independently and in a team setting. Good interpersonal and communication skills are still key. The literature in library and information science, as well as other service professions, identifies numerous methods by which librarians can communicate and market their professional skills, services, and resources. These methods include Internet-based dissemination, online chats, RSS feeds, blogs, "meet a librarian" sessions, features in local news sources, to name just a few. Margulies (2006) suggests that librarians "must continue to meet their objectives of supporting and in fact enhancing search, retrieval and understanding through the synthesis of information, and must also participate in active outreach and communication" (¶4). She also suggests that it is critical that librarians quantify value and effectively market "their solutions in order to raise their skill, talent and offerings' perceived value" (Margulies, 2006, ¶4). Communication skills may also include a technical component. It is not enough to understand software programs, network and online languages and technologies, digital and image technologies, programming and scripting languages, XML standards and technologies, or basic systems administration. One must be able to communicate with other technical staff or to communicate complex information to non-technical users.

Important Components in SLIS/GIS Curricula

Critical Thinking

A key issue in library education is the development of critical thinking and critical reflection, with transferable concepts and skills, rather than rote skills (Doskatsch, 2003; Harvey, 2001). Critical thinking is not a *"surface learning approach* (simply memorising and reproducing the content as presented with limited attention to application or transformation to new contexts), rather than a *deep approach* (in which the student intends to make sense of the content and develop a personal understanding)" (Johnston & Webber, 2003, p. 342). There is a seemingly limitless amount of library literature addressing critical thinking and information literacy. However, two problems must be addressed. First, many library organizations and librarians assume that once an individual has mastered a skill, one can "tick the box" and move on (Johnston & Webber, 2003), leading to a fragmented learning situation that does not encourage advanced problem solving, much less critical thinking. The second problem is that librarians may need more education about learning and teaching, as well as how best to evaluate student learning when assessing critical thinking skills and knowledge (Johnston & Webber, 2003; Pausch & Popp, 1997). It is not professional when "What is shown in the literature, for the most part, is user satisfaction with the one-shot session, when it is possible that the patrons do not know enough to be dissatisfied" (Pausch & Popp, 1997, ¶22).

With the emergence of critical GIS, information literacy will not be enough. The librarian must also understand and integrate issues of ontology and epistemology into the provision of GIS reference and research services. Formal ontologies, which form a logical universe, become the basis for communicating diverse epistemologies, which are ways of knowing the world (Schuurman, 2006). If an epistemology is like a category, then context shapes the formation and selection of categories (Fonseca, Davis, & Cmara, 2003). Therefore, changes in context or point of view lead to a shift in perceived and/or enumerated categories, which can lead to an alternative ontology that can be created in an environment (Fonseca et al., 2003; Schuurman, 2006). As librarians become more involved with researchers, especially in GIS services, understanding "GIScience ontology research and social theory perspectives on spatial relations, events, and processes" will be an essential skill (Schuurman, 2006), p.736). This is an important knowledge base and skill since critical geographers and other disciplines focus on how the production of knowledge reveals and reinforces certain relations of power (Sparke, 2000). With the increase in the use of surrounding community as "laboratory" for academic coursework and research, such "public scholarship" requires, as well as fosters, "critically informed inquiry, analysis and interpretation within community-based research venues for both students and teachers" (Jarosz, 2004, p.919). Library science education must produce graduates who can assist and participate in this type of research and classroom teaching.

The Importance of Pedagogical Models

Preservice librarianship training often does not provide a pedagogic grounding for designing and delivering training. Further, as discipline-based silos are slowly superseded by a relatively seamless culture of collaborative approaches to learning and teaching, the definition of educator is broadening in academic settings. Traditional teaching activities may mean "that an individual academic no longer has sole responsibility for curriculum decisions, materials and delivery design, student services and support, interaction with students, marking assignments and quality assurance of both course and the teaching and learning process" (Doskatsch, 2003, p. 112). Librarians are already involved in the reengineering of the teaching and learning environment through provision of resource discovery to support curriculum development, incorporation of information literacy into curricula, and participation on curriculum committees (Doskatsch, 2003; Varalakshmi, 2003). However, to teach GIS concepts, such as data modeling, remote sensing, and data collection, librarians will need to be conversant with not only the application of math and science concepts but how to teach them effectively to new users who may not have a science background (Bruce, 2003). Doskatsch (2003) reminds us that "...Effectiveness in this role requires the convergence of pedagogical knowledge, information expertise, technological competence, strategic skills and professionalism" (p. 113).

Doskatsch (2003) also argues, "the metamorphosis from librarian to educator and learning facilitator will not occur unless librarians are provided with opportunities to develop teaching competencies and pedagogic understandings" (p. 119). It is the responsibility of library educators, as well as individual librarians, library managers, and professional associations, to ensure that librarians have credibility in the role of educator. To gain credibility, performance indicators focused on staff expertise and quality assurance mechanisms to measure the effectiveness of what we teach are essential, not just the use of satisfaction measures. Librarians must also publish research evidence to "substantiate our claims that the educative role of librarians benefits teaching and learning outcomes" (Doskatsch, 2003). In short, "librarians need to become conversant as early as their preservice education with pedagogical concepts and how people learn. They may also need to develop the capacity to teach" (Bundy, 2001, p. 4).

Spatial Concepts

To work effectively with geospatial data in answering research questions to training new users in the use of resources and applications requires the librarian to understand spatial concepts, tools used to create representations of geospatial data, and cognitive processes using geospatial data. Parihar (2002) suggests that GIS is ubiquitous in academic settings. Further, that as students progress to graduate, doctoral, and postdoctoral programs, the need to use GIS increases across all disciplines, from the "hard" sciences (Geology, Engineering, etc.) to the "soft" sciences (Biology, Zoology, Botany etc.) to even Architecture and Public Health. Table 1 addresses many of the spatially based questions encountered in academic settings (The reader is also referred to Chapter VII, which addresses spatial concepts and competencies in the provision of reference and research services).

Much of the newer literature defining GIS services in libraries, especially digital library environments, have been written by individuals from outside of the library community. However, the "ideals and beliefs held by librarians (curators and archivists as well) however, are not shared by geographers and GI scientists, and the reverse may also be true" (Boxall, 2002, p. 1)—and that any differences between researchers can be attributed to changes in respective disciplines and academic focus (Boxall, 2002).

With the development of digital information networks and ubiquitous distribution of computing, researchers from disciplines outside of library information science, such as computer science and engineering, have borrowed its organizational concepts to organize growing collections of data. Even with the differences in academic foci, the shared goal of facilitating access to data can be a framework for collaboration between disciplines. Geographers and other individuals from related academic disciplines, who work in the various forms of geolibraries, can become partners with librarians in an effort to "increase access, use and preservation of cartographic materials and geospatial information" (Boxall, 2002, p. 1). With their long history of providing information services to the public and to the academic community, librarians can contribute to the collection, cataloging, and facilitation of geospatial data in a digital library setting.

An integrative approach to designing educational programs should be taken in training professional librarians to provide GIS services in a digital library setting, and in training new librarians to meet the challenges of providing GIS services in academic libraries. Focused technical programs may suffice for experienced librarians but, for beginning librarians, geographic information coursework can be designed

Table 1. Spatial information in academia

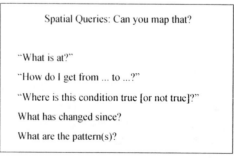

Spatial Queries: Can you map that?
"What is at?"
"How do I get from ... to ...?"
"Where is this condition true [or not true]?"
What has changed since?
What are the pattern(s)?

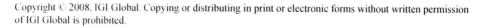

for library graduate educational programs. In order to obtain the necessary skills needed to work with the technical, service, and managerial aspects of delivering GIS services, LIS education could include subjects from other academic discipline, such as geography and epidemiology. The following section is a brief review of the ARL Geographic Information Systems Literacy Project as an example of the diverse GIS environments in academic/research libraries.

The ARL Geographic Information Systems Literacy Project

In 1992, the ARL, in partnership with Environmental Systems Research Institute, Inc. (ESRI), launched the GIS Literacy Project. Member libraries were invited to send librarians to ESRI for free training in using its GIS software. Seven years later, a study was conducted by the ARL to measure the impact of the program upon participating libraries, especially in their design of GIS delivery programs. Surveys distributed to participating libraries covered in four main categories: (1) general information about the library's role in delivering GIS services; (2) the number, level, and academic preparation or other training of staff involved; (3) the amount and kind of equipment, software, and data files that support GIS in the library; and (4) the kind of service offered and by whom it is used. Sent to 121 member libraries, the response rate was 60%, or 72 institutions. Of the 72 libraries, 64 reported that they provide GIS services. These services were administered by both the library (53 universities; 83%) and by academic departments offering GIS courses (45 universities; 70%). Among libraries that offer GIS services but did not administer them, the most common activity was offering guidance in finding appropriate data sets.

Data from the survey indicated that GIS services in responding libraries were mostly facilitated by the government documents center (48%) or the map library (52%). Subject bibliographers offered the service at 23% of responding institutions. Only three libraries (5%) reported having a discrete GIS unit and only seven (11%) provide the service at the general reference desk (Association of Research Libraries, 1999).

Data from the survey also indicated that most staff members in charge of a library's GIS services were librarians with an MLS degree (81%). In addition to the MLS, 54% of the GIS librarians held at least one additional graduate degree. The "typical" ARL library devoted the following staff resources to GIS services: a librarian, a support staff member, a graduate assistant (10 hours per week), and a student worker (10 hours per week), with librarians and support staff having other duties. The most common GIS training among respondents was the ARL GIS Literacy Project, to which 37 libraries sent librarians. GIS librarians at 31 institutions had training by GIS software providers, GIS librarians at 28 libraries learned GIS in coursework. Library staff provided technical support for GIS hardware or software at 51 institutions (80%). GIS services also varied widely across the ARL Libraries due to the technical knowledge and interdisciplinarity required to use GIS. Except for the specialist in the discipline, self-service in GIS was usually not adequate. GIS users also required more time and effort on the part of the librarian to produce satisfactory results. What this survey does indicate is that, in addition to the requisite skills needed by librarians in today's hybrid libraries, additional skill sets are basic for librarians who want to work with geospatial data.

Promoting Geographic Literacy: Skills Needed by Librarians, Educators, and Students

Howser and Callahan (2003) identify three levels of geographic information literacy that are an essential part of GIS instruction. They suggest that, in order to promote geographic literacy efforts, "libraries should provide workshops that balance background information with GIS hands-on exercises. In order to address every potential participant's skill level, GIS instruction must be provided in a three-tier approach: introductory, intermediate and advanced" (Howser & Callahan, 2004, p. 3).

In addition to Howser and Callahan, Hyland (2002) also advocates a three-tiered approach to the delivery of GIS services in libraries: a basic, or "bare-bones" GIS service, a physical GIS collection, and the digital library (or clearinghouse node) (p. 209). Similar to Howser and Callahan, Hyland also identifies a number of skills

Table 2. Levels of instruction

Levels	Instruction
Introductory	Essentials of geographic literacy: • the basics of map projections, scale, legend construction, data classification, and color usage. Critical analysis: • participants are shown three maps (same data, different classification methods) to show how maps can be misleading Hands-on training and exercises using ESRI's ArcGIS software Upon completion of this workshop, the foundation of geographic literacy has been established and allows for further GIS instruction.
Intermediate	Guided in-depth instruction on ArcGIS • georeferencing, data preparation, and shapefile creation • using ArcGIS extensions. Each session includes hands-on exercises and time for questions.
Advanced	One-on-one instruction • preparing a GIS map for printing • modifying datasets and boundary files.

needed by library staff to deliver the GIS services according to identified level. The first tier, a "bare-bones" service, requires only the time commitment of a public services librarian who can help patrons navigate various data nodes and has a familiarity with free GIS software resources, such as ESRI's ArcExplorer (Hyland, 2002). The library, in this role, is simply as an intermediary, helping the user to find the data and providing only the most basic of support for Web-based mapping systems. Therefore, any public services librarian with an interest and aptitude "should be able to provide a minimum-level GIS service after about 20 hours of work in either a hands-on workshop or through a self-paced tutorial" (Hyland, 2002, p. 209). The second tier, supporting a physical GIS collection, requires more monetary and staff resources. In addition to the hardware and software costs, considerable staff time needs to be allotted to cataloging, providing public service support, and software training. For the third tier, a geolibrary, Hyland uses the Cornell University Geospatial Information Repository (CUGIR) at the Mann Library as an example (Hyland, 2002). In this geolibrary, librarians from technical services, collection development, information technology, and public services all participate in providing GIS services. The technical services librarian supports metadata services within CUGIR, for example, data are described with FGDC metadata standards and converted into XML, SGML, HTML, Dublin Core, and MARC to allow for the broadest possible access to the data. The information-technology team member supports the server on which CUGIR is housed, designs new relational databases for better access, and provides all programming needed for the Web and the Z39.50 interface. Collection development staff assist in refining the preservation and collection policies. The public services librarian is responsible for end-user support, and is the primary contact and negotiator with data partners (Hyland, 2002).

In reflecting upon the delivery of services in libraries, library and information science educators should ensure that new generations of librarians are equipped to "facilitate the most effective use of vast amounts and kinds of information" (Varalakshmi, 2003, p. 44). Library and information science (LIS) graduates are expected to have critical thinking, intellectual, and technical skills needed for the profession. Varalakshmi (2003) also advocates flexible learning styles wherein LIS students are prepared for library environments that have been altered by emerging information technologies. In order to illustrate her viewpoint, she cites a sample survey, conducted to seek the opinions of academic, special, and public librarians on the existing LIS educational programs in the State of Andhra Pradesh (India). Survey respondents highlighted the following as critical to LIS education:

- "Focus on producing knowledge managers than mere librarians."
- "Focus on imparting knowledge on Web-based services as the future belongs to the Internet."

- "Organize more fields of study and interact with working librarians to gain knowledge on real time situations and enables them to blend the learning with working skills."
- "Have a more practical component in curriculum as IT is practice based."
- "Offer specialization in the areas of Knowledge Management, Multimedia systems, Web design and development and Digital libraries."
- "The expectation is for a high-level performance of the fresh librarian with professional knowledge, technological skills, communication skills, managerial capabilities and attitudinal flexibilities, or in simple term a skilled digital librarian." (Varalakshmi, 2003, p. 45)

Varalakshmi (2003) also emphasizes an LIS curriculum that is open to integrating new technologies and ideas to meet professional challenges in library environments. Areas that should be addressed in the development of educational and training programs in GIS and library and information science include the nature of information environment, de-institutionalization of education, levels of IT component in curriculum, assessment of job market, professionalization of librarianship, knowledge base, teaching practices and technological competencies of faculty (Varalakshmi, 2003, p. 48).

Another survey of information services delivery analyzed the success and failure of GIS services in two libraries that had participated in the ARL/GIS Literacy Project: the library at SUNY Albany and the New York State Library (Shawa, 1998). The SUNY Albany library was not successful in expanding their small desktop GIS service to their library users, while the New York State Library managed to make GIS a successful library tool. A review of the library at SUNY Albany indicated several major problems. The first major problem was the lack of staff preparation and training for the service. One staff librarian had training in GIS, but had no regularly assigned hours to provide the service. When she was not available, none of the other staff could assist patrons. A second major problem was the lack of planning for the service. Without understanding the service parameters, technology, and staff preparation, GIS service delivery will fail (Shawa, 1998).

The discussion of information service delivery in libraries illustrates the need of extensive and continuing training for library staff. Surveys of information delivery in libraries indicate a staff training need that extends far beyond the instructional benefits of a workshop into more extensive coursework, or possible certification. Based upon the need for skills to provide GIS services in any of the three tiers described so far, new approaches to graduate school curriculum can be advocated. There should also be numerous training options available. Online training courses and self-paced tutorials from vendors, such as ESRI and MapInfo, may suffice for the bare-bones services discussed earlier by Hyland (2002). Ultimately, for a thorough preparation of librarians to enable a more effective delivery of information service to patrons, geographic information literacy components must be integrated into graduate library science programs. Beginning with core competencies, as defined by ALA, curriculum can be developed that would integrate new information technologies with long-standing tenets of professional librarianship.

Developing a GIS-Integrated Curriculum for Library and Information Science

Planning for coursework for beginning librarians with developing technologies, such as GIS, can begin with examining some core competencies as advocated by the American Library Association (ALA). ALA has identified a number of skills and knowledge bases that are basic to a librarian or information professional in contemporary society (American Library Association, 2005). The list of competencies also provide a rationale for working with other disciplines in the building of new geographic information services, especially in terms of working with computer technologies.

The development of curriculum from core competencies of librarianship, such as knowledge organization and knowledge dissemination, create a base of knowledge from which outside subjects provided by courses outside of the department may be integrated. Often, parallels are drawn between educational priorities between other disciplines and Library and Information Science. The suggested areas of LIS study as advocated by Varalakshmi (2003) mirror the research-based GIS graduate educational priorities advocated by the University Consortium for Geographic Information Science (UCGIS). Composed of research universities whose mission is not only to monitor, but also to advance emerging technologies as they apply to geographic information science, the UCGIS provides a forum for sharing knowledge gained by individual research efforts involving advanced technology. By examining, formalizing, and combining approaches to research and to education that use new and innovative techniques, the UCGIS works to codify GIScience. Educational priorities of research-based GIS education of the UCGIS are:

- "**Supporting infrastructure:** Science advances more rapidly with institutional and outside support for research activities.
- **Emerging technologies:** Creating as well as exploiting technologies.
- **Professional Education:** Research plays an important supporting role for data, tools, and course content provided to the professional student.
- **Learning with GIS:** Access to research-based instruction promotes learning in all settings, thereby building the connections that cause out-of-the-classroom experiences to contribute to in-the-classroom achievement. There needs to

Table 3. American Library Association competency statements

Knowledge Organization	Can apply the basic principles involved in the organization and representation of knowledge and information structures. Understands the system of standards and methods used to control and create information structures.
Technological Knowledge	Demonstrates a comprehension of current information and communication technologies, and other related technologies, as they affect the resources and uses of libraries and other types of information providing entities. Has basic knowledge of the concepts and processes related to the assessment and evaluation of the specifications, economic impact and efficacy of technology-based products and services. Understands and can apply the principles of techniques used to continuously track and analyze emerging technologies to recognize relevant innovations. Demonstrates proficiency in the use of standard information and communication technology and tools consistent with prevailing service norms and professional applications.
Knowledge Dissemination: Service	Knows and demonstrates service concepts, principles and techniques that facilitate information access, relevance, and accuracy for individuals or groups of users. Can retrieve, evaluate and synthesize information from diverse sources for use by individuals or groups. Can interact with individuals or groups of users to provide consultation, mediation or guidance in their use of information resources. Can recognize and respond to diversity in user needs and preferences for resources and services. Can relate assessments of emerging or chronic situations, circumstances or conditions to the design and implementation of appropriate service and resource responses.
Knowledge Accumulation: Education and Lifelong Learning	Can interact with individuals or groups of users to provide consultation, mediation or guidance in their use of information resources. Knows basic learning theories, instructional methods, and achievement measures, and can apply them to learning situations within libraries and other information providing entities. Understands the principles related to the teaching and learning of information seeking, evaluating and using concepts, processes and skills. Appreciates the importance of continuing education and lifelong learning as principles of good service, and as personal guidelines for continuous professional development.
Knowledge Inquiry: Research	Understands the nature of research, research methods and research findings within the library and information fields and has an awareness of current literature in these and related areas. Is familiar with the fundamentals of research, survey and data collection designs of current or potential value to library and information settings.
Institution Management	Knows the fundamental principles of planning, management and the evaluation of libraries or other information providing entities. Is aware of the currently prevailing types of library and information professions, and information providing settings. Displays a knowledge of how change occurs, and how institutional and individual change strategies and options are developed. Demonstrates oral and written communication skills necessary for group work, collaborations, and professional level presentations. Has a grasp of concepts behind, and methods for, developing partnerships, collaborations, networks and other structures within a community of stakeholders. Demonstrates the ability to serve a diversity of stakeholders. Understands the basic principles related to reaching specific audiences and promoting concepts or services.

Source: http://www.ala.org/ala/accreditiationb/Draft_Core_Competencies_07_05.pdf

Table 4. The NCGIA core curriculum in GIScience

1. FUNDAMENTAL GEOGRAPHIC CONCEPTS FOR GISCIENCE
1.1.	The World in Spatial Terms
1.1.1.	Human Cognition of the Spatial World
1.1.2.	Asking Geographic Questions
1.2.	Representing the earth digitally
	features, pictures, variables; points, lines, areas, fields, 3D; processes and time
1.3.	Position on the earth
1.3.1.	Coordinate Systems Overview
1.3.2.	Latitude and Longitude
1.3.3.	The Shape of the Earth
1.3.4.	Discrete Georeferencing
1.3.5.	Global Positioning Systems Overview
1.4.	Mapping the earth
1.4.1.	Projections and transformations
1.4.2.	Maps as Representations of the World
1.5.	Spatial relationships
	connections and topology; networks; distance and direction; flow and diffusion; spatial hierarchies; boundaries; spatial patterns; attributes of relationships
1.6.	Abstraction and incompleteness
1.6.1.	Sampling the World
1.6.2.	Line Generalization
2. IMPLEMENTING GEOGRAPHIC CONCEPTS IN GISYSTEMS
2.1.	Defining characteristics of computing technology
2.1.1.	Fundamentals of Data Storage - Carol Jacobson
2.1.2.	Algorithms
2.1.2.1.	Simple Algorithms for GIS I: Intersection of Lines
2..1.2.2.	Simple Algorithms for GIS II: Operations on Polygons
2.1.2.3.	The Polygon Overlay Operation
2.2.	Fundamentals of computing systems
	operating systems; programming languages and software engineering; developing algorithms; user interfaces; computer networks; hardware for GISystems
2.3.	Fundamentals of information science
2.3.1.	Information Organization and Data Structure
2.3.2.	Non-spatial Database Models
	data modeling
2.4.	Representing fields
2.4.1.	Rasters
2.4.2.	TINs
2.4.3.	Quadtrees and Scan Orders
	polygon coverages
2.5.	Representing discrete objects
	storing relationships; computing relationships; topology for geodata; object hierarchies
2.6.	Representing networks
2.7.	Representing time and storing temporal data
2.8.	Populating the GISystem
	creating digital data - sampling the world; remote sensing; GPS as a data source; digitizing and scanning; editing accessing existing data - data exchange; open GIS; finding data; data conversion; transfer standards; distributed networked databases; generating data from existing data metadata
2.9.	Kinds of geospatial data

continued on next page

Table 4. continued

2.9.1.	Transportation Networks
2.9.2	Natural Resources Data
2.9.2.1.	Soil Data for GIS

hydrography; land cover and vegetation; geology; climate; terrain

2.9.3.	Land Records

administrative boundary data; demographic and health data; global data

2.10.	Handling uncertainty
2.10.1.	Managing Uncertainty in GIS2.10.2. Uncertainty Propagation in GIS
2.10.3.	Detecting and Evaluating Errors by Graphical Methods
2.10.4.	Data Quality Measurement and Assessment

storing uncertainty information

2.11.	Visualization and cartography
2.11.1.	cartographic fundamentals

principles of graphic design; digital output options; scientific visualization; animation and virtual worlds; cognitive basis of visualization

2.12.	User interaction

user interfaces; forms of user interaction with GIS

2.13.	Spatial analysis

combining data; map algebra; terrain modeling; finding and quantifying relationships; generalization; spatial statistics; geostatistics; spatial econometrics; spatial interpolation; spatial search; location/allocation; districting; spatial interaction modeling; cellular automata; distance modeling; neighborhood filtering; pattern recognition; genetic algorithms

2.14.	Implementation paradigms
2.14.1.	Spatial Decision Support Systems
2.14.2	Exploratory Spatial Data Analysis
2.14.3	Process Modeling and Simulation
2.14.4.	Multimedia and Virtual Reality
2.14.5.	WebGIS
2.14.6.	Artificial Neural Networks for Spatial Data Analysis

interoperability; object oriented GIS; knowledge based and expert systems; collaborative spatial decision making

3. GEOGRAPHIC INFORMATION TECHNOLOGY IN SOCIETY

3.1.	Making it work

needs assessment; conceptual design of the GIS; survey of available data; evaluating hardware and software; database planning and design; database construction; pilot studies and benchmark tests; acquisition of GIS hardware and software; GIS system integration; GIS application development; GIS use and maintenance

3.2	Supplying the data
3.2.1.	Public access to geographic information
3.2.2.	WWW Basics
3.2.3.	Digital Libraries
3.2.4.	Legal Issues

standards; national and international data infrastructures; marketing data

3.3	The social context

digital democracy; geographic information in decision making; human resources and education; ethics of GIS use

3.4	The industry

history and trends; current products and services; careers in GIS

3.5.	Teaching GIS
3.5.1.	Curriculum Design for GIS
3.5.2.	Teaching and Learning GIS in Laboratories

4. APPLICATION AREAS AND CASE STUDIES

4.1.	Land Information Systems and Cadastral Applications
4.2.	Precision Agriculture

also: facilities management; network applications; emergency response and E911; recreation; resource management (agriculture, forestry); urban planning and management; environmental health; environmental modeling; emergency management; studying and learning geography; business and marketing (real estate)

Source: http://www.ncgia.ucsb.edu/education/curricula/giscc/

be an assessment of how research can support (and equally important, where it cannot provide) an environment conducive to the development of spatial information skills.

- **Educational policy:** Research coupled with cutting-edge education can transform the way we learn and teach, if not the entire educational framework. Bad research will not attract or keep the best graduate students. Therefore, research plays an important role when establishing educational policies" (Research-based Education Working Group, 1997).

An expansion of geographic information literacy components from the UCGIS priorities is outlined in Table 4, National Center for Geographic Information & Analysis (NCGIA) Core Curriculum. The table outlines how geographic information concepts integrate into applied aspects of GIS. Section 3 "Geographic Information Technology in Society" is especially relevant to library and information science.

Current SLIS Curricula

There can be many variations on the amount of coursework offered in Library Science programs to enable students to "hit the ground running" as they enter the profession. Opportunities range from introductory coursework or special topic course to certificates to dual degree programs. For example, the University of Missouri offers a seminar in map librarianship introducing cartographic resources, the basics of cartography and map reading, issues in reference, collection development, and preservation peculiar to cartographic material, and electronic mapping. Simmons College offers a course, "Managing spatial information," that covers the principles and fundamentals of spatial information librarianship, an overview of geographic information systems, and spatial visualization and modeling of government produced and commercially distributed digital data. The University of Pittsburgh's School of Information Sciences has a track in geoinformatics[2] that has a strong emphasis on information science.

Currently, not only two SLIS programs offer a formal degree program integrating geography and library science. The University of Wisconsin-Milwaukee Coordinated Degree program is comprised of 48 credits, of which 18 are required in Geography. The program adds that the program "normally builds upon an undergraduate background in that discipline" http://www.uwm.edu/Dept/SLIS/academics/coordinatedMLIS/geography.htm. In addition to the 30 credit hour courses in LIS, the required courses in Geography include an introduction to techniques of research and presentation, quantitative analysis, and cartography. Students then have a choice of Growth in Geographic Thought or Theory and Methodology in Geography. Only the A summer field course and two geography seminars round out the program. However, the University of Maryland Geography/Library & Information Systems

Table 5. Mapping current LIS coursework to UCGIS guidelines

SUPPORTING INFRASTRUCTURE	
"Science advances more rapidly with institutional and outside support for research activities."	
U W-Milwaukee	U Maryland
The Academic Library *	Information Environments #
The Public Library *	Information Access in Electronic Environments
The Special Library and Information Center *	Information Access in the Humanities
Introduction to Information Science *	Information Access in the Arts
Theory and Methodology in Geography *	Information Access in the Social Sciences
Introduction to Reference Services and Resources *	Information Access in Science and Technology
Ethics and the Information Society *	Information Access in the Health Sciences
Information Policy *	Access to Legal Information
Information Technology & Organizational Context	Access to Business Information
	Access to Federal Government Information
	Library Service to the Disadvantaged
	User Instruction
	Seminar in International and Comparative Librarianship and Information Science
	Planning and Evaluating Library Services
	Knowledge Management
	Libraries and Information Services in the Social Process
	Public Library Seminar
	Library Services for Client Groups with Disabilities
	Seminar in Library and Information Networks
	Seminar in the Academic Library
	Legal Issues in Managing Information
	Designing Information Products and Services
	Seminar in the Special Library and Information Center

EMERGING TECHNOLOGIES	
"Creating as well as exploiting technologies."	
U W-Milwaukee	U Maryland
Library Automation *	Principles of Human-Computer Communication
Information Systems: Analysis and Design*	Principles of Software Evaluation
Advanced Topics in Information Retrieval *	Database Design
Multimedia*	Introduction to Expert Systems
Seminar: Geographic Techniques*	Building the Human-Computer Interface
Contemporary Geographic Approaches*	Information Policy
Microcomputers for Information Resources Management *	Information Technology
Digital Libraries*	Library Systems Analysis

continued on next page

Table 5. continued

Information Marketing*	Principles of Records and Information Management
Electronic Networking and Information Services*	Seminar in Information for Decision-Making
Advanced Use of Microcomputers in Library and Information Management*	Recent Trends and Issues in Library and Information Services

PROFESSIONAL EDUCATION

"Research plays an important supporting role for data, tools, and course content provided to the professional student."

U W-Milwaukee	U Maryland
Indexing and Abstracting	Information Use
Cataloging and Classification	Cataloging and Classification
Techniques of Research and Presentation	Library and Archives Preservation
Introduction to Techniques of Research and Presentation	Government Information Sources and Services
Internship in Geography Seminar: Geographic Techniques	Seminar in Technical Services
Research Methods in Library & Information Science	Research Methods in Library and Information Studies
	Documentation, Collection, and Appraisal of Records
	Seminar in Archives, Records, and Information Management
	Advanced Archival Administration
	Archival Principles, Practices, and Programs
	Bibliographic Control
	Seminar in the Organization of Knowledge
	Classification Theory
	Seminar in Linguistic Topic
	Construction and Maintenance of Index Languages and Thesauri

LEARNING WITH GIS

"Access to research-based instruction promotes learning in all settings, thereby building the connections that cause out-of-the-classroom experiences to contribute to in-the-classroom achievement. There needs to be an assessment of how research can support (and equally important, where it cannot provide) an environment conducive to the development of spatial information skills."

U W-Milwaukee	U Maryland
Introduction to the History of European Cartography Before 1500 A.D*	Advanced Computer Cartography #
Advanced Geographic Information Science: Geographic Modeling*	Geographic Information Systems and Spatial Analysis #
GIS/Cartography Internship*	Design for Geographic Information Systems#
Instructional Technologies*	Quantitative Spatial Analysis#

continued on next page

Table 5. continued

Intermediate Geographic Information*	Computerized Map Projections and Transformations#
Geographic Information Science*	Information Retrieval Systems
Cartographic Design*	Seminar in Research Methods and Data Analysis
Problems in the History of Cartography*	Access Techniques and Systems for Archives
GIS and Society*	Collaborative Instructional Design and Evaluation
Cartographic Design*	Information Access
	Information Structure

EDUCATIONAL POLICY

"Research coupled with cutting edge education can transform the way we learn and teach, if not the entire educational framework. Bad research will not attract or keep the best graduate students. Therefore, research plays an important role when establishing educational policies."

U W-Milwaukee	U Maryland
Managing Library Collections*	Culture and Natural Resource Management#
Organization of Information *	Information Architecture I #
Management of Libraries and Information: Information and Records*	Principles of Records Management
An Introduction to Modern Archives Administration *	Selection and Evaluation of Resources for Learning
Competitive Intelligence and Business Information*	Integrating Technology into Learning and Teaching
Information Sources and Services in the Health *	Management and Administration for the Information Professional
Legal Issues for Library and Information Managers*	
Seminar in International and Multicultural Information Services*	
Seminar in Library Administration*	

Source: University of Wisconsin-Milwaukee Geography MA/MLS Coordinated Degree Program from http://www.uwm.edu/Dept/SLIS/academics/coordinatedMLIS/geography.htma

program is a dual degree program, resulting in two Master's degrees: the Master of Library Science (MLS) and the Master of Arts in Geography. With a minimum of 54 graduate credit hours, students must apply separately, and be admitted to both the College of Library and Information Services and to the Geography Department.

Table 5 illustrates the integration of curricula in Geography and Library and Information Science by examining the course offerings from the degree programs of the University of Wisconsin-Milwaukee and the University of Maryland.

1. University of Wisconsin-Milwaukee Geography MA/MLS Coordinated Degree Program from http://www.uwm.edu/Dept/SLIS/academics/coordinatedMLIS/geography.htm.
2. University of Maryland Geography/Library Science Dual Degree Program from http://www.gradschool.umd.edu/catalog/programs/printable.cfm?code=223.

What is important to take away from this comparison is that elements of GIS may well be in many curricula or easily integrated, may be incorporated into a variety of course offerings, and may be offered jointly for those students wishing to receive a second degree or minor in a subject area. The Guide to Geographic Information Systems lists over 200 U.S. schools offering GIS certificates and over 100 schools offering associate or bachelor GIS degrees (http://www.gis.com/education/formal.html). A 2006 publication by the Association of American Geographers includes 10 knowledge areas, 73 units, 329 topics, and over 1,600 formal educational objectives. The *GIS&T Body of Knowledge* serves as a resource for course and curriculum planning for academic and professional programs at four-year and two-year institutions, as a basis for professional certification and program accreditation, and as a resource in continuing professional development (DiBiase, DeMers, Johnson, Kemp, Taylor Luck, et al. 2006). Both of these resources have postgraduation competencies should enable library and information educators to build in those competencies or to find additional or complementary niche areas to possibly start their own joint programs.

To start such a program, it is critical to focus on what a student is expected to be able after completing the certificate, such as acquiring a higher-level position, transferring career paths, achieving X in continuing education credits, enabling a move organizationally, and so forth. It is also worthwhile to determine what credentialing or postdegree involvement is required from professional organizations. Once what possible outcomes are determined, design of the certificate or supplemental curriculum can proceed more readily. It is important to eliminate redundancy in the coursework and to not overburden the student. Finally, required courses may be modularized to incorporate their content into other department's courses, broadening the opportunities to reach new students interested in GIS and in library and information science.

Linking Preservice to In-Service Training

Unlike other professional schools, such as social work and education, librarianship does not have a requirement for its curriculum to link student-based curriculum (preservice) to continuing education or certification programs post graduation (in-service). Often, librarians struggle to mediate and balance what they have learned in their graduate programs to real-world practice. Consider too that working conditions, and institutional and organizational practices that make digital libraries most usable continue to change, often by exponential, not arithmetic, measures. Further, the emerging cultural model for today's library include focal technologies and numerous interrelationships with users, other groups, and organizations using these technologies, expecting new delivery methods of services and resources, and requiring novel, and sometimes immediate, instruction and training.

Continuing Education

A tiered approach in delivering GIS information services is advocated, depending on need, from the most basic locational question to more intense research support (Hyland, 2002). Even the most minimal of GIS services requires significant investment in training programs and resources for staff to acquire GIS-related skills (Dereksen, Sweetkind, & Williams, 2000; Kinikin & Hench, 2005; Strasser, 1995; Wikle, 1998; Wikle & Finchum, 2003). In a review of geolibraries, data infrastructures, and their impact on field of map librarianship, librarians are encouraged to take a more holistic approach to training and to work with other disciplines in creating development opportunities for librarians with GIS technologies. Boxall (2002) writes "And what of training and professional development? Because of changes in GIS and cartography, not to mention other disciplines, we need to make sure that our skills are current" (p. 12). A study done in Canada about human resources issues with geographic information systems illustrates changes in other disciplines: in "Canada, a broader study of the human resource issues, jointly funded by the Geomatics Industry Association of Canada and the Canadian Institute of Geomatics, and the Canadian Association of Land Surveyors called the HAL, suggests that the demand from government and industry for highly qualified GIScientists and technicians will continue to grow" (Boxall, 2002, p. 12).

Besides emphasizing the need to collaborate with other disciplines in creating GIS education programs, the study also describes a need for professional and mid-career training. Boxall (2002) envisions that a closer alliance with disciplines, such as Geography, Cartography, and Geographic Information Science, will integrate the field of librarianship into the cartographic visualization process as described by (Kraak & Ormeling, 1996). As part of this process, librarians would become involved in all aspects of geographic visualization rather than associated with just the output of mapping software programs (Kraak & Ormeling, 1996). Further, collaboration with other disciplines that use GIS software and information management technology enables librarians to quickly adapt to changes in service demand from their community of users (Boxall, 2002). After all, information literacy, analytical reasoning, and model-based reasoning, skills important in librarianship and geographic visualization, begin with the ability to gather data about an environment, to understand cause and effect relationships, and to do deductive reasoning within an environment (Vitolo & Coulston, 2002). Parallel to the evolvement of mapping technology, the move from descriptive mapping to prescriptive mapping has also increased the value of GIS to decision making and management. These are all areas that can be seen as potential professional development and continuing education opportunities for library and information science educators.

Mentoring

Simply defined, mentoring is the process by which one individual assists another individual to learn something that the latter may or may not have learned by herself or himself. Mentoring is often considered "… a professionally supportive relationship between an experienced, successful mid- career employee and a beginner. It is a time-honored method of encouraging new talent, of sharing expertise and connections, and of providing rapid, upward mobility" (Lary, 1998, pp.23). There are a number of benefits from mentoring (Munde, 2000). Mentoring provides new knowledge and skill sets that can result in higher salary, promotion, and overall satisfaction with their career choice. There is a also a renewed sense of what it means to be a professional, possible recognition and impact on the profession, and satisfaction with a career choice. For the organisation, there is better staff retention, reduced turnover, better staff orientation to existing cultural and organisational norms and expectations (Munde, 2000). Mentoring can also improve the integration of leadership at all levels of the organisation (Munde, 2000), increase professional and personal empowerment/development, open new partnerships for research and scholarship, or create new service opportunities in the university-community venue (Nofsinger & Lee, 1994).

There are a number of examples of formal mentoring programs in librarianship. Kuyper-Rushing (2001) describes elements of a successful formal mentoring program for tenure-track librarians at Louisiana State University Libraries. The mentor, not the supervisor, serves as advisor and advocate for the protégé. Also, participants were told "mentors serve in an advisory capacity only and the establishment of a mentoring relationship should not be seen as a guarantee of advancement or career success for either protégé or mentor" (Kuyper-Rushing, 2001, p. 43). The program was successful in ensuring participation in professional service and in identifying research areas appropriate for the individual and was, at the time of publication, addressing the implications of posttenure review mentoring process (Kuyper-Rushing, 2001). Ricker (2006) describes a mentoring program designed for GIS librarians based on the key differences between GIS librarianship and other forms of librarianship. She believes that there is a variety of roles for a mentor: from "addressing job and career development – including mentoring and coaching" to being a "process advisor and consultant" (p. 356). Her 10-point agenda includes "set goals.., have regular meetings with an agenda…, challenge the protégé…, when you learn, teach them…, take the protégé to meetings….,be patient …, give protégés room to develop their own areas…, provide group training sessions…, be available for questions…, be actively involved" (pp. 357-359).

Although most mentoring programs stress the development of a professional relationship and a focus on professional goals, mentoring can also create reflective practitioners of both the protégé and the mentor. To best understand professional practice, it is essential to understand the professional. Simply, the sum of what a professional knows is greater than the sum of what he is aware he knows, much less all that he or she can articulate (Schön, 1983). This creates a hidden world of practitioner competence that allows the practitioner the ability to accommodate to new, unique, or uncertainty. Or, "…We have to fall back on routines in which previous thought and sentiment has been sedimented. It is here that the full importance of reflection-on-action becomes revealed. As we think and act, questions arise that cannot be answered in the present. The space afforded by recording, supervision and conversation with our peers allows us to approach these. Reflection requires space in the present and the promise of space in the future" (Smith, 1994, p. 150). Reflective learning is modeled on Lave and Wenger's (1991) situated learning model, which suggests that all learning is contextual, embedded in a social and physical environment. Reflection allows the practitioner to better understand the process of his or her learning within this environment.

Two reflective, or cognitive, mentoring approaches seem particularly appropriate to facilitate learning in a mentoring situation. The first, cognitive coaching, is a peer coaching/mentoring model based on the understanding that metacognition, that is, being aware of one's own thinking processes leads to independent learning (Costa & Garmston, 1994; Ellison & Hayes, 2003). A planning conference with the mentor and the protégé helps the protégé clarify the goals and objectives of the activity, evaluate strategies to achieve those goals and objectives, and determine evaluation methods that demonstrate the goal was achieved (Costa & Garmston, 1994). Under the observation of the mentor, the protégé performs the activity. After the completion of the activity, there is a reflective conference, which addresses the activity and its performance, relating it to current, tangential, and overlapping practice (Costa & Garmston, 1994). The effectiveness of the cognitive coaching model lies in the reflective process that leads to the development of metacognitive skills. Throughout the process, the protégé learns how to assess her/his own thinking through reflection, which builds effective problem solving and increases creativity and innovation. The second, cognitive apprenticeship, is a mentoring/internship model ideally suited for practica, fieldwork, or internships. A pedagogical model of support also developed within the situated learning paradigm (Lave & Wenger, 1991), it is a good model for students who are trying to connect their "book/classroom knowledge" into a real-life setting. Using the cognitive apprenticeship approach, the student is actually working under the guidance of a professional librarian. Ideally, the professional scaffolds the novice practitioner's learning and practice, with conferences and problem-solving of actual experiences.

As mentioned earlier, effective mentoring can have many positive benefits to the protégé, the mentor, and the organisation. Reflective practice and engagement with other professionals also provides new opportunities for peripheral learning and formation of formal and informal networks, that is, "the purpose is not to learn from talk as a substitute for legitimate peripheral participation; it is to learn *to* talk as a key to legitimate peripheral participation" (Lave & Wenger, 1991, pp. 108-109). Situated learning, as a pedagogical model, does not acknowledge talk of knowledge that is decontextualized, abstract, or general. Further, it squarely situates new knowledge and learning as being located *in situ*, in existing communities of practice (Tennant, 1997, p. 77).

Communities of Practice

Relatively new to the field of librarianship (Abram, 2005; Hanson, 2000; Margulies, 2006), the notion of communities of practice emerged from work done on situated learning in 1991 (Lave & Wenger, 1991). A community of practice is often described as a process of social learning that occurs when individuals with common interests collaborate over an extended period, sharing ideas, problem solving, or creating new innovations or innovative practices (Wenger, 1998). Libraries, seen as a community and social entity, operate much on a "less is more" philosophy, through the contextualization of information within that community (Margulies, 2006). The essence of communities of practice evolves from three characteristics: the valuation of work roles, the degree of participation in "peripheral" learning permitted under working conditions such as conferences, workshops, and networking opportunities, and opportunities for participation in innovative implementations (Bourdieu, 1977; Brown & Duguid, 1991; Lave & Wenger, 1991). The real value of a community of practice lies in its ability to connect one's personal development and professional identity of practitioners to an organization's mission and goals (Wenger, McDermott, & Synder, 2002). While knowledge is often thought to be the property of individuals, a great deal of knowledge is both produced and held collectively, readily generated when people work together in tightly knit groups (Hanson, 2000). Therefore, for organizations, "… learning is an issue of sustaining the interconnected communities of practice through which an organization knows what it knows and thus becomes effective and valuable as an organisation" (Wenger, 1998, p. 8).

Graduate schools of library and information science and academic libraries need to work together to support one another in development of librarians, especially those students planning to enter specialty areas, such as GIS librarianship. This requires graduate schools to link preservice education to in-service training. Novice librarians will find that their educational knowledge transfers easily to the practice setting, and practicing librarians have the benefit of incorporating current best practices in librarianship with cutting-edge, state-of-the-art technology and pedagogical methods. This philosophy of graduate education also provides rich opportunities to establish and become part of communities of practice, as peers and mentors. Support is provided to new librarians, professional development is enhanced for both new and practicing librarians, and new opportunities for scholarship are available as preservice and in-service programs reach across disciplines, and create or blend diverse communities. Further, participation in innovative technologies and implementations, such as GIS, emphasizes an important factor in transformation, that is, "The central issue in learning is becoming a practitioner not learning about practice" (Brown & Duguid, 1991, p.48, original emphasis).

Distance Learning in GIS

The concept of remote access to the contents and services of libraries and other information resources provides the user technology that brings access to the resources of multiple libraries and information services. Distance learning is now defined as taking courses by teleconferencing or using the Internet as a method of communication. Many colleges and universities offer online or distance learning programs in GIS that count toward formal degrees or certificates. Vendors, such as ESRI, also often provide online training, which also may count toward degrees or certificates. A number of Web sites provide information about distance GIS programs, such as gis.com, which is part of ESRI's outreach for GIS education. Due to the interdisciplinary nature of GIS, the academic background of online students is very diverse, ranging from geosciences, archaeology, military, civil engineering, to business administration. Online programs appeal to non-traditional students who may vary in age from recent undergraduates to older students returning for graduate work after a successful career or raising a family. Online GIS classes use both synchronous and asynchronous technologies, including one-to-one real-time synchronous communication using messaging technologies, one-to-one asynchronous communication via standard e-mail, group asynchronous communication using customized message board facilities, and other applications available in course authoring software. With the use of Web-based GIS applications, students are able to work on the Web and create deliverables for the classroom based on real-time and real-life data.

Conclusion/Summary

GIS is a distinct service apart from those traditionally offered by libraries and requires a complex set of skills beyond those typically taught in the library science curriculum. Delivery of GIS services in libraries is a tiered approach; the training and development needs of librarians are also tiered. By focusing on the core competencies of professional librarianship and by working more closely with other discipline such as geography, librarians can create dynamic training programs to meet their varied needs. To ensure that future needs are met, graduate library education should include core competencies from related disciplines. Further, continuing education should include new services, such as Web-mapping, and e-commerce, and location-based services. Professional education in GIS for librarians may include information technology, geographic information systems, location-based services, database systems, electronic commerce, Web-mapping, mathematics and statistics, mapping science, remote sensing, visualisation, spatial analysis, computer science, administration & management, and professional development (Hunter & Ogleby, 2002). All of these areas can be framed within the context of librarianship. Only this holistic approach will help to bridge some of the differences between librarianship and other disciplines as described by Boxall (2002).

References

Abram, S. (2005). The Google opportunity: Google's new initiatives are rocking our world. Here's how to rock back. *Library Journal, 130*(2), 34-36.

American Library Association. (2005). *Draft: Statement of core competencies: Competency statements.* Chicago, IL: American Library Association. Retrieved July 2006, from http://www.ala.org/ala/accreditationb/Draft_Core_Competencies_07_05.pdf

Arms, W. Y. (2001). *Digital libraries.* Cambridge, MA: MIT Press.

Association of Research Libraries. (1999). *The ARL Geographic Information Systems Literacy Project: A SPEC kit* (ARL SPEC Kit No. 238). Washington, DC: Association of Research Libraries, Office of Leadership and Management Services.

Bourdieu, P. (1977). *Outline of a theory of practice.* Cambridge, England: Cambridge University Press.

Boxall, J. (2002). Geolibraries, the global spatial data infrastructure and digital Earth: A time for map librarians to reflect upon the moonshot. *INSPEL, 36*(1), 1-21.

Brown, J. S., & Duguid, P. (1991). Organizational learning and communities-in-practice: Toward a unified view of working, learning, and innovation. *Organization Science, 2*(1), 40-57.

Bruce, D. (2003). *Geography matters to higher education: Using GIS science to meet state required learning objectives in high schools.* Retrieved 15 February 2006, from http://env1.kangwon.ac.kr/sdwr%202003/Literature%20Survey/International%20Web%20Sites/Esri/industries/university/stateobjectives.html

Bundy, A. (2001). *The 21st century profession: Collectors, managers or educators?* Paper presented to Information Studies Program students University of Canberra *Professional perspectives series* Canberra, Australia: University of Canberra, Information Studies Program. Retrieved December 2006, from http://www.library.unisa.edu.au/about/papers/collectors.pdf

Choi, Y., & Rasmussen, E. (2006). What is needed to educate future digital librarians: A study of current practice and staffing patterns in academic and research libraries. *D-Lib, 12*(9). Retrieved December 2006, from http://www.dlib.org/dlib/september06/choi/09choi.html

Chowdhury, G. G. (2002). Digital libraries and reference services: Present and future. *Journal of Documentation, 58*(3), 258-283 .

Chowdhury, G. G., & Chowdhury, S. (2003). *Introduction to digital libraries.* London, England: Facet.

Costa, A. L., & Garmston, R. J. (1994). *Cognitive coaching: A foundation for renaissance schools.* Norwood, MA: Christopher-Gordon Publishers.

DeMers, M. N. (1997). *Fundamentals of geographic information systems.* New York, NY: J. Wiley & Sons.

Dereksen, C. R. M., Sweetkind, J. K., & Williams, M. J. (2000). The place of geographic information system services in a geoscience information center. *Proceedings of the Geoscience Information Society, 31,* 29-46.

DiBiase, D., DeMers, M., Johnson, A., Kemp, K., Taylor Luck, A., Plewe, B., & Wentz, E. (2006). *Geographic information science & technology body of knowledge.* Washington, DC: Association of American Geographers.

Doskatsch, I. (2003). Perceptions and perplexities of the faculty-librarian partnership: An Australian perspective. *Reference Services Review, 31*(2), 111-121. Retrieved November 2006, from http://www.library.unisa.edu.au/about/papers/RSRarticle.pdf

Ellison, J., & Hayes, C. (2003). *Cognitive coaching: Weaving threads of learning and change into the culture of an organization.* Norwood, MA: Christopher-Gordon Publishers.

Fonseca, F., Davis, C., & Cmara, G. (2003). Bridging ontologies and conceptual schemas in geographic information integration. *Geoinformatica, 7*(4), 355-378.

Hanson, A. (2000). Overcoming barriers in the planning of a virtual library: Recognising organisational and cultural change agents. In M. Khosrowpour (Ed.), *Challenges of information technology management in the 21st century* (pp. 299-302). Hershey, PA: Idea Group Publishing.

Harvey, R. (2001). Losing the quality battle in Australian education for librarianship. *Australian Library Journal, 50*(1), 15-22. Retrieved December 2006, from http://alia.org.au/publishing/alj/50.1/full.text/quality.battle.html

Howser, M., & Callahan, J. (2004). *Beyond locating data: Academic libraries role in providing GIS services.* Paper presented at the annual meeting of the *ESRI EdUC conference* (p. [Online]). San Diego, CA: ESRI. Retrieved July 2006, from http://gis.esri.com/library/userconf/educ04/papers/pap5127.pdf

Hunter, G., & Ogleby, C. (2002). New trends in geographic information technology education . *Geospatial information portal: Papers and articles on education.* Retrieved from http://www.gisdevelopment.net/education/papers/edpa0016.htm

Hyland, N. C. (2002). GIS and data sharing in libraries: Considerations for digital libraries. *INSPEL, 36*(3), 207-215.

Jarosz, L. (2004). Political ecology as ethical practice. *Political Geography, 23*(7), 917-927.

Johnston, B., & Webber, S. (2003). *Studies in Higher Education, 28*(3), 335-352.

Kinikin, J., & Hench, K. (2005). Survey of GIS implementation and use within smaller academic libraries. *Issues in Science and Technology Librarianship,* (42), [n].

Kraak, M. J., & Ormeling, F. J. (1996a). *Cartography: Visualization of spatial data.* Essex, UK: Addison Wesley Longman Limited.

Kraak, M. J., & Ormeling, F. J. (1996b). *Cartography: Visualization of spatial data.* Essex, UK: Addison Wesley Longman Limited.

Kuyper-Rushing, L. (2001). A formal mentoring program in a university library: Components of a successful experiment. *The Journal of Academic Librarianship, 27*(6), 440-446.

Lary, M. S. (1998). Mentoring: A gift for professional growth. *The Southeastern Librarian, 47*(4), 23-26.

Lave, J., & Wenger, E. (1991). *Situated learning: Legitimate peripheral participation.* Cambridge, England; New York: Cambridge University Press.

Margulies, P. (2006). Leveraging the skills of the corporate special librarian to enhance the perceived value of information and sustain communities of practice. *Electronic Journal of Academic and Special Librarianship, 7*(1).Retrieved December 2006, from http://southernlibrarianship.icaap.org/content/v07n01/margulies_p01.htm

Munde, G. (2000). Beyond mentoring: Toward the rejuvenation of academic libraries. *The Journal of Academic Librarianship, 26*(3), 171-176.

Nofsinger, M. M., & Lee, A. S. W. (1994). Beyond orientation: The roles of senior librarians in training entry-level reference colleagues. *College & Research Libraries, 55*(3), 161-170.

Parihar, S. M. (2002). Higher education in spatial information industry: A case for promoting mutual interest in geography at University level. *Geospatial information portal: Papers and articles on education.* Retrieved from http://www.gisdevelopment.net/education/papers/edpa0014pf.htm

Pausch, L. M., & Popp, M. P. (1997). *Assessment of information literacy: Lessons from the Higher Education Assessment Movement.* Paper presented at the annual meeting of the Association of College and Research Libraries Nashville, TN: ACRL. Retrieved December 2006, from http://www.ala.org/ala/acrlbucket/nashville1997pap/pauschpopp.htm

Research-based Education Working Group. (1997). *UCGIS education white paper: Research-based GIScience graduate education.* Santa Barbara, CA: National Center for Geographic Information & Analysis, University of California, Santa Barbara. Retrieved July 2006, from http://www.ncgia.ucsb.edu/other/ucgis/ed_priorities/research.html

Ricker, K. M. (2006). GIS mentoring. *Library Trends, 55*(2), 349-360.

Schön, D. A. (1983). *The reflective practitioner: How professionals think in action.* New York, NY: Basic Books.

Schuurman, N. (2006). Formalization matters: Critical GIS and ontology research. *Annals of the Association of American Geographers, 96*(4), 726-739.

Shawa, T. W. ([1998]). *A critical study of GIS services in research libraries: A case study of SUNY Albany and New York State libraries.* Retrieved August 2006, from http://www.princeton.edu/~shawatw/gis.html

Smith, M. K. (1994). *Local education.* Buckingham, England: Open University Press.

Sparke, M. (2000). Graphing the geo in geo-political: Critical geopolitics and the re-visioning of responsibility. *Political Geography, 19*(3), 373-380.

Strasser, T. C. (1995). Desktop GIS in libraries, technology and costs: A view from New York State. *Journal of Academic Librarianship, 21*(4), 278-282.

Tennant, M. (1997). *Psychology and adult learning.* London, England: Routledge.

Varalakshmi, R. S. R. (2003). Educating digital librarians: Expectations, realities and future perspectives. *Malaysian Journal of Library & Information Science, 8*(2), 43.

Vitolo, T. M., & Coulston, C. (2002). Taxonomy of information literacy competencies. *Journal of Information Technology Education, 1*(1), 43-51. Retrieved 01/12/02, from http://jite.org/documents/Vol1/v1n1p043-052.pdf

Wenger, E. (1998). *Communities of practice: Learning, meaning, and identity.* Cambridge, England: Cambridge University Press.

Wenger, E., McDermott, R., & Synder, W. (2002). *Cultivating communities of practice.* Boston, MA: Harvard Business School Press.

Wikle, T. A. (1998). Continuing education and competency programmes in GIS. *International Journal of Geographical Information Science, 12*(5), 491-507.

Wikle, T. A., & Finchum, G. A. (2003). The emerging GIS degree landscape. *Computers, Environment and Urban Systems, 27*(2), 107-122.

Endnotes

[1]   A 'digital librarian' was someone who is responsible for and involved in technology-based projects to deliver digital information resources in non-public service areas.

[2]   http://www.sis.pitt.edu/~dist/academics/specializations/geoinformatics.htm

Chapter X

What the Future Holds:
Trends in GIS and
Academic Libraries

John Abresch, University of South Florida Libraries, USA

Ardis Hanson, University of South Florida Libraries, USA

Susan Heron, University of South Florida Libraries, USA

Peter Reehling, University of South Florida Libraries, USA

Introduction

Geographic information is ubiquitous, from MapQuest in Google to the use of global positioning systems on PDAs and automobiles. More people use geographic information on a daily basis, from directions and a review of a local restaurant to building new infrastructures for communities. Therefore, libraries and librarians should be planning on how best to obtain, market, and provide this type of information for their users' personal and professional needs. What are some of the emerging themes in geographic information systems, particularly for libraries? In the convergence of services and resources, emergent themes are cartography; platform/network development; "geoweb" services and resources; geodata management trends; and societal impacts. Sui (2004) postulates that GIScience research will be involved in "computational, spatial, social, environmental, and aesthetic dimensions" (p. 65), therefore "geocomputation, spatially integrated social sciences, social informatics, information ecology, and humanistic GIScience" are areas of research to watch (p. 65). This chapter will address these themes from both a GIS and libraries perspective.

Getting From Then to Now

GIS has changed considerably since its introduction in the 1960s. Map data moved from a tangible medium, lines drawn on paper, to intangible media, digital values stored electronically. The early cartographic and spatial design vocabulary developed to automate the drafting of maps in the 1970s established today's geospatial conceptual and theoretical constructs. As computing evolved, so did GIS. The emergence of relational database management systems allowed numeric data to link with geospatial data. Although dual encoding data models (vector and raster) brought debate into the GIS community as to which was best, it was determined that the "nature of the data and the processing determines the appropriate data structure" (Berry, 2006, ¶ 10). As mentioned in previous chapters, the improvement in hardware and software applications, as well as the emerging information infrastructure, played an important role in moving GIS out of the back room into an everyday activity. Spatial statistics, which describe the geographic patterns of mapped data, are a direct extension of traditional non-spatial statistics. These are often used in data mining large quantitative datasets (LeSage & Pace, 2001) as well as in meteorological, geophysical, and public health analyses (Gould & Arnone, 2004; Härdle, Mori, & Vieu, 2007; Waller & Gotway, 2004). Spatial analysis uses statistical models that represent life or social phenomena in a mathematical or statistical way. Modeling real-life phenomena allows researchers to determine factors or variables that influence the behavior of the phenomena. It also allows prediction or forecasting of long-term behavior of the phenomena, by changing factors that influence them or by noting historical events. Anything that has a contextual component to it can be enhanced through the use of spatial analysis (Berry, Marble, & Joint Comp, 1968; Chou, 1997; Maguire, Batty, & Goodchild, 2005; Paulston, 1996; Worrall, 1991).

Berry (2006) notes that spatial mathematics has extended conventional mathematical concepts. Although "map algebra" uses logical sequencing of basic operations (e.g., addition, subtraction, exponentiation) to perform complex, multifactor map analyses, "mapematics" allows new operations specific to geographical applications, such as distance and optimal path routing (Berry, 2006). Further, geotechnology (comprising GIS, GPS, and remote sensing) was identified as one of the three fastest growing fields (the other two being biotechnology and nanotechnology) (Gewin, 2004). Since geographic information (GI) science research plays a large role in the growth of geotechnology, it is important to review how the theoretical constructs of GI science may influence the development of GIS.

GIScience Research

GIScience is about representations of space and time and how to implement them in a digital environment. A number of persons have contributed to the emergence of this new field, including Chrisman (1978), Couclelis (1998), Goodchild (1992), Peuquet (1988, 1994, 2002), and Raper (2000). Skov-Petersen (2003) notes that GIS borrows much of its basic theory and methodology from the traditions of quantitative geography, causing many academicians to consider it within the positivist research tradition. Positivism essentially is the idea that science alone presents reliable knowledge of nature, therefore, having clear principles for and descriptions of events, with a focus on quantification, measurement, and observable phenomena, is critical. Sense making, in short, is contained in what can be derived from scientific observation. Characteristics of positivism focus on realism, demarcation between scientific theories and other types of beliefs, cummulative scientific beliefs, deductive theories, theoretical postulates, and precise scientific concepts and vocabularies (Hacking, 1981). Initially, researchers saw GIS as a quantitative-, technology- and data-driven discipline: "It is seemly self-evident that GIS is quantitative and empiricist, given its computational roots in Boolean mathematics and its use for manipulating empirical spatial databases" (Sheppard, 2001). Other researchers feel that the tremendous growth in the applications of GIS in both the public and private sector is creating a significant positivist influence (Lake, 1993; Schuurman, 1999).

The geography and planning community fears being subsumed by GIS. To many non-geographers, GIS is the most visible component of geography and considered the most useful (Skov-Petersen, 2003). However, it is important to remember that GIS is not just a method or mechanism (Skov-Petersen, 2003), but that it may "transform the planning process itself by focusing attention on technical issues at the expense of political or ethical questions and by narrowing analytical attention to questions answerable via available technology" (Lake, 1993, p. 406). Further, there is a question of whether GIS is a "fair representation of a reality," since reality changes through time and in space (Skov-Petersen, 2003). Although any technology can be considered neutral, analytical shortcomings can affect how information is derived; analysis, model design, and verification of results can only take into account measurable data. However, there is a movement to make GIS more holistic, further refuting the positivist claims for GIS. Sheppard (2001) argues two points. First, since GIS has the ability to process qualitative data ranging from text to audiovisual, its use and background cannot be entirely positivist (Sheppard, 2001). Second, GIS can be used for scenario building, as in the case of prediction of hurricane trajectory or, in the recent case of Hurricane Katrina, the extent of damage with a massive storm surge (Sheppard, 2001). Both qualitative information and scenario building are interpretative in nature, limiting the positivist argument. Skov-Petersen (2003) adds that, with a more holistic approach to GIS, the research process involves more than just the mechanistic and quantitative aspects of data gathering, it also must address the environmental and social aspects of the data gathering process. He reinforces Lake's premise that "the important thing is not what one is doing but on how one is doing it" (p. 406), especially if "the consequences of the analytical concepts behind … GIS … [are] unknown or undiscussed," then "not much can be gained by discussing or adjusting the method as it is implemented and used in practice" (Skov-Petersen, 2003, p. 9). In examining the research environment involved with GIS, Skov-Petersen (2003) cites six arguments from the academic literature on the nature of techniques and methodology in GIS derived from the works of several scholars (Lake, 1993; Taylor & Overton, 1991; Wack, 1985).

a.   "The data-background in question has to be measurable and quantifiable.
b.   A method or a science like GIS, based on technological development, which again is a consequence of commercial and governmental strategies, cannot claim political or social neutrality.
c.   When dealing with people as one obviously does in human geography and planning, subject and object are interlinked. Accordingly the GIS-analyst being a human being him- or herself will - aware or unaware - be influenced by the object and the related processes when analyzing it.
d.   The basic approach to the world as constituted in terms of lawful regularities is incompatible with a respect of the individual human being of the society.
e.   GIS can be accused of merely being technology- and power-driven is questioned. Accordingly it can from an ethical position be asked if it is right at all to participate in the development of a technology eventually used in warfare? And again, can a technology developed for war be used in the name of peace.
f.   Since data is the ultimate background for GIS, it is claimed that data-rich regions of the world will be favored as technology ascends" (Skov-Petersen, 2003, p. 9).

However, Couclelis and Golledge (1983) suggest that there is room enough for multiple approaches. After all, they suggest, positivism has "set standards of clarity, consistency, and rigor in the development of argument and in the conduct of inquiry that are unparalleled in the history of human thought" (p. 334). Further, positivism insists upon "open, public, intersubjective tests of knowledge by continuous reference to experience" (p. 334), which assists in keeping GIS research and GIScience free from "intellectualism, apriorism, armchair theorizing, unbridled speculation, and anything that resembles dogma or that could block the road to free inquiry and the progress" (p. 334). What does emerge from positivism is "a distinct mode of discourse, a space of possibilities for theoretical languages that meet the criteria of clarity, coherence, intersubjective validity, and a concern never to lose sight of experience" (Couclelis & Golledge, 1983, p. 334). Therefore, practitioners and theorists must engage in discourse on the nature of reality, the essence of knowledge, and the relationship between and among the inner, physical, and virtual worlds. There are other ways of "knowing" than only the ways of science. It is not for any of us to say which way of "knowing" is the best.

Social Constructs, Media, and Communication

Broader conceptual frameworks from which to conceptualize social and environmental issues within GIS research include GIS as a social construct and GIS as media. Social constructionism (social constructivism) is a sociological theory of knowledge that seeks to uncover the ways in which individuals and groups participate in the creation of their perceived reality, that is, how individuals create, institutionalize, and make traditions out of social phenomena. Socially constructed reality is seen as an ongoing, dynamic process in which reality is reproduced by people acting on their interpretations and their knowledge of it. Social constructionism links the sociology of science, technology, and knowledge. GIS can, therefore, be seen as a "social construct" (Sheppard, Couclelis, Graham, Harrington, & Onsrud, 1999; Sheppard, 2001). The National Science Foundation's Project Varenius incorporated computational, cognitive, and societal components in the hope of advancing GIS research. Sheppard *et al.* (1999) explored the societal component, introducing key research initiatives and also to set "a benchmark by which to assess, a few years from now, the specific contributions of the Varenius project to that increasingly vital research area" (p. 798). Sheppard and colleagues suggest that quantitative geography is associated with empiricism, positivism, and social *status quo* while qualitative geography is non- or post-empiricist, contextual, and social empowerment. Sheppard further argues that this dualism can be broken down by deconstructing the underlying representation. The social construction of geographical information system (GIS) technology requires two-way relationships between technology and people, connecting "different social groups in the construction of new localized social arrangements" (Harvey & Chrisman, 1998, p. 1683).

One framework for analyzing the design and implementation of technological innovations uses concepts of interpretive flexibility, technological frames, and boundary objects. Interpretive flexibility describes the many ways in which a specific technology is interpreted by different social groups (Berger & Luckmann, 1967). Technological frames are how members of a social group understand specific uses, conditions and consequences of a specific technology (Orlikowski & Gash, 1994, p. 178). Boundary objects simultaneously separate different social groups and delineate important points of reference between them, creating flexible and dynamic relationships between social groups, time, and place. Harvey and Chrisman (1998) use the concept of boundary objects to understand how GIS technology "exists as part of an intricate web of social relations" (p. 1693). Since GIS is a "technology [that] successfully connects multiple, even opposing perspectives ... through stabilizations of facts [, relations,] and artifacts" (Harvey & Chrisman, 1998, p. 1683), "boundary objects" may be seen as the place where power relations intersect, power being a major focal area in social constructionism.

Sui and Goodchild (2001, 2003) assert that the relationships between GIS and society would be better understood if GIS could be conceptualized as a medium. After all, "Media are generally understood as means of sending messages or communicating information to the general public, and mass media are the instruments by which mass communication takes place in modern societies" (Sui & Goodchild, 2001, p. 387; Sui & Goodchild, 2003). As GIS has become more ubiquitous through Internet access, interactive maps on PDAs, or sitting on the dashboards of our cars, GIS has increasingly become a means to communicate certain aspects of the real world to the general public.

In addition, communication issues, such as the enduring interest in issues, place and identity, empowerment, marginalization, and public participation, and how we select and transform information, have influenced GIScience and GIS research. Several theories, including Shannon and Weaver's communication theory and Chomsky's linguistic theory, have been applied to both cartography and to GIS (Martin, 1996; Moellering, 1984; Nyerges, 1980; Robinson & Petchenick, 1975; Tobler, 1979). Shannon and Weaver (1949) postulate that information is transmitted in a linear fashion from sender to receiver via a defined channel through the surrounding "noise" environment where information exists, focused on reducing the amount of equivocality (ambiguity) in the transmission and receipt of the message. Their work on the invariance of a message under transformation greatly influenced Tobler's notion of geographic filters (1969). Chomsky's (1957) transformation theory that each sentence in a language has two levels representing the core semantic relations of a sentence mapped on to the surface structure of speech via transformations influenced Tobler's 1979 work on cartographic transformations: "Cartographic transformations are applied to locative geographic data and to substantive geographic data. … Substantive transformations occur in map interpolation, filtering, and generalization, and in map reading" (p. 101). How information is transformed is an important issue in any study of social construction, media, and communication in GIS.

Communication theory also influenced the change from Shannon and Weaver's transmissional model to a ritual model that addresses how information (communication) is preserved across time and in place. The introduction of systems theory, specifically Bateson's notion of recursivity, "the re-entry of the whole into the part" (1972), erased the artificial distinction when observing natural and human systems. Therefore, human systems can be studied as natural systems and natural systems can be studied as social constructs. It also follows that, if all we have is our knowledge (measurable and observable events) of the natural system and whatever construct we have layered on it, *how* we interpret is critical. Vicker's *appreciative system* is "an unique interpretive screen, yielding one of many possible ways of interpreting and valuing experience" (Vickers, 1965, p. 69). The appreciative system, which serves to guide action, is "…a mental construct, partly subjective, largely inter-subjective … based on a shared subjective judgment and constantly challenged or confirmed by experience" (Vickers, 1983, p. 55). One's mental construct should do two things well. First, the construct should correspond with reality sufficiently to guide action. Second, the construct should be created in such a manner that it is able to mediate communication (in whatever media) between individuals. Checkland and Scholes (1990) extend Vickers' individual construct perspective to the system as the "process of enquiry itself" (p. 277). Simply, communication helps us to understand recurrent design problems in X, identify extant patterns in our own designs of X, and use these patterns to create a language to share our discoveries. This language also helps form our identity as we "class" ourselves individually, as a group, or systemically, or as we tease out new ways of knowing. Blending communication and GIS theory creates complicated discourses in both mental and physical environments.

However, there is a caveat to accepting GIS as social construct, media, or as communication, that of authoritativeness of the information, since "it is obvious that the information is formed and filtered by the organization and technology behind the media" (Skov-Petersen, 2003, p. 278). Professional ethics and citizen education become critical components to ensure that as information travels across networks, "what is communicated how it is communicated, and what supporting information is communicated along with the message" retains integrity of the sender across the transmission (Skov-Petersen, 2003, p. 278). For GIS to avoid the taint of biased data that has touched all other forms of media, it will be key that there are objective results, with "no doubt about the fidelity of the value of the data and methods behind" so that user could accept it as "a true and valid representation of reality" (Skov-Petersen, 2003, p. 278).

Literacy issues again come into play as different individuals require differing levels of representation depending on their skill level and cognitive abilities. Historically, geodata and results from GIS analyses have focused on how accurate the geospatial information (place, attributes, and time), with little importance placed in how the information is used. As the user-oriented development of GIS continues, user interfaces and presentation of results require more easily communicated information to assess the quality of data, models, and conclusions.

In 1995, a seminal book, the *Ground Truth: The Social Implications of Geographic Information Systems* was published. According to the preface, *Ground Truth* was:

"first, a book about the transformation of data handling and mapping capabilities that have emerged in the past two decades, and the impact they have had within the discipline of geography. Second, it is a book about the constellation of ideas, ideologies, and social practices that have emerged with the development of new forms of data handling and spatial representation. Third, it situates GIS as a tool and an approach to geographical information within wider transformations of capitalism in

the late 20th century: as a tool to protect disciplinary power and access to funding; as a way of organizing more efficient systems of production; and as a reworking (and rewriting) of cultural codes – the creation of new visual imaginaries, new conceptions of earth, new modalities of commodity and consumer, and new visions of what constitutes market, territory and empire." (Pickles, 1995, p. viii)

Ten years later, the discipline of Geography has become inter- and transdisciplinary. Further, researchers envision the discipline as contributing an essential role in an emerging "transdisciplinary synthesis science" (Skole, 2004, p. 739) as new research and education directions for many global environmental, development, and health issues will extend beyond current disciplinary frameworks within geography. Skole (2004) argues that "while the important disciplinary work must continue to be strengthened, present and future challenges will focus on connecting across disciplines and scales, supporting synthesis studies and activities, more tightly linking science, technology, and decision making, and achieving predictive capability where possible" (p. 739). The interrelatedness of the disciplines is clearly illustrated in the following examples. Research on global climate change and its effect on natural and man-made ecosystems moves across the disciplines of the "natural" and "hard" sciences into urban planning. Urban growth and the attendant problems of poverty concentration, social and political fragmentation, and reduced environmental quality uses demographic and population research. Tracking population movement, biosocial linkages and health, and family structure and changes requires researchers to examine public health issues (Demographic and Behavioral Sciences Branch, 2006). Vulnerability to politically destabilizing environmental disasters, water contamination, infectious diseases, and human health require public health and environmental engineering, risk and safety perspectives (Skole, 2004) that bring us back to global perspectives. Clearly, innovation in research for effective, applied solutions requires interdisciplinary research, data collection and data sharing, translation from research to practice, and training of staff to answer complex and integrative questions. To do so will require large-scale data are available to use for long-term studies. These data will contain nationally representative collections, ethnographies, and microsimulation models. Long-term longitudinal studies should address space, place, biomarkers, and time for research on transitional populations and environmental changes.

Integral to the research process in investigating the challenges will be technological applications, such as geospatial information technologies. Applications of GIS and the compilation of spatial datasets in interoperable formats are creating many sharing opportunities among researchers. Skole (2004) notes that "sharing of data is in turn creating new opportunities for collaboration among scientists, with the potential to move across disciplinary boundaries, and even the prospect for developing entirely new methodologies and fields of knowledge" (p. 741). He also describes how the "rise in GIScience as an intellectual domain of geography that has advanced the requisite spatial analytical methodologies, and is fostering cross-discipline collaborations" (p. 741).

As disciplines outside of the field of geography and cartography integrate geographic information systems into their disciplines' techniques and methodologies, the breadth and extent of applications are extended beyond the traditional research areas as defined by geographers and cartographers. This situation has a variety of effects, such as in the technology and in the research process. One effect of the growth in the use of GIS as a research tool is in its research application. Gold (2006) emphasizes that in many contemporary applications of GIS, users are interested in moving away from the traditional two-dimensional vertical overlay of spatial information and need more complex data visualizations. Users often need more advanced techniques to assist in the analysis of spatial information, such as "three-dimensional visualizations of data" (Gold, 2006, p. 506) as well as contemporary interest in Web-based map display, availability of data files, and metadata.

Social Informatics and GIS

Online activities of searching or surfing the Internet can be an anonymous and individual experience as users view many impermanent graphic images and read temporary postings of textual data. The experience of interacting in an online environment can also be a social one. The Internet is composed of many networked information nodes, which are often accessed simultaneously by many users. Communication software using e-mail and messaging in chat rooms promote the exchange of ideas and real-time dialogue between users. The use of social networking Web sites such as MySpace and Facebook allow for the posting of textual information and graphics on customizable personal spaces on the Internet.

Shumar and Renninger (2002) describe how the discursive interactions between individuals in particular contexts can create unique social spaces, which are defined in ongoing narratives as individuals add information or commentary about the specific environment or topic being discussed. The narrative process of communicating in an online environment can thus build a shared knowledge of values and opinions within defined boundaries of a social space similar to a community space (Shumar & Renninger, 2002, pp. 6-12). As the distances between physical places continue to shrink, permitting almost instantaneous communication between individuals, the boundaries of the Internet community space are increasingly flexible across space and time. While this allows a variety of hierarchal online relationships, over time online users prefer to interact with online community members with stronger ties to real physical places, such as family members and colleagues (Shumar & Renninger, 2002, p. 10).

The concept of community in online social spaces can be also be strengthened by building upon existing personal networks in existing urban neighborhoods (Gaved & Mulholland, 2005). The ubiquitous availability of Internet mapping software programs, such as Google Earth and MapQuest, enables the end user to examine the spatial aspects of their neighborhood communities, such as location of specific items, or the route or layout of particular roads. The Internet GIS works as a facilitator of various spatial data and organizes them in a manner that would allow the end user to make spatial queries, such as distance between two distinct locations. End users may also have access to detailed socioeconomic data and numeric data that local governments may have made available on their community Web site. Using software, such as ESRI's ARC/IMS, many local governments in the United States have implemented interactive Web sites to offer accessibility to digital geospatial data. An effect of the development of the Internet is that the large amount of information available is often unorganized and difficult to comprehend or interpret without software tools and substantial computer use skills.

Spatially Integrated Social Sciences

As with any discipline, philosophical perspectives affect how a science or methodology is viewed and used. In geography, one prevailing perspective is the separation of physical geography from human geography, advocating "the separation of nature and society in geographic discourses" (Kwan, 2004, p. 756). A second perspective is the separation of spatial-analytical geographies from social-cultural geographies, which "separate[s] spatial patterns and relations from social, cultural, and political processes (Kwan, 2004, p. 756). However, Kwan (2004) and Sui (2004) suggest that social-cultural and spatial-analytical geographies can be richer in content and context if viewed from a hybrid, or "third culture" perspective. These hybrid perspectives "cut across the divides between the social-cultural and the spatial-analytical, the qualitative and the quantitative, the critical and the technical, and the social-scientific and the arts-and-humanities. It is a future not of "either/or" but of "both-and" (Kwan, 2004).

The development of spatial decision support systems is one example how networking, analysis, and artificial intelligence techniques use geographic information and analytics procedures to assist in decision making and strategic planning. As GIS is integrated within other information technology systems and enterprise-wide operations, horizontally and vertically integrated systems linked to workflow will ensure automatic maintenance of data. However, success in the development and use of these systems will require systematic creation of metadata in standardized formats to insure the interoperability of different databases, models, applications, and other tools for better decision making. Sui (2004) notes that the "diffusion of spatial analytical tools" and their integration with "visualization tools" will lead to the use of geographic metaphors important in describing political-economic activities across contemporary social and cultural regions (p. 66).

Information Ecology

Information ecology includes a conceptualization of information and information systems as life forms and life support systems in the context of a broader perspective of unity with nature. A new perspective, that is, one that sees the entire material world, including technology and the fruits of technology, as part of nature emerges. It examines the dynamics and properties of dense, complex digital information within our distributed environments. For example, "[i]nformation architecture, in the broadest sense, is simply a set of aids that match information needs with information resources. A well implemented architectural design structures information in an organization through specific formats, categories, and relationships" (Davenport & Prusak, 1997, p. 156). Information ecology is regard by some as "an expansion of geography's human-environment interaction tradition during the Information Age" (Sui, 2004, p. 66).

One significant trend in the information ecology of geographic information systems has been the effect of the contributions of disciplines outside of geography and planning. Advances in areas such as mapping hardware and software, computer graphics, portable global positioning systems, and visualization techniques, have extended GIS beyond the digital cartography laboratory and into the hands of users in disciplines, such as public health, law, and even primary education.

Spatial Visualization

Although one of the first instances of the term visualization in the cartographic literature can be traced back to the early 1950s (Philbrick, 1953), visualization today emphasizes facilitating thinking and problem solving as well as the generation of ideas and hypotheses through the use of visual displays (Fisher, Dykes, & Wood, 1993; McCormick, DeFanti, Brown, & Zaritsky, 1987). In fact, cartographic visualization systems may represent the principal technology for the scientific visualization of digital spatial information (Dykes, 1996).

Skupin and Fabrikant (2003) note that "[s]ome cartographers are engaged in the interpretation and transformation of specific computational approaches in the light of cartographic tradition and informed by geographic information science, while other cartographers are using a cognitive approach that emphasizes the user side of spatialization" (pp. 99-100). The focus of this approach is to provide "understanding on how human perception and human cognition of geographic spaces interact with visual representations" (pp. 99-100). They also contend that the two perspectives are complementary, with geographic information science providing "a synthesis that matches geometric primitives against the cognitive categories that underlie our understanding of space" (Skupin & Fabrikant, 2003, p. 100).

Influences behind cartographic visualization include geography, linguistics, information science, cognitive science, and human-computer interaction (Couclelis, 1998; Edwards, 2001; Gahegan & Pike, 2006; Griffin, MacEachren, Hardisty, Steiner, & Li, 2006; Miller & Han, 2001) . When looking at the historical development of visualization, the influence of geography clearly dominates. Numerous American and European authors have affected the field, including Tobler (Tobler, 1962, 1965, 1979, 2004), Olson (Olson, 1975), Monmonier (Monmonier, 1965, 1980, 1981), Jenks (Jenks, 1953, 1973; Jenks & Brown, 1966), Muller (Muller, 1975, 1979), Rhind (Longley, Goodchild, Maguire, & Rhind, 2001; Rhind, 2000), Taylor (Taylor, 1978, 1991), MacEachren (Griffin et al., 2006; MacEachren, 1982, 1992, 1995; Maceachren, Buttenfield, et. al., 1992; MacEachren, Edsall, Haug, D., Baxter, Otto, Masters, et al., 1999), Kraak (Kraak, 1999, 2000, 2003; Kraak & Ormeling, 1996; MacEachren & Kraak, 1997; MacEachren & Kraak, 2001), and Batty (Batty, 1976, 1987; Batty & Longley, 1994; Longley & Batty, 1996; Maguire et al., 2005), to name just a few.

In illustrating various types of visualization techniques, such as multidimensional scaling, spring models, tree maps, and cognitive visualizations, organization of data is an integral component in building retrieval systems for the visualizations (Skupin & Fabrikant, 2003). The process of spatialization visually summarizes and describes large data repositories, and also provides opportunities for visual query and sense making of large data collections. However, information seekers tend to "prefer navigation within clearly defined hierarchal semantic spaces" (Chang, Smith, Beigi, & Benitez, 1997, p. 67). In this regard, researchers working with GIS to create new spatializations of data can work with professionals in the fields of library and information science in devising semantic ontologies and indexes that would offer efficient data mining opportunities to users.

Geocomputation

A map of data "concisely communicates spatial distribution, enabling the viewer to better understand patterns and relationships ... and also among other potential map features such as populated areas, roads, and physiographic regions. Spatial analytical techniques alone are inadequate to study spatial data; mapping spatial data is necessary to understand the data fully" (Hallisey, 2005, p. 350). With Hallisey's assertion in mind, it is easy to see how new cartographic techniques and applications allow quick and easy interactive visualization of complex science, such as ability to graphically display mathematical equations of nonlinear dynamics and chaos theory (O'Sullivan & Unwin, 2003). Further, as the dominant paradigm in cartography changes from the map communication model to the cartographic visualization model, the map's real power to analyse, explore data, and "visualize spatial datasets to understand patterns better" (Crampton, 2001, p. 235) will expand to "generate hypotheses, develop problem solutions and construct knowledge" (Kraak, 2003, p. 390). As the "third culture" in geography, using state-of-the-art technology to creatively generate and meld ideas from the arts and sciences (Sui, 2004) expands MacEachren's iterative and comparative concept of visualization, "there is a continual give-and-take between vision and visual cognition through the intermediary of knowledge schema" (MacEachren, 1995, p. 366–367).

The Role(s) of Libraries

The emergence of the many new applications of digital geospatial data and GIS Web services available on the Internet is creating a demand for guidance in the access and use of digital geospatial data. Librarians, with their traditional skills of information collection, description, organization, and dissemination, can provide a more holistic learning experience for the community of digital library users. Librarians can prepare a learning environment of Internet resources that is easier to navigate by classifying different online information sources. Librarians can also offer bibliographic instruction and technical assistance in helping patrons use Internet resources. However, much of the work that libraries do is considered invisible (Paisley, 1980). Borgman (2003) suggests the "the invisibility is partly due to the successes of the institution. Good library design means that people can find what they need, when they need it, in a form they want" (p. 656). Researching how visible libraries are to their users and stakeholders and how well stakeholders' goals are represented in library plans and policies is critical to determine the library's role in the emerging cyberinfrastructure (Borgman, 2003a, p. 657). Other critical research questions address collections, preservation and access, and institutional boundaries (Borgman, 2003a). These include defining the concept and coherence of a "collection" in an online environment, where access is to content libraries may or may not own. Preservation and access focuses on the stability of access to online resources and user persistence in discovery and acquisition of information. Institutional boundaries are blurring between types of information institutions, as well as blurring the boundaries between services and collections (Borgman, 2003a). Marcum (2003) suggests that research is also needed into the needs and behaviors of library resource users. She sees the data from this research informing preservation options for digital materials as well as redefining the requisite skills for professional librarianship and informing LIS education and post-MLS training (Marcum, 2003). Bertot and McClure (2003) suggest relating traditional evaluation components and terminology to the networked environment to assist libraries in their decisionmaking processes regarding the provision of information services and resources. Further, framework development for outcomes assessment requires more complex analyses of the operating environment of the library and the impact of situational factors on library services and resources outcomes. They also suggest that, in reality, libraries may not always be able to anticipate and/or predict the outcomes of their services/resources on users (Bertot & McClure, 2003; Bertot, Snead, Jaeger, & McClure, 2006). Five "grand challenges" for library research are proposed: "Library service: Could library services be made more meaningful? Library theory: Who knew what when? Library design: Have digital libraries been designed backwards? Library values: How neutral can libraries be? and, Library communities: How do communities differ?" (Buckland, 2003, p. 675). Borgman (2003) sees these challenges as "intertwined and research on each of them will inform the others" (p. 672) as "libraries find their best fit in the information infrastructure of our networked world" (p. 672).

Effective planning for GIS services and digital geospatial collections would involve faculty and staff from public and technical services, since the application of new computer technology often creates much environmental change in libraries. The use of such strategies in the planning of digital libraries would counter the lack of guidance that often accompanies information available on computer networks (Brown & Duguid, 2002). In closing, Sui (2004) states "almost every aspect of natural and social reality that geographers study has, implicitly or explicitly, become computational. This is fully embodied in the recent concepts of digital individuals, digital communities, digital government, digital cities, digital terrain models, and all the way to a digital earth" (pp. 65-66). What we hope is that readers of this monograph will see not only the computational component of geospatial data but also the informational and societal aspects of geography and librarianship.

As discussed in Chapter II, the development of telecommunications networks and of the Internet tends to be spatially uneven. The uneven distribution of technology, of computer networks and of the information they transmit, has been characterized as contributing to a digital divide between those who do have access to the Internet and those who do not (Norris, 2001). The digital divide has been described by researchers in a variety of dichotomies in both contemporary American and international settings (Burkett, 2000). In contemporary American society, researchers apply econometrics to socioeconomic data to portray how particular income groups have sufficient capital to purchase computer hardware and software for Internet access, while other income groups do not. Researchers also look at other factors, such as race, ethnicity, and educational attainment, to further illustrate differences in technology access (Ferrigno-Stack, Robinson, Kestnbaum, Neustadtl, & Alvarez, 2003). Different technological standards are also identified within contemporary American society that can contribute to uneven access to information technologies. In international settings, the literature on the digital divide includes many of the factors described in articles about technology and information access in America, though other factors are emphasized, for example, political ones (Fahmi, 2002).

The literature also emphasizes, in both American and international settings, the importance of libraries as nodes across emerging international information networks (Hull, 2001). Libraries are natural facilitators of communication technologies. Further, they are also the logical choice to facilitate the transmission of information to local and remote users. Information often flows across the new medium of the Internet in unique ontologies and communicative forms between online communities with different characteristics and perspectives (Kent, 2000). Socioeconomic factors may combine with differing technological capabilities between communities to influence accessibility trends associated with the digital divide. Libraries can provide a digital environment that facilitates interoperability and virtual knowledge exchange between online communities and individual users. Libraries can also provide information access to the individual user as well as the larger community in which they are located physically or virtually.

Despite attention to GIS as a medium for integrating data and communicating information, GIS as a system remains a black box for many librarians (Schuurman, 2003). The term "black box" was used by Bruno Latour, to describe technologies in which the inner workings are hidden from the user (Latour, 1987). For all but specialists, GIS's intensive, multiapplication computing environment is not only a mystery, it is non-intuitive. Poore (2003) investigates the intersection between data, integration strategies and systems development. She argues that, in order for GIS to truly accommodate user needs, it must account for a network of systems that includes the natural software (Poore, 2003). After all, "[u]nderstanding user practices is the basis for re-engineering the black box" (Schuurman, 2003, p. 3). This focus on practice illustrates some of the ways in which people actually use geospatial data and GIS.

In another study of end-user issues, this time focusing on access, the United States Federal Geographic Data Committee (FGDC) conducted a nationwide survey of likely framework data producers (Tulloch & Shapiro, 2003). The survey queried data producers' questions on the different ways in which they allow access to data. The responses indicated a progressing spectrum of access activities. While the vast majority of producers allowed data sharing, very few actively advertised their data in data clearinghouses or catalogs. In between the producers who actively facilitated the distribution of their data and those who did not share data at all were other levels of access: data sharing (88%), limited data redistribution (75%), and participation in a coordinating council (42%). Since these three activities often require minimal exertion on the part of the data producer, we group them as "casual" access practices. The three less common access activities were implementation of a policy on data dissemination (40%), unrestricted data redistribution (30%), and advertising data in a clearinghouse or catalog (9%). The latter three activities demonstrate a greater commitment to access, since they often require more work and can involve additional exposure to risk (Harvey & Tulloch, 2006).

The use of GIS by nongovernmental organizations (NGOs), and environmental groups for social activism, points to the power and scope of the technology to operate with multiple epistemologies and social visions. Public participation GIS (PPGIS) is a way of extending decision-making processes to include groups that may not otherwise be heard in the context of policy development. Recently, NGOs have extended their reach across borders as they join to protest against the World Trade Organization and the Free Trade Agreement of the Americas. Understanding user issues, such as access and resource discovery, is also addressed in PPGIS, which is being adopted as a transnational tool with the potential to empower multiple communities in different cultural contexts (Sieber, 2003).

The real promise of digital libraries is the potential for transformation of disciplinary research and trans-boundary research of current and emerging disciplines (Atkins, Droegemeier, Feldman, Garcia-Molina, Klein, Messerschmitt, et al., 2003; Henry, 2003). After all, "[t]he classic two approaches to scientific research, theoretical/analytical and experimental/observational, have been extended to *in silico* simulation to explore a larger number of possibilities at new levels of temporal and spatial fidelity" (Atkins et al. 2003, p. 4). Further, the notion of infrastructure has also evolved. As discussed in Chapter II, the move from an industrial society and economy to a knowledge society and economy requires a change in our infrastructure. "Cyberinfrastructure" refers to today's distributed technologies in computing, information, and communication (Atkins et al., 2003). A cyberinfrastructure is well suited to support new ways of working, such as collaboratories, which are a "new networked organizational form that also includes social processes; collaboration techniques; formal and informal communication; and agreement on norms, principles, values, and rules" (Cogburn, 2003, p. 86). These innovative forms require "flexible tools and services to gather information from multiple sources, including digital libraries, and to manipulate them for their own purposes" (Borgman, 2003b, p. 1). The cyberinfrastructure, which can be seen as both an object of research and as a platform in service of research, will contain "grids of computational centers, some with computing power second to none; comprehensive libraries of digital objects including programs and literature; multidisciplinary, well-curated federated collections of scientific data; thousands of online instruments and vast sensor arrays; convenient software toolkits for resource discovery, modeling, and interactive visualization; and the ability to collaborate with physically distributed teams of people using all of these capabilities" (Atkins et al., 2003, p. 7).

Libraries are well suited to this endeavor, with their long history of collecting, preserving, and archiving information. Without a doubt, the need for long-term, distributed, and stable data and metadata repositories is critical, as research communities struggle with access and discovery issues using federated data from multiple sources and disciplines. Stewardship, an increasingly important discussion in the scientific world, is also part of librarianship. Libraries are used to handling convertability of data as obsolesced hardware and software are translated to newer forms of storage, with documentation and the obsolesced items stored. Digitization and stewardship of legacy data are critical to resource discovery. Data archaeology, coined in 1993, identifies critical computer data encrypted in now obsolete media or formats, and analyzes data entities and attributes to ensure that historical data is successfully migrated (O'Donnell, 1998; Ravindranathan, Shen, Gonçalves, Fan, Fox, & Flanagan, 2004). Libraries, to paraphrase Ranganathan, are living repositories, not data mortuaries.

Stewardship also includes ongoing creation and improvement of the metadata of these digital objects. The traditional activities of cataloging, authority control, and classification have evolved to rich metadata and semantic relationships, creating new paradigms for information classification, manipulation, and visualization of complex, distributed information and data systems. Metadata quality and interoperability is also required to support search and retrieval, administration and preservation, and evaluation and use. Metadata quality issues include "(1) specificity, (2) completeness of fields, (3) syntactic correctness, (4) semantic correctness, and (5) consistency, as implemented through authority control" (Larsen, 2002, p. 15). Representation of information, through standards, protocols, formats, and languages, is the natural provenance of librarians as we create new information frameworks, crosswalks, and ontologies to increase resource discovery of contextual information. New data-mining techniques and applications allow the discovery of new knowledge "in problem areas never intended at the time of the original data acquisition" (Atkins et al., 2003, p. 42).

Collaboration and partnerships with other disciplines producing data will require understanding and development of middleware, standard or interoperable formats, and related data storage strategies (Atkins et al., 2003). Libraries continue to integrate standards and techniques, such as WSDL, UDDI, and SOAP, to deal with differences among systems, ontologies, and data formats in library systems and catalogues (Alonso, Casati, & Machiraju, 2004; Gonçalves, France, & Fox, 2001; Ravindranathan et al., 2004). These solutions may integrate data-harvesting techniques and frameworks to address data quality and scalability to create architectures based on object-oriented ontologies of search modules and metadata (Jordan, 2006; Smart, Abdelmoty, & Jones, 2004). An emerging area is how to make data reconciliation within harvested data joined to a single collection view for the user (Gonçalves et al., 2001). Authority control may assist in handling construction of conceptual and contextual maps in these settings (Smith & Crane, 2001; Weaver, Delcambre, & Tolle, 2003; Weaver, Delcambre, Shapiro, Brewster, Gutema, & Tolle, 2003; Weinheimer & Caprazli, 2004). The literature suggests that abstracting and indexing services, with more browsable interfaces, are critical to the discovery and gathering processes (Atkins et al., 2003; Makedon, Ford, Shen, Steinberg, Saykin, Wishart et al., 2002). Consider that about 250 megabytes of data, or approximately 250 books per year per person, is produced yearly, of which only 0.003 percent is in printed form (Lyman & Varian, 2003). Therefore, contextual and conceptual mapping and selection, as well as how to best maximize cognitive load, may well highlight the importance of these functions of librarianship.

The Breakout Group on Evaluating Digital Library Users and Interfaces (Borgman, Griffiths, Kovacs, Mostapha, Sfakakis, & Banwell, 2002) suggests that digital libraries need new evaluation approaches, including reviews of users and uses of digital libraries; language and culture; multiple content formats; multimodal environments; non-textual interfaces; and evaluation granularity (p. 138). Criteria include the "cost of evaluation; cost benefit of evaluation (e.g., how much money is saved through productivity improvements as a ratio of the development cost of the system); the ability to share methods, instruments, and testbeds; and validity and reliability (e.g., tradeoffs between evaluation in situ and evaluation in laboratory settings)" (Borgman et al., 2002, p. 138).

How these trends and forecasts affect libraries and their staff is still unknown. As addressed in Chapters VII and IX, core competencies for library staff continue to emerge. Librarians will need to demonstrate "interpersonal competence, personal integrity, and the capacity to think systemically, innovatively, and integratively about how work systems and people need to collaborate. Combine these competencies with the librarian's traditional role of connecting users to information; designing and managing complex, interconnected systems and organizations; selecting and organizing information resources; teaching and consulting; creating logical and intuitive insights about information; and formulating and articulating information policy, and you have the competencies framework for the 21st century research librarian" (Hanson, 2004, p. 2).

Conclusion/Summary

The emergence and development of advanced computer technology in the United States has altered the contemporary economic landscape. Both public agencies and private firms rely heavily upon computerized information technologies. In the information economy, data is an important commodity as it flows between individuals and organizations across communication networks like the Internet. The development of such networks tends to be spatially uneven, often concentrating in one sector of society while marginalizing other sectors of society. The marginalized population in society would lack access to the mechanisms and tools of communication and of the information economy, like the Internet, and of the information to which they provide access. The uneven distribution of technology, computer networks, and the information they transmit has been characterized as contributing to a digital divide between those who do have access to the Internet and those who do not. With over 16,000 public libraries, 98,000 school library and media centers, and 4,000 academic libraries located across the country with the necessary resources to provide information access and services, libraries are in a position "to overcome the social and technological barriers to access" (Hull, 2001). The American Library Association also asserts that libraries can alter the social effects of the digital divide by strengthening the collaborative work within the library community ,and increasing collaborative efforts with others in the government, corporate, and non-profit sectors. It is up to individual libraries and librarians to initiate programs that are relevant to their community of users.

Digital libraries of discipline-specific collections and archives are key components of the cyberinfrastructure. Their collaborative technologies and interoperable, distributed databases require coordinated, large, and long-term investment in four areas: basic research to advance the evolution of the cyberinfrastructure; development activities targeted to build advanced operational structures; institutions to provide operational support and services; and high-impact applications of advanced cyberinfrastructure in science, technology, engineering, medicine, and allied education (Atkins et al., 2003).

Borgman (2003) predicts the rise of personal digital libraries. These created collections are much more than repositories. They allow researchers the ability to work with malleable, mutable resources, creating and sharing new knowledge in an easily transportable and interoperable form. Designing tools and services that enable individuals to create and manage their own personal digital libraries is the next research frontier (Borgman, 2003b).

Other emerging issues for librarians surface in the intersection of the emerging conceptions of virtual space and the traditional conceptions of geographic space. For example, what future is there for borders and boundaries in a world where "there is no there" (Sheppard *et al.*, 1999)? What topologies, ontologies, and vocabularies need to be created to accommodate the notion of "space-time" in both the physical and virtual worlds (Latour, 2005)? How do we handle the interface of "virtual and physical" analytically?

We would like to end this chapter with a quote from the NSF cyberinfrastructure report reminding us of what challenges, both good and bad, may be present in the digital, information/knowledge environment we find ourselves in.

"A vast opportunity exists for creating new research environments based upon cyberinfrastructure, but there are also real dangers of disappointing results and wasted investment for a variety of reasons including underfunding in amount and

duration, lack of understanding of technological futures, excessively redundant activities between science fields or between science fields and industry, lack of appreciation of social/cultural barriers, lack of appropriate organizational structures, inadequate related educational activities, and increased technological ("not invented here") balkanizations rather than interoperability among multiple disciplines. The opportunity is enormous, but also enormously complex, and must be approached in a long-term, comprehensive way." (Atkins et al., 2003, p. 4).=

It is through the collaboration among those of us in library, information, and GIS sciences that these questions may be answered. To paraphrase Korzybski, the map is not the territory, but maps are not all we possess. We hope that this volume clearly illustrate how academic libraries and GIS can be successfully explored with a variety of techniques, tools, theoretical frameworks, and processes.

References

Alonso, G., Casati, F. K. H., & Machiraju, V. (2004). *Web services: Concepts, architectures and applications*. New York, NY: Springer.

Atkins, D. E., Droegemeier, K. K., Feldman, S. I., Garcia-Molina, H., Klein, M. L., Messerschmitt, D. G. et al. (2003). *Revolutionizing science and engineering through cyberinfrastructure: Report of the National Science Foundation Blue-Ribbon Advisory Panel on Cyberinfrastructure*. Washington, DC: National Science Foundation. Retrieved from http://www.nsf.gov/cise/sci/reports/atkins. pdf [Also known as the Atkins Report].

Bateson, G. (1972). *Steps to an ecology of mind*. Chicago, IL. University of Chicago Press.

Batty, M. (1976). *Urban modelling: Algorithms calibrations, predictions*. Cambridge, England: Cambridge University Press.

Batty, M. (1987). *Microcomputer graphics: Art, design, and creative modelling*. London, England: Chapman and Hall.

Batty, M., & Longley, P. (1994). *Fractal cities: A geometry of form and function*. London, England: Academic Press.

Berger, P., & Luckmann, L. (1967). *The social construction of reality: A treatise in the sociology of knowledge*. New York, NY: New York.

Berry, B. J. L., Marble, D. F., & Joint Comp. (1968). *Spatial analysis: A reader in statistical geography*. Englewood Cliffs, N.J: Prentice-Hall.

Berry, J. K. (2006). Early GIS technology and its expression. *GeoWorld, October*. Retrieved January 2007, from http://www.innovativegis.com/basis/MapAnalysis/Topic27/Topic27.htm#Early

Bertot, J. C., & McClure, C. R. (2003). Outcomes assessment in the networked environment: Research questions, issues, considerations, and moving forward. *Library Trends, 51*(4), 590-614.

Bertot, J. C., Snead, J. T., Jaeger, P. T., & McClure, C. R. (2006). Functionality, usability, and accessibility: Iterative user-centered evaluation strategies for digital libraries. *Performance Measurement and Metrics, 7*(1), 17-28.

Borgman, C., Griffiths, J., Kovacs, L., Mostapha, J., Sfakakis, M., & Banwell, L. (2002). Report on breakout group: Evaluating digital library users and interfaces, June 12, 2002. In C. Borgman, I. Sølvberg, & L. Kovács (Eds.), *Fourth DELOS workshop: Evaluation of digital libraries: Testbeds, measurements, and metrics* (pp. 137-138). Budapest, Hungary: Hungarian Academy of Sciences, Computer and Automation Research Institute (MTA SZTAKI). Retrieved December 2006, from http://www.sztaki.hu/conferences/deval/presentations/DELOSWorkshop4OnEval_report.pdf

Borgman, C. L. (2003a). The invisible library: Paradox of the global information infrastructure. *Library Trends, 51*(4), 652-675.

Borgman, C. L. (2003b). *Personal digital libraries: Creating individual spaces for innovation.* Paper presented at the meeting of the NSF Post Digital Library Futures Workshop Wave of the future, June 15 - 17, 2003, Wequassett Inn, Cape Cod, Chatham, Massachusetts Alexandria, VA: National Science Foundation. Retrieved January 2007, from http://www.sis.pitt.edu/~dlwkshop/paper_borgman.pdf

Brown, J. S., & Duguid, P. (2002). *The social life of information.* Boston, MA: Harvard Business School Press.

Buckland, M. K. (2003). Five grand challenges for library research. *Library Trends, 51*(4), 675-687.

Burkett, I. (2000). Beyond the "information rich and poor": Futures understandings of inequality in globalizing informational economies. *Futures, 32,* 679-694.

Chang, S.-F., Smith, J. R., Beigi, M., & Benitez, A. (1997). Visual information retrieval from large distributed online repositories. *Communications of the ACM, 40*(12), 63-71.

Chou, Y.-H. (1997). *Exploring spatial analysis in geographic information systems* (1st ed.). Santa Fe, NM: OnWord Press.

Chrisman, N. R. (1978). Concepts of space as a guide to cartographic data structures. *Harvard Papers on Geographic Information Systems, 5,* 1-19.

Cogburn, D. L. (2003). HCI in the so-called developing world: What's in it for everyone. *Interactions, 10*(2), 80-87.

Couclelis, H. (1998). Worlds of information: The geographic metaphor in the visualization of complex information. *Cartography and Geographic Information Systems, 25*(4), 209-220.

Couclelis, H., & Golledge, R. (1983) Analytic research, positivism, and behavioral geography. *Annals of the Association of American Geographers, 73*(3), 331-339.

Crampton, J. W. (2001). Maps as social constructions: Power, communication and visualization. *Progress in Human Geography, 25*(2), 253-260.

Davenport, T. H., & Prusak, L. (1997). *Information ecology: Mastering the information and knowledge environment.* London, England: Oxford University Press.

Demographic and Behavioral Sciences (DBS) Branch (2006). *Demographic and Behavioral Sciences (DBS) Branch Long-Range Planning 2006-2007: Highlights from a panel discussion 2006.* Bethesda, MD: Center for Population Research, National Institute of Child Health and Human Development. Retrieved June 12, 2007, from http://www.nichd.nih.gov/publications/pubs/upload/dbsb_panel_highlights_2006_2007.pdf

Dykes, J. (1996). Dynamic maps for spatial science: A unified approach to cartographic visualization. In D. Parker (Ed.), *Innovations in GIS 3* (pp. 177-187). London, England: Taylor & Francis.

Edwards, G. (2001). A virtual test bed in support of cognitively-aware geomatics technologies. In D. R. Montello (Ed.), *Spatial information theory: Foundations of geographic information science: International Conference, COSIT 2001 Morro Bay, CA, USA, September 19-23, 2001, Proceedings* (pp. 140-155). Berlin, Germany: Springer-Verlag GmbH.

Fahmi, I. (2002). The Indonesian Digital Library Network is born to struggle with the digital divide. *International Information & Library Review, 34*(2), 153-174.

Ferrigno-Stack, J., Robinson, J. P., Kestnbaum, M., Neustadtl, A., & Alvarez, A. (2003). Internet and society: A summary of research reported at Webshop 2001. *Social Science Computer Review, 21*(1), 73-117.

Fisher, P., Dykes, J., & Wood, J. (1993). Map design and visualization. *Cartographic Journal, 30*(2), 136-142.

Gahegan, M., & Pike, W. (2006). A situated knowledge representation of geographical information. *Transactions in GIS, 10*(5), 727-749.

Gaved, M., & Mulholland, P. (2005). Grassroots initiated networked communities: A study of hybrid physical/virtual communities. In *HICSS '05. Proceedings of the 38th Annual Hawaii International Conference on System Sciences* (p. 191-193). Retrieved January 2007, from http://doi.ieeecomputersociety.org/10.1109/HICSS.2005.288

Gewin, V. (2004). Mapping opportunities. *Nature, 427*(Jan 22), 376-377.

Gold, C. M. (2006). What is GIS and what is not? *Transactions in GIS, 10*(4), 505-519.

Gonçalves, M. A., France, R. K., & Fox, E. A. (2001). MARIAN: Flexible interoperability for federated digital libraries. In P. Constantopoulos, & I. T. Sølvberg (Eds.), *Research and advanced technology for digital libraries: 5th European conference, ECDL 2001, Darmstadt, Germany, September 4-9, 2001, proceedings* Vol. 2163 (pp. 173-187). Berlin, Germany: Springer-Verlag GmbH.

Goodchild, M. F. (1992). Geographical information science. *International Journal of Geographical Information Systems, 6*(1), 31-45.

Gould, R. W. Jr., & Arnone, R. A. (2004). Temporal and spatial variability of satellite sea surface temperature and ocean colour in the Japan/East Sea. *International Journal of Remote Sensing, 25*(7-8), 1377-1382.

Griffin, A. L., MacEachren, A. M., Hardisty, F., Steiner, E., & Li, B. (2006). A comparison of animated maps with static small-multiple maps for visually identifying space-time clusters. *Annals of the Association of American Geographers, 96*(4), 740-753.

Hacking, I. (1981). *Scientific revolutions.* Oxford, England: Oxford University Press.

Hallisey, E. J. (2005). Cartographic visualization: An assessment and epistemological review. *The Professional Geographer, 57*(3), 350-364.

Hanson, A. (2004). *Research library: Annual report, 2004.* Tampa, FL: University of South Florida, The Louis de la Parte Florida Mental Health Institute.

Härdle, W., Mori, Y., & Vieu, P. (2007). *Statistical methods for biostatistics and related fields.* Berlin, Germany: Springer.

Harvey, F., & Chrisman, N. (1998). Boundary objects and the social construction of GIS technology. *Environment and Planning A, 30*, 1683-1694.

Harvey, F., & Tulloch, D. (2006). Local-government data sharing: Evaluating the foundations of spatial data infrastructures. *International Journal of Geographical Information Science, 20*(7), 743-768.

Henry, C. H. M. (2003). *Alliance for the 21st century library: Proposal to monitor and analyze transformations in academic disciplines. Proposal to the Council on Library and Information Resources*: Retrieved November 2006, from http://cohesion.rice.edu/library/alliance/emplibrary/NewProject.doc

Hull, B. (2001). Can librarians help to overcome the social barriers to access? *New Library World, 102*(10), 382-388.

Jenks, G. (1953). "Pointillism" as a cartographic technique. *Professional Geographer, 5*(5), 4-6.

Jenks, G. (1973). Visual integration in thematic mapping, fact or fiction. *The International Yearbook of Cartography, 13*, 27-35.

Jenks, G., & Brown, D. (1966). Three dimensional map construction. *Science, 3750*(857-864).

Jordan, M. (2006). The CARL metadata harvester and search service. *Library Hi Tech, 24*(2), 197-210.

Kent, R. (2000). The information flow foundation for conceptual knowledge organization. In *Dynamism and stability in knowledge organization: Proceedings of the sixth international ISKO conference,* (pp. 111-117,). Ergon Verlag: Würzburg, Germany.

Kraak, M.-J. (2003). Geovisualization illustrated. *ISPRS Journal of Photogrammetry & Remote Sensing, 57*, 390-399.

Kraak, M.-J. (1999). *Dealing with time.* Bethesda, MD: American Congress on Surveying and Mapping.

Kraak, M.-J. (2000). *Web cartography.* London: Taylor & Francis.

Kraak, M.-J., & Ormeling, F. J. (1996). *Cartography: Visualization of spatial data.* Essex, UK: Addison Wesley Longman Limited.

Kwan, M.-P. (2004). Beyond difference: From canonical geography to hybrid geographies. *Annals of the Association of American Geographers, 94*(4), 756-763.

Lake, R. W. (1993). Planning and applied geography: Positivism, ethics and GIS. *Progress in Human Geography, 17*(3), 404-413.

Larsen, R. L. (2002). The DLib Test Suite and Metrics Working Group: Harvesting the experience from the digital library initiative. In C. Borgman, I. Sølvberg, & L. Kovács (Eds.), *Fourth DELOS workshop: Evaluation of digital libraries: Testbeds, measurements, and metrics* (pp. 15-19). Budapest, Hungary: Hungarian Academy of Sciences, Computer and Automation Research Institute (MTA SZTAKI). Retrieved December 2006, from http://www.sztaki.hu/conferences/deval/presentations/DELOSWorkshop4OnEval_report.pdf

Latour, B. (1987). *Science in action: How to follow scientists and engineers through society.* Cambridge, MA: Harvard University Press.

Latour, B. (2005). *Reassembling the social: An introduction to actor-network theory.* Oxford, England: Oxford University Press.

LeSage, J. P., & Pace, R. K. (2001). Spatial dependence in data mining. In R. L. Grossman, C. Kamath, P. Kegelmeyer, V. Kumar, & R. R. Namburu (Eds.), *Data mining for scientific and engineering applications* (pp. 439-460). Boston, MA: Kluwer Academic.

Longley, P., & Batty, M. (1996). *Spatial analysis: Modelling in a GIS environment.* Cambridge, England: GeoInformation International.

Longley, P. A., Goodchild, M. F., Maguire, D. J., & Rhind, D. W. (2001). *Geographic information systems and science.* Chichester, England: John Wiley & Sons Ltd.

Lyman, P., & Varian, H. R. (2003). *How much information.* Retrieved from http://www.sims.berkeley.edu/how-much-info-2003

MacEachren, A. M. (1982). The role of complexity and symbolization method in thematic map effectiveness. *Annals of the Association of American Geographers, 72*(4), 495-513.

MacEachren, A. M. (1992). Application of environmental learning theory to spatial knowledge acquisition from maps. *Annals of the Association of American Geographers: Annals of the Association of American Geographers, 82*(2), 245-274.

MacEachren, A. M. (1995). *How maps work: Representation, visualization, and design.* New York: Guilford Press.

MacEachren, A. M., Buttenfield, B. P., & et al. (1992). Visualization. In R. F. Abler, M. G. Marcus, & J. M. Olson *Geography's inner worlds: Pervasive themes in contemporary American geography* (1st ed., pp. 99-137). New Brunswick, N.J.: Rutgers University Press.

MacEachren, A. M., Edsall, R., Haug, D., Baxter, R., Otto, G., Masters, R. et al. (1999). Virtual environments for geographic visualization: Potential and challenges. In *Proceedings of the 1999 workshop on new paradigms in information visualization and manipulation in conjunction with the eighth ACM international conference on information and knowledge management.* New York, NY: ACM. Retrieved November 2006, from http://citeseer.ist.psu.edu/cache/papers/cs2/422/http:zSzzSzwww.geovista.psu.eduzSzpublicationszSzN-PIVM99zSzammNPIVM.pdf/maceachren99virtual.pdf

MacEachren, A. M., & Kraak, M.-J. (1997). *Exploratory cartographic visualization.* Oxford, England: Elsevier Science.

MacEachren, A. M., & Kraak, M.-J. (2001). *Research challenges in geovisualization: Special issue on geovisualization.* Gaithersburg, MD: American Congress on Surveying and Mapping.

Maguire, D. J., Batty, M., & Goodchild, M. F. (2005). *GIS, spatial analysis, and modeling* (1st ed.). Redlands, CA: ESRI Press.

Makedon, F., Ford, J., Shen, L., Steinberg, T., Saykin, A., Wishart, H. et al. (2002). MetaDL: A digital library of metadata for sensitive or complex research data. In M. Agosti, & C. Thanos (Eds.), *Research and advances technology for digital technology: 6th European conference, ECDL 2002, Rome, Italy, September 16-18, 2002, proceedings* (pp. 374-389). Berlin, Germany: Springer-Verlag GmbH.

Marcum, D. B. (2003). Research questions for the digital era library. *Library Trends, 51*(4), 636-651.

Martin, D. (1996). *Geographic information systems: Socioeconomic applications* (2nd ed.). London, England: Routledge.

McCormick, B. H., DeFanti, T. A., Brown, M. D., & Zaritsky, R. (1987). Visualization in scientific computing. *Computer Graphics, 21*(6). New York, NY: ACM SIGGRAPH.

Miller, H. J., & Han, J. (2001). *Geographic data mining and knowledge discovery. Research monographs in geographic information systems.* London, England: Taylor & Francis.

Moellering, H. (1984). Real maps, virtual maps and interactive cartography. In G. L. Gaile, & C. J. Willmott (Eds.), *Spatial statistics and models* (pp. 109-131). Dordrecht: Reidel Publishing.

Monmonier, M. S. (1965). The production of shaded maps on the digital computer. *The Professional Geographer, 17*(5), 13-14.

Monmonier, M. S. (1980). The hopeless pursuit of purification in cartographic communication, a comparison of graphic-art and perceptual distortions of graytone symbols. *Cartographica, 17*(1), 24-39.

Monmonier, M. S. (1981). Map-text coordination in geographic writing. *The Professional Geographer, 33*(4), 406-412.

Muller, J.-C. (1975). Associations in choropleth map comparison. *Annals of the Association of American Geographers, 65*(3), 403-413.

Muller, J.-C. (1979). Perception of continuously shaded maps. *Annals of the Association of American Geographers, 69*(2), 240-249.

Norris, P. (2001). *Digital divide: Civic engagement, information poverty, and the Internet worldwide.* Cambridge, England: Cambridge University Press.

Nyerges, T. L. (1980). *Modeling the structure of cartographic information for query processing.* Unpublished doctoral dissertation, Ohio State University, Columbus, OH.

O'Donnell, J. J. (1998). *Avatars of the word: From papyrus to cyberspace.* Boston, MA: Harvard University Press.

Olson, J. (1975). Autocorrelation and visual map complexity. *Annals of the Association of American Geographers, 65*(2), 189-204.

Orlikowski, W., & Gash, D. (1994). Technological frames: Making sense of information technology in organizations. *ACM Transactions on Information Systems, 12,* 174-207.

O'Sullivan, D., & Unwin, D. (2003). *Geographic information analysis.* Hoboken, NJ: John Wiley and Sons.

Paisley, W. J. (1980). Information and work. In B. Dervin, & M. J. Voigt (Eds.), *Progress in the communication sciences, 2* (pp. 114-165). Norwood, NJ: Ablex.

Paulston, R. G. (1996). *Social cartography: Mapping ways of seeing social and educational change.* New York, NY: Garland Pub.

Peuquet, D. J. (1988). Representations of geographic space: Toward a conceptual synthesis. *Annals of the Association of American Geographers, 78*(3), 375-394.

Peuquet, D. J. (1994). It's about time: A conceptual framework for the representation of temporal dynamics in geographic information systems. *Annals of the Association of American Geographers, 84*(3), 441-461.

Peuquet, D. J. (2002). *Representations of space and time.* New York, NY: Guilford Press.

Philbrick, A. K. (1953). Toward a unity of cartographical forms and geographical content. *Professional Geographer, 5*(5), 11-15.

Pickles, J. (1995). Representations in an electronic age: Geography, GIS, and democracy. In J. Pickles (ed.), *Ground truth: The social implications of geographic information systems* (pp. 1-30). New York, NY: The Guildford Press.

Pickles, J. (1999). Social and cultural cartographies and the spatial turn in social theory. *Journal of Historical Geography, 25*(1), 93–98.

Poore, B. S. (2003). The open black box: The role of the end-user in GIS integration. *The Canadian Geographer, 47*(1), 62-74.

Raper, J. (2000). *Multidimensional geographic information science.* New York, NY: Taylor and Francis.

Ravindranathan, U., Shen, R., Gonçalves, M. A., Fan, W., Fox, E. A., & Flanagan, J. W. (2004). Prototyping digital libraries handling heterogeneous data sources: The ETANA-DL case study. In R. Heery, & L. Lyon (Eds.), *Research and advanced technology for digital libraries: 8th European conference, ECDL 2004, Bath, UK, September 12-17, 2004, proceedings* (pp. 186-197). Berlin, Germany: Springer-Verlag GmbH.

Rhind, D. (2000). Current shortcomings of global mapping and the creation of a new geographical framework for the world. *The Geographical Journal, 166*(4), 295-305.

Robinson, A. H., & Petchenick, B. B. (1975). The map as a communication system. *The Cartographic Journal, 12*, 7-14.

Schuurman, N. (1999). Speaking with the enemy? A conversation with Michael Goodchild. *Environment and Planning D: Society & Space, 17*(1), 1-2.

Schuurman, N. (2003). The ghost in the machine: Spatial data, information and knowledge in GIS. *The Canadian Geographer / Le Géographe Canadien, 47*(1), 1-4.

Shannon, C. E., & Weaver, W. (1949). *The mathematical theory of communication.* Urbana, IL: University of Illinois Press.

Sheppard E., Couclelis H., Graham S., Harrington J. W., & Onsrud H. (1999). Geographies of information society. *International Journal of Geographical Information Science, 13*(8), 797-823.

Sheppard, E. (2001). Quantitative geography: Representations, practices, and possibilities. *Environment and Planning D: Society & Space, 19*(5), 535-554.

Shumar, W., & Renninger, A. K. (2002). On conceptualizing community. K. A. S. W. Renninger *Building virtual communities: Learning and change in cyberspace* (pp. 1-17). New York: Cambridge University Press.

Sieber, R. E. (2003). Public participation geographic information systems across borders. *The Canadian Geographer/Le Géographe Canadien, 47*(1), 50-61.

Skole, D. L. (2004). Geography as a great intellectual melting pot and the preeminent interdisciplinary environmental discipline. *Annals of the Association of American Geographers, 94*(4), 739-743.

Skov-Petersen, H. (2003). Defining GIS: Assessment of ScanGIS. In K. Virrantaus, & H. and Tveite (Eds.), *ScanGIS 2003 - Proceedings of the 9th Scandinavian Research Conference on Geographical Information Science, 4-6 June 2003, Espoo, Finland* (pp. 269-284). Helsinki, Finland: Department of Surveying, Helsinki University of Technology. Retrieved January 2007, from http://www.scangis.org/scangis2003/papers/31.pdf

Skupin, A., & Fabrikant, S. I. (2003). Spatialization methods: A cartographic research agenda for non-geographic information visualization. *Cartography and Geographic Information Science, 30*(2), 95-115.

Smart, P. D., Abdelmoty, A. I., & Jones, C. B. (2004). An evaluation of geo-ontology representation languages for supporting Web retrieval of geographical information. In *Proceedings of the GIS research UK (GISRUK) 12th annual conference* (pp. 175-178). Norwich, UK: University of East Anglia. Retrieved February 2006, from http://www.geo-spirit.org/publications/ psmart-gisruk04-final.pdf

Smith, D. A., & Crane, G. (2001). Disambiguating geographic names in a historical digital library. In P. Constantopoulos, & I. T. Sølvberg (Eds.), *Research and advanced technology for digital libraries: 5th European conference, ECDL*

2001, Darmstadt, Germany, September 4-9, 2001, proceedings (pp. 127-136). Berlin, Germany: Springer-Verlag GmbH.

Sui, D. Z. (2004). GIS, cartography, and the "third culture": Geographic imaginations in the computer age. *The Professional Geographer, 56*(1), 62-72.

Sui, D. Z., & Goodchild, M. (2001). GIS as media? *International Journal of Geographical Information Science, 15*(5), 387-390.

Sui, D. Z., & Goodchild, M. F. (2003). A tetradic analysis of GIS and society using McLuhan's law of the media. *The Canadian Geographer/Le Géographe Canadien, 47*(1), 5-17.

Taylor, D. R. F. (1978). *Recent trends in geographic information processing in the National Capital Region*. Ottawa, Canada: Carleton University.

Taylor, D. R. F. (1991). *Geographic information systems: The microcomputer and modern cartography* (1st ed.). Oxford, England: Pergamon.

Taylor, P. J., & Overton, M. (1991). Further thoughts on geography and GIS: A preemptive strike? *Environment and Planning A, 23*, 1087-1094.

Tobler, W. R. (1962). A classification of map projections. *Annals of the Association of American Geographers, 52*(2), 167-175.

Tobler, W. R. (1965). Computation of the correspondence of geographical patterns. *Papers in Regional Science, 15*(1), 131-139.

Tobler, W. R. (1969). Geographical filters and their inverses. *Geographical Analysis, 1*(July), 234-253.

Tobler, W. R. (1979). A transformational view of cartography. *The American Cartographer, 6*, 101-106.

Tobler, W. R. (2004). Thirty five years of computer cartograms. *Annals of the Association of American Geographers: Annals of the Association of American Geographers, 94*(1), 58-73.

Tulloch, D. L., & Shapiro, T. (2003). The intersection of data access and public participation: Impacting GIS users' success? *URISA Journal, 15*(APA II), 55-60.

Vickers, G. (1965). *The art of judgment: A study of policy making*. New York, NY: Basic Books, Inc.

Vickers, G (1983). *Human systems are different*. London, England: Harper & Row.

Wack, P. (1985). Scenarios: Uncharted waters ahead. *Harvard Business Review, 63*(5), 73-89.

Waller, L. A., & Gotway, C. A. (2004). *Applied spatial statistics for public health data*. Hoboken, NJ: John Wiley & Sons.

Weaver, M., Delcambre, L., Shapiro, L., Brewster, J., Gutema, A., & Tolle, T. (2003). A digital geolibrary: Integrating keywords and place names. In T. Koch, & I. T. Sølvberg (Eds.), *Proceedings: Research and advanced technology for digital libraries: 7th European conference, ECDL 2003 Trondheim, Norway, August 17-22, 2003* (pp. 422-433). Trondheim, Norway: ECDL.

Weaver, M., Delcambre, L., & Tolle, T. (2003). Metadata++: A scalable hierarchical framework for digital libraries. In T. M. T. Sembok, H. B. Zaman, H. Chen, S. R. Urs, & S. H. Myaeng (Eds.), *Digital libraries: Technology and management of indigenous knowledge for global access* (pp. 696-696). Berlin, Germany: Springer-Verlag GmbH.

Weinheimer, J., & Caprazli, K. (2004). A framework for unified authority files: A case study of corporate body names in the FAO catalogue. In T. Koch, & I. T. Sølvberg (Eds.), *Research and advanced technology for digital libraries: 7th European Conference, ECDL 2003 Trondheim, Norway, August 17-22, 2003 Proceedings* (pp. 374-386). Berlin, Germany: Springer-Verlag GmbH.

Worrall, L. (1991). *Spatial analysis and spatial policy using geographic information systems.* London, England: Belhaven Press.

About the Contributors

John Abresch has a Master of Arts in geography and in library and information science. He has worked as a geographic information systems analyst in both the private and public sectors on many research projects, using both urban and environmental applications of GIS. Currently employed as an instructor librarian, Research Services and Collections, in the Tampa Library of the University of South Florida, Mr. Abresch is responsible for providing specialised research and reference services to social sciences faculty and graduate students. He also has collection development responsibility that encompasses geographic and GIS resources.

Ardis Hanson, MLS, is the director of the research library at the Louis de la Parte Florida Mental Health Institute at the University of South Florida (USF). Interested in the use of technology to enhance research, she has presented at the USF Symposium on 21st Century Teaching Technologies and Internet2 showcasing innovative software applications. She is an adjunct instructor in the School of Library and Information Science and the College of Public Health at USF. Ms. Hanson was a member of the USF Virtual Library Planning Committee, the Implementation Team, the Interface Design Project Group, and the Metadata Team. In addition to being a member of the executive council of ACURIL (Association of Caribbean, University, Research, and Institutional Libraries), Ms. Hanson is pursuing a PhD in communication.

Susan Jane Heron, MLS, is the associate director of the Collection Analysis and Technical Services Department at the University of South Florida Libraries (USF). Previously, Ms. Heron was head of Technical Services at the University of San Diego, head of Cataloging at San Diego State University, project coordinator for Database Conversion at Temple University, and a library liaison with the Research Libraries Group. Ms. Heron was a member of the USF Virtual Library Implementation Team as well as the Metadata, Digitization, Thesis and Dissertation, and Electronic Reserve Teams. Currently, Ms. Heron is on the State University System Technical Services Policy Committee, which is implementing ALEPH v.18 and examining metadata issues for geographic data and electronic resources.

Peter J. Reehling received a Master of Arts in geography from Ball State University and a second master's degree in library and information science from Indiana University. He is currently geographic information librarian at the University of South Florida. His duties range from the provision of GIS services to managing the ESRI Site License and coordinating the GIS Help Desk Coordinator. His collection development duties include the Geography Department and the Military Sciences. Previously, Mr. Reehling served as a geospatial analyst for the National Imagery and Mapping Agency and a geographer for the U.S. Census Bureau. Mr. Reehling is currently the vice-chair/chair-elect chair of MAGERT (Map and Geography Round Table of the American Library Association).

Index